CHROMATOGRAPHY

CHROMATOGRAPHY
METHODS AND INDUSTRIAL APPLICATIONS

Chris Dayton
George Witmore
Michal Witmore

Kruger Brentt
Publishers
2023

Kruger Brentt Publishers UK. LTD.
Company Number 9728962

Regd. Office: 68 St Margarets Road, Edgware, Middlesex HA8 9UU

ISBN: 9781787150850

Disclaimer:

Every possible effort has been made to ensure that the information contained in this book is accurate at the time of going to press, and the publisher and author cannot accept responsibility for any errors or omissions, however caused. No responsibility for loss or damage occasioned to any person acting, or refraining from action, as a result of the material in this publication can be accepted by the editor, the publisher or the author. The Publisher is not associated with any product or vendor mentioned in the book. The contents of this work are intended to further general scientific research, understanding and discussion only. Readers should consult with a specialist where appropriate.

Every effort has been made to trace the owners of copyright material used in this book, if any. The author and the publisher will be grateful for any omission brought to their notice for acknowledgement in the future editions of the book.

For information on all our publications visit our website at http://krugerbrentt.com/

PREFACE

Chromatography is an important biophysical technique that enables the separation, identification, and purification of the components of a mixture for qualitative and quantitative analysis. Proteins can be purified based on characteristics such as size and shape, total charge, hydrophobic groups present on the surface, and binding capacity with the stationary phase. Four separation techniques based on molecular characteristics and interaction type use mechanisms of ion exchange, surface adsorption, partition, and size exclusion. Other chromatography techniques are based on the stationary bed, including column, thin layer, and paper chromatography. Column chromatography is one of the most common methods of protein purification. The purpose of applying chromatography which is used as a method of quantitative analysis apart from its separation, is to achieve a satisfactory separation within a suitable time interval. Various chromatography methods have been developed to that end. Some of them include column chromatography, thin-layer chromatography (TLC), paper chromatography, gas chromatography, ion exchange chromatography, gel permeation chromatography, high-pressure liquid chromatography, and affinity chromatography.

Initially chromatographic techniques were used to separate substances based on their color as was the case with herbal pigments. With time its application area was extended considerably. Nowadays, chromatography is accepted as an extremely sensitive, and effective separation method. Column chromatography is one of the useful separation, and determination methods. Column chromatography is a protein purification method realized especially based on one of the characteristic features of proteins. Besides, these methods are used to control purity of a protein. HPLC technique which has many superior features including especially its higher sensitivity, rapid turnover rate, its use as a quantitative method, can purify amino acids, proteins, nucleic acids, hydrocarbons, carbohydrates, drugs, antibiotics, and steroids. Chromatography is the separation method for biological molecules from the analytical to the preparative scale. Few of the recent advances in life science, whether in "proteomics" or in the industrial production of recombinant therapeutics, would have been possible without the help of chromatographic separations. Concomitantly, chromatography nowadays stands for much more than a batch column packed with porous particles.

Chromatography: Methods and Industrial Applications

The present book contains *Advanced Chromatography: Methods and Industrial Applications* chapters covering all related disciplines. These chapters include introduction to chromatography, basic principles of chromatography, separation techniques-chromatography, fundamentals of gas chromatography, fundamentals of high performance liquid chromatography, fundamentals of paper chromatography, chromatography and mass spectrometer, theory and practice of UHPLC and UHPLC-MS, recent advances in hydrophilic interaction: liquid chromatography, advanced chromatographic techniques in the analysis of luteolin and its derivatives, principles of ion exchange, application of thin layer chromatography for qualitative analysis of gunpowder, role of standard flow liquid chromatography for shotgun proteomics in bioenergy research, supercritical fluid chromatography, application of gas and liquid chromatography coupled tandem mass spectrometry systems for pesticide quantification in exported fruit and vegetables, liquid chromatography, instrumental analysis of poison-by chromatographic techniques, ion chromatography an advantageous technique in soil analysis, chromatographic applications in medicine, paper chromatography of antifungal antibiotics, metabolomics of gastric cancer metastasis detected by gas chromatography and mass spectrometry and role of chromatography for monitoring of breast cancer biomarkers. Related terminology is given at the end for ready reference. It is well suited for advanced analytical instrumentation students as well as for analysts seeking additional knowledge or a deeper understanding of familiar techniques. It will cater to the requirements of the Upper-level undergraduate students and graduate students, researchers, technicians, and scientists.

We are grateful to all those persons as well as various books, manuals, periodicals, magazines, journals etc. that helped in the preparation of this book. In spite of the best efforts, it is possible that some errors may have occurred into the compilation and editing of the book. Further queries, constructive suggestions and criticisms for the improvement of the book are always welcome and shall be thankfully acknowledged.

Chris Dayton

George Witmore

Michal Witmore

Contents

1

INTRODUCTION TO CHROMATOGRAPHY

1.1 INTRODUCTION: AN OVERVIEW

Chromatography is an important biophysical technique that enables the separation, identification, and purification of the components of a mixture for qualitative and quantitative analysis.

Chromatography is a separation method where the analyte is combined within a liquid or gaseous mobile phase., which is pumped through a stationary phase. Usually one phase is hydrophilic and the other lipophilic. The components of the analyte interact differently with these y=two phases. Depending on of their polarity they spend more or less time interacting with the stationary phase and are thus retarded to a greater or lesser extent. This leads to the separation of the different components present in the sample. Each sample component elutes from the stationary phase at a specific time, its retention time. As the components pass through the detector their signal is recorded and plotted in the form of a chromatogram.

- The Russian botanist Mikhail Tswett coined the term chromatography in 1906.
- The first analytical use of chromatography was described by James and Martin in 1952, for the use of gas chromatography for the analysis of fatty acid mixtures.
- A wide range of chromatographic procedures makes use of differences in size, binding affinities, charge, and other properties to separate materials.
- It is a powerful separation tool that is used in all branches of science and is often the only means of separating components from complex mixtures.

Chromatography is a biophysical technique that creates separation, identification, and purification of a mixture to then analyze the components of the mixture. To do so, chromatography relies specifically on two different phases, the mobile and stationary phases. To separate the components of the mixture, it is dissolved in a solvent during the mobile phase. The mobile phase then carries the mixture through the stationary phase where the components of the mixture interact with the stationary phase. The components all interact with the stationary phase at their own pace, creating the separation of the components located within the mixture. The components of the mixture will move differently through this phase due to differences in absorption, partition, and molecular weights. Chromatography can be done in many ways to adhere to the different substances being analyzed. Some of the most common types of chromatography are gas chromatography, liquid chromatography, thin-layer chromatography, and column chromatography.

TERM	DEFINITION
Mobile phase or carrier	solvent moving through the column
Stationary phase or adsorbent	substance that stays fixed inside the column
Eluent	fluid entering the column
Eluate	fluid exiting the column (that is collected in flasks)
Elution	the process of washing out a compound through a column using a suitable solvent
Analyte	mixture whose individual components have to be separated and analyzed

2.2 PRINCIPLE OF CHROMATOGRAPHY

Chromatography is based on the principle where molecules in mixture applied onto the surface or into the solid, and fluid stationary phase (stable phase) is separating from each other while moving with the aid of a mobile phase.

- The factors effective on this separation process include molecular characteristics related to adsorption (liquid-solid), partition (liquid-solid), and affinity or differences among their molecular weights.
- Because of these differences, some components of the mixture stay longer in the stationary phase, and they move slowly in the chromatography system, while others pass rapidly into the mobile phase, and leave the system faster

.Three components thus form the basis of the chromatography technique.

1. **Stationary phase:** This phase is always composed of a "solid" phase or "a layer of a liquid adsorbed on the surface solid support".
2. **Mobile phase:** This phase is always composed of "liquid" or a "gaseous component."
3. **Separated molecules**

The type of interaction between the stationary phase, mobile phase, and substances contained in the mixture is the basic component effective on the separation of molecules from each other.

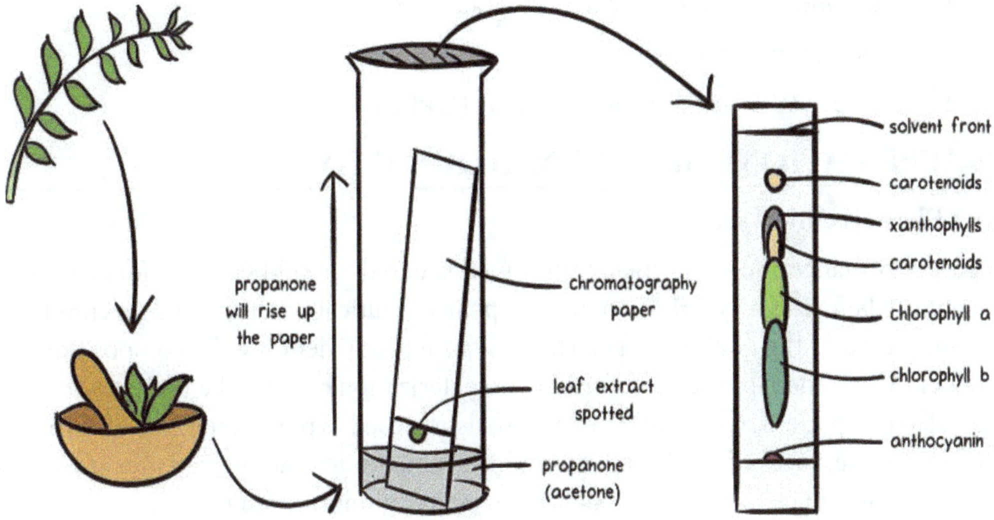

2.3 TYPES OF CHROMATOGRAPHY

- Substances can be separated on the basis of a variety of methods and the presence of characteristics such as size and shape, total charge, hydrophobic groups present on the surface, and binding capacity with the stationary phase.
- This leads to different types of chromatography techniques, each with their own instrumentation and working principle.
- For instance, four separation techniques based on molecular characteristics and interaction type use mechanisms of ion exchange, surface adsorption, partition, and size exclusion.
- Other chromatography techniques are based on the stationary bed, including column, thin layer, and paper chromatography.

2.4 COMMONLY EMPLOYED CHROMATOGRAPHY TECHNIQUES

- Column chromatography
- Ion-exchange chromatography
- Gel-permeation (molecular sieve) chromatography
- Affinity chromatography
- Paper chromatography
- Thin-layer chromatography
- Gas chromatography (GS)
- Dye-ligand chromatography

- Hydrophobic interaction chromatography
- Pseudo-affinity chromatography
- High-pressure liquid chromatography (HPLC)

2.5 APPLICATIONS OF CHROMATOGRAPHY

2.5.1 Pharmaceutical sector

High-performance liquid chromatography is considered the gold standard for compound separation before characterization in the pharmaceutical industry. The technique is generally used at the quality control stage, quantitatively defining the composition of a product. As in other applications, further characterization can then be undertaken of the separated compounds. Chromatography may also be used to purify intermediate products throughout the stages of synthesis, and sophisticated modern automated processes allow crude products to be refined continuously on an optimum timescale.

2.5.2 Vaccine Industry

As in the synthesis of other pharmaceutical products, many vaccines undergo purification by chromatography during or in the final stages of production. The antigen of interest to the production of the vaccine also may be obtained by chromatography, for example, the SARS coronavirus spike protein has been isolated by liquid chromatography allowing it to be produced in good quantities. This has enabled researchers to perform a variety of tests on the isolated protein, revealing many structural and biochemical characteristics that explain the high infectivity of the virus, and has allowed the development of vaccines against it. Vaccines may contain the inactivated or attenuated free virus, or just the antigen that the body's immune defenses recognize. In any case, chromatography is often used to separate the components of interest from by-products and compounds present in the growth medium of the cells within which the virus or proteins are cultured.

2.5.3 Chemical Industry

The chemical industry must adhere to numerous environmental safety precautions. Perfluoroalkyl substances, also known as PFAS, have become a persistent threat to the human body and the environment. PFAS can be found in items such as protective coatings on shoes and other fabrics, electronics, and even firefighting foams. While these substances benefit products by making them extremely durable, they pose an environmental concern as they continue to accumulate. PFAS in our drinking water can also lead to damaging health concerns such as reproductive and developmental setbacks. By using solid-phase extraction, liquid chromatography, and mass spectrometry, we can detect PFAS in the environment and our drinking water, even at very low limits.

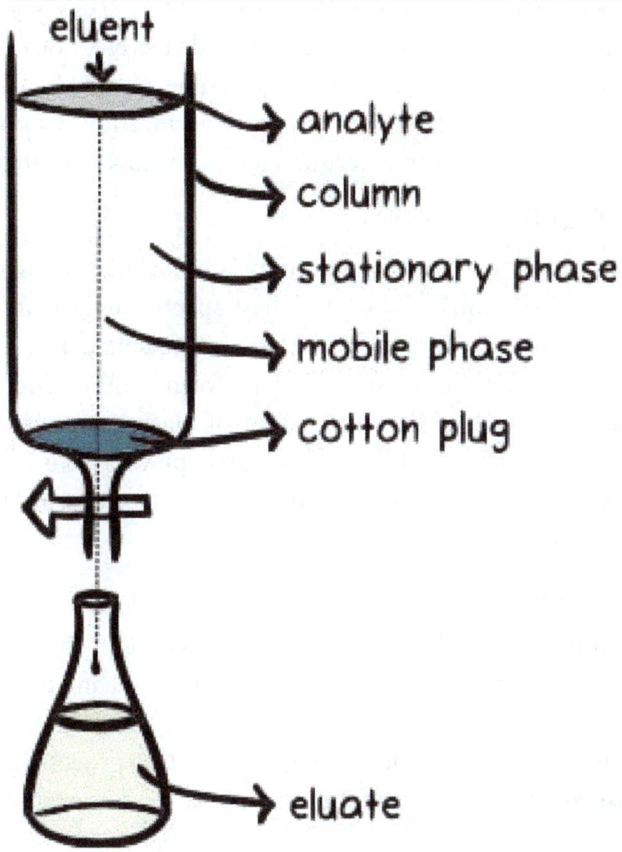

2.5.4 Food Industry

Food products contain a large and complex mixture of compounds, which may be naturally occurring or imparted at a later stage of the manufacturing or harvesting process. Many such compounds may be a cause for concern when present in food products above a particular concentration, and several types of chromatography are generally used to separate the numerous compounds from one another for deeper analysis. For example, plant-based products may bear lingering traces of pesticides, while meat may contain veterinary drugs. In addition to detecting potentially toxic impurities in food products, chromatography is used to standardize and authenticate the nutritional profile of many goods at the quality control stage.

2.5.5 Forensic Science

Similar to security precautions, Forensics is a unique application of chromatography. Gas chromatography can be used for more in-depth forensics procedures, for example, crime scene analysis to test evidence such as blood, hair, and fabric samples to further understand what may have happened at the scene. Chromatography is massively important to forensic pathology work. Gas chromatography is widely used to identify the types of fluids and compounds that exist in a body postmortem. In such cases, a possible cause of death and motive can be determined based on finding drugs, alcohol, or toxic substances in the body. Another unique form of forensics that can be assisted by chromatography is arson verification. By using chromatography in arson verification, it's possible to identify flammable substances in fire debris to determine the exact substance that created the fire.

2.5.6 Molecular Biology Studies

One of the most complex uses of chromatography is molecular biology studies. Hybrid techniques between electrochemistry (EC) and mass spectrometry with chromatography are often applied to studies of proteins, peptides, and nucleic acids. This combination is largely used for metabolomics such as biotransformation reactions like oxidative reactions and proteomics such as the purification of plasma proteins, hormones, and antibodies. Chromatography in nucleic acid research plays a role in accelerating the identification process of nucleobases, nucleotides, and nucleosides, as well as identifying their oxidization process.

2.5.7 Environmental Studies

Aspects of environmental quality can also be assessed and monitored using chromatography. The concentration of pollutants in water, soil, and air are each regularly determined firstly by mixture separation by chromatography in tandem with other characterization techniques such as mass spectrometry. This process has proven vital in city planning and assessing the impact of industry on particular locations and the planet in general.

2.5.8 Beverage Testing

Food isn't the only thing you consume which has been tested using chromatography. Many drinks manufacturers use this technique to ensure each bottle of their product is exactly the same, so you can rely on a consistent taste.

2.5.9 Security

Security practices are a unique industry where chromatography can be utilized. Gas chromatography can be used to determine volatile gases, furthering safety measures at locations such as airports and large gatherings with similar safety precautions like concerts and sporting events to eliminate deadly threats.

2

BASIC PRINCIPLES OF CHROMATOGRAPHY

2.1 INTRODUCTION

Chromatography has a great impact on all areas of analysis and, therefore, on the progress of science in general. Chromatography differs from other methods of separation in that a wide variety of materials, equipment, and techniques can be used. This chapter will focus on the principles of chromatography, mainly liquid chromatography (LC). The general principles of extraction are first described as a basis for understanding chromatography.

Chromatography is a separation method based on the partitioning of a solute between a mobile phase and a stationary phase. The mobile phase may be liquid, gas, or a supercritical fluid. The stationary phase may be an immobilized liquid or a solid, in either a planar or column form. Based on the physicochemical characteristics of the analyte, and the availability of instrumentation, a chromatographic system is chosen to separate, indentify, and quantify the analyte. Chromatographic modes include adsorption, partition, ion exchange, size exclusion, and affinity chromatography. Factors to be considered when developing a separation include mobile phase variables (strength, pH, temperature, and flow rate), and column efficiency, selectivity, and capacity.

2. 2 EXTRACTION

In its simplest form, extraction refers to the transfer of a solute from one liquid phase to another.

Extraction in myriad forms is integral to food analysis - whether used for preliminary sample cleanup, concentration of the component of interest, or as the actual means of analysis. Extractions may be categorized as batch, continuous, or counter current processes.

2.2.1 Batch Extraction

In batch extraction the solute is extracted from one solvent by shaking it with a second, immiscible solvent. The solute partitions, or distributes, itself between the two phases and, when equilibrium has been reached, the partition coefficient, K, is a constant.

$$K = \frac{\text{(Concentration of solute in phase 1)}}{\text{(Concentration of solute in phase 2)}} \quad [1]$$

After shaking, the phases are allowed to separate, and the layer containing the desired constituent is removed, for example, in a separatory funnel. In batch extraction, it is often difficult to obtain a clean separation of phases, owing to emulsion formation. Moreover, partition implies that a single extraction is usually incomplete.

2.2.2 Continuous Extraction

Continuous liquid–liquid extraction requires special apparatus, but is more efficient than batch separation. One example is the use of a Soxhlet extractor for extracting materials from solids. Solvent is recycled so that the solid is repeatedly extracted with fresh solvent. Other pieces of equipment have been designed for the continuous extraction of substances from liquids, and different extractors are used for solvents that are heavier or lighter than water.

2.2.3 Counter Current Extraction

Counter current distribution refers to a serial extraction process. It separates two or more solutes with different partition coefficients from each other by a series of partitions between two immiscible liquid phases.

Liquid–liquid partition chromatography also known as counter current chromatography, is a direct extension of counter current extraction. Years ago the counter current extraction was done with a "Craig apparatus" consisting of a series of glass tubes designed such that the lighter liquid phase (mobile phase) was transferred from one tube to the next, while the heavy phase (stationary phase) remained in the first tube . The liquid–liquid extractions took place simultaneously in all tubes of the apparatus, which was usually driven electromechanically. Each tube in which a complete equilibration took place corresponded to one theoretical plate of the chromatographic column. The greater the difference in the partition coefficients of various substances, the better was the separation. A much larger number of tubes were required to separate mixtures of substances with close partition coefficients, which made this type of counter current extraction very tedious. Modern liquid–liquid partition chromatography that developed from this concept is much more efficient and convenient.

2.3 CHROMATOGRAPHY

2.3.1 Historical Perspective

Modern chromatography originated in the late nineteenth and early twentieth centuries from independent work by David T. Day, a distinguished American geologist and mining engineer, and Mikhail Tsvet, a Russian botanist. Day developed procedures for fractionating crude petroleum by passing it through Fuller's earth, and Tsvet used a column packed with chalk to separate leaf pigments into colored bands. Because Tsvet recognized and correctly interpreted the chromatographic processes and named the phenomenon chromatography, he is generally credited with its discovery.

After languishing in oblivion for years, chromatography began to evolve in the 1940s due to the development of column partition chromatography by Martin and Synge and the invention of paper chromatography.

The first publication on GC appeared in 1952. By the late 1960s, GC, because of its importance to the petroleum industry, had developed into a sophisticated instrumental technique, which was the first instrumental chromatography to be available commercially. Since early applications in the mid 1960s, HPLC, profiting fromthe theoretical and instrumental advances of GC, has extended the area of liquid chromatography into an equally sophisticated and useful method. SFC, first demonstrated in 1962, is finally gaining popularity. Modern chromatographic techniques, including automated systems, are widely utilized in the characterization and quality control of food raw materials and food products.

2.3.2 General Terminology

Chromatography is a general term applied to a wide variety of separation techniques based on the partitioning or distribution of a sample (solute) between a moving or mobile phase and a fixed or stationary phase. Chromatography may be viewed as a series of equilibrations between the mobile and stationary phase. The relative interaction of a solute with these two phases is described by the partition (K) or distribution (D) coefficient (ratio of concentration of solute in stationary phase to concentration of solute inmobile phase). The mobile phase may be either a gas (GC) or liquid (LC) or a supercritical fluid (SFC). The stationary phase may be a liquid or, more usually, a solid. The field of chromatography can be subdivided according to the various techniques applied, or according to the physico-chemical principles involved in the separation. Table 1 summarizes some of the chromatographic procedures or methods that have been developed on the basis of different mobile–stationary phase combinations. Inasmuch as the nature of interactions between solute molecules and the mobile or stationary phases differ, these methods have the ability to separate different kinds of molecules.

2.3.3 Gas Chromatography

Gas chromatography is a column chromatography technique, in which the mobile phase is gas and the stationary phase is either an immobilized liquid or a solid packed in a closed tube. GC is used to separate thermally stable volatile components of a mixture. Gas chromatography, specifically gas–liquid chromatography, involves vaporizing a sample and injecting it onto the head of the column. Under a controlled temperature gradient, the sample is transported through the column by the flow of an inert, gaseous mobile phase.

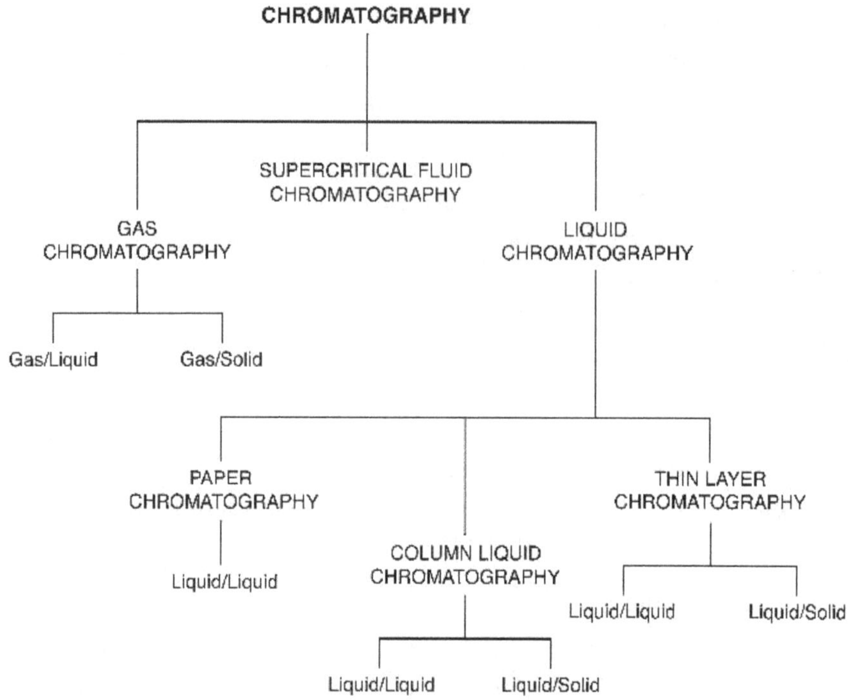

A scheme for subdividing the field of chromatography, according to various applied techniques]

Characteristics of Different Chromatographic Methods

Method	Mobile/Stationary phase	Retention Varies with
Gas-liquid chromatography	Gas/liquid	Molecular size/polarity
Gas-solid chromatography	Gas/solid	Molecular size/polarity
Supercritical fluid chromatography	Supercritical fluid/soild	Molecular size/polarity
Reversed-phase chromatography	Polar liquid/nonpolar liquid or solid	Molecular size/polarity
Normal-phase chromatography	Less polar liquid/more polar liquid or solid	Molecular size/polarity
Ion-exchange chromatography	Polar liquid/Ionic solid	Molecular charge
Size-exclusion chromatography	Liquid/solid	Molecular size
Hydrophobic-interaction chromatography	Polar liquid/nonpolar liquid or solid	Molecular size/polarity
Affinity chromatography	Water/binding sites	Specific structure

Reprint from(8).p. A21, with kind permission from elsevier Science-NL, Sara Burgerhartstraat 25, 1055 KVAmsterdam,

Volatiles are then separated based on several properties, including boiling point, molecular size, and polarity.

2.3.4 Liquid Chromatography

There are several liquid chromatography techniques applied in food analysis, namely paper chromatography, thin layer chromatography (TLC) (both of these techniques may be referred to as planar chromatography), and column liquid chromatography, all of which involve a liquid mobile phase and either a solid or a liquid stationary phase. However, the physical form of the stationary phase is quite different in each case.

2.3.4.1 Paper Chromatography

Paper chromatography was introduced in 1944. In paper chromatography the stationary phase and the mobile phase are both liquid. Paper generally serves as a support for the liquid stationary phase. The dissolved sample is applied as a small spot or streak one half inch or more from the edge of a strip or square of filter paper (usually cellulose), which is then allowed to dry. The dry strip is suspended in a closed container in which the atmosphere is saturated with the developing solvent (mobile phase), and the paper chromatogram is developed. The end closer to the sample is placed in contact with the solvent, which then travels up or down the paper by capillary action (depending on whether ascending or descending development is used), separating the sample components in the process. When the solvent front has traveled the length of the paper, the strip is removed from the developing chamber and the separated zones are detected by an appropriate method.

The stationary phase in paper partition chromatography is usually water. However, the support may be impregnated with a nonpolar organic solvent and developed with water or other polar solvents or water (reversed-phase paper chromatography). In the case of complex sample mixtures, a two-dimensional technique may be used. The sample is spotted in one corner of a square sheet of paper, and one solvent is used to develop the paper in one direction. The chromatogram is then dried, turned 90°, and developed again, using a second solvent of different polarity.

Another means of improving resolution is the use of ion-exchange papers, that is, paper that has been impregnated with ion-exchange resin or paper, with derivatized cellulose hydroxyl groups (with acidic or basic moieties).

In paper and thin-layer chromatography, components of a mixture are characterized by their relative mobility (*R*f) value, where:

$$R\_f= \frac{\text{(Distance moved by component)}}{\text{(Distance moved by solvent)}} \qquad [2]$$

Unfortunately, *R*f values are not always constant for a given solute/sorbent/solvent, but depend on many factors, such as the quality of the stationary phase, layer thickness, humidity, development distance, and temperature.

2.3.4.2 Thin-Layer Chromatography

Thin-layer chromatography (TLC), first described in 1938, has largely replaced paper chromatography because it is faster, more sensitive, and more reproducible.

The resolution in TLC is greater than in paper chromatography because the particles on the plate are smaller and more regular than paper fibers. Experimental conditions can be easily varied to achieve separation and can be scaled up for use in column chromatography, although thin-layer and column procedures are not necessarily interchangeable, due to differences such as the use of binders with TLC plates, vapor-phase equilibria in a TLC tank, etc. There are several distinct advantages to TLC: high sample throughput, low cost, the possibility to analyze several samples and standards simultaneously, minimal sample preparation, and that a plate may be stored for later identification and quantification.

TLC is applied in many fields, including environmental, clinical, forensic, pharmaceutical, food, flavors, and cosmetics. Within the food industry, TLC may be used for quality control. For example, corn and peanuts are tested for aflatoxins/mycotoxins prior to their processing into corn meal and peanut butter, respectively.

2.3.4.2.1 General Procedures:

TLC utilizes a thin (ca. 250 μm thick) layer of sorbent or stationary phase bound to an inert support in a planar configuration.

The support is often a glass plate (traditionally, 20 cm × 20 cm), but plastic sheets and aluminum foil also are used. Precoated plates, of different layer thicknesses, are commercially available in a wide variety of sorbents, including chemically modified silicas.

Four frequently used TLC sorbents are silica gel, alumina, diatomaceous earth, and cellulose. Modified silicas for TLC may contain polar or nonpolar groups, so both normal and reversed-phase thin-layer separations may be carried out. High-performance thin-layer chromatography (HPTLC) simply refers to TLC performed using plates coated with smaller, more uniform particles. This permits better separations in shorter times.

If adsorption TLC is to be performed, the sorbent is first activated by drying for a specified time and temperature. Sample (in carrier solvent) is applied as a spot or streak 1–2 cm from one end of the plate. After evaporation of the carrier solvent, the TLC plate is placed in a closed developing chamber with the end of the plate nearest the spot in the solvent at the bottom of the chamber. Traditionally, solvent migrates up the plate (ascending development) by capillary action and sample components are separated.

After the TLC plate has been removed from the chamber and solvent allowed to evaporate, the separated bands are made visible or detected by other means. Specific chemical reactions (derivatization), which may be carried out either before or after chromatography, often are used for this purpose. Two examples are reaction with sulfuric acid to produce a dark charred area (a **destructive chemical method**) and the use of iodine vapor to form a colored complex (a **nondestructive method** inasmuch as the colored complex is usually not permanent). Common **physical detection methods** include the measurement of absorbed or emitted electromagnetic radiation (e.g., fluorescence) by means of autoradiography and the measurement of b-radiation from radioactively labeled compounds. Biological methods or biochemical inhibition tests can be used to detect toxicologically active substances. An example is measuring the inhibition of cholinesterase activity by organophosphate pesticides.

Quantitative evaluation of thin-layer chromatograms may be performed (1) in situ (directly on the layer) by using a densitometer or (2) after scraping a zone off the plate, eluting compound from the sorbent, and analyzing the resultant solution (e.g., by liquid scintillation counting).

2.3.4.2.2 Factors Affecting Thin-Layer Separations

In both planar and column liquid chromatography, the nature of the compounds to be separated determines what type of stationary phase is used. Separation can occur by adsorption, partition, ion-exchange, size-exclusion, or multiple mechanisms.

In simple adsorption TLC, the higher the solvent strength, the greater the Rf value of the solute. An Rf value of 0.3–0.7 is typical. Mobile phases have been developed for the separation of various compound classes on the different sorbents.

In addition to the sorbent and solvent, several other factors must be considered when performing planar chromatography. These include the type of developing chamber used, vapor phase conditions (saturated vs. unsaturated), development mode (ascending, descending, horizontal, radial, etc.), and development distance.

Sorbent	Chromatographic Mechanism	Typical Application
Silica gel	Adsorption	Steroids, amino acids, alcohols, hydrocarbons, lipids, aflatoxins, bile acids, vitamins, alkaloids
Silica gel RP	Reversed phase	Fatty acids, vitamins, steroids, hormones, carotenoids
Cellulose, kieselguhr	Partition	Carbohydrates, sugars, alcohols, amino acids, carboxylic acids, fatty acids
Aluminum oxide	Adsorption	Amines, alcohols, steroids, lipids, aflatoxins, bile acids, vitamins, alkaloids
PEI cellulose[a]	Ion exchange	Nucleic acids, nucleotides, nucleosides, purines, pyrimidnes
Magnesium silicate	Adsorption	Steroids, pesticides, lipids, alkaloids

Reprinted from (15) by permission of Wiley, New York. [a]PEI cellulose refers to cellulose derivatized with polyethyleneimine (PEI)

2.3.4.3 Column Liquid Chromatography

Column chromatography is the most useful method of separating compounds in a mixture. Fractionation of solutes occurs as a result of differential migration through a closed tube of stationary phase, and analytes can be monitored while the separation is in progress.

In column liquid chromatography, the mobile phase is liquid and the stationary phase can be either solid or liquid supported by an inert solid. A system for low-pressure (i.e., performed at or near atmospheric pressure) column liquid chromatography is illustrated in Fig. 2.

Having selected a stationary and mobile phase suitable for the separation problem at hand, the analyst must first prepare the stationary phase (resin, gel, or packing material) for use according to the supplier's instructions. (For example, the stationary phase often must be hydrated or pres welled in the mobile phase). The prepared stationary phase then is packed into a column (usually glass), the length and diameter of which are determined by the amount of sample to be loaded, the separation mode to be used, and the degree of resolution required. Longer and narrower columns usually enhance resolution and separation. Adsorption columns may be either dry or wet packed; other types of columns are wet packed. The most common technique for wet packing involves making a slurry of the adsorbent with the solvent and pouring this into the column. As the sorbent settles, excess solvent is drained off and additional slurry is added.

This process is repeated until the desired bed height is obtained. (There is a certain art to pouring uniform columns and no attempt is made to give details here.) If the packing solvent is different from the initial eluting solvent, the column must be thoroughly washed (equilibrated) with the starting mobile phase. The sample to be fractionated, dissolved in a minimum volume of mobile phase, is applied in a layer at the top (or head) of the column. Classical or low-pressure chromatography utilizes only gravity flow or a peristaltic pump to maintain a flow of mobile phase (eluent or eluting solvent) through the column. In the case of a gravity-fed system, eluent is simply siphoned from a reservoir into the column. The flow rate is governed by the hydrostatic pressure, measured as the distance between the level of liquid in the reservoir and the level of the column outlet. If eluent is fed to the column by a peristaltic pump (see Fig. 2), then the flow rate is determined by the pump speed and, thus, regulation of hydrostatic pressure is not necessary.

The process of passing the mobile phase through the column is called elution, and the portion that emerges from the outlet end of the column is sometimes called the eluate (or effluent). Elution may be isocratic (constant mobile-phase composition) or a gradient (changing the mobile phase, e.g., increasing solvent strength or pH) during elution in order to enhance resolution and decrease analysis time. As elution proceeds, components of the sample are selectively retarded by the stationary phase based on the strength of interaction with the stationary phase, and thus they are eluted at different times.

The column eluate may be directed through a detector and then into tubes, changed at intervals by

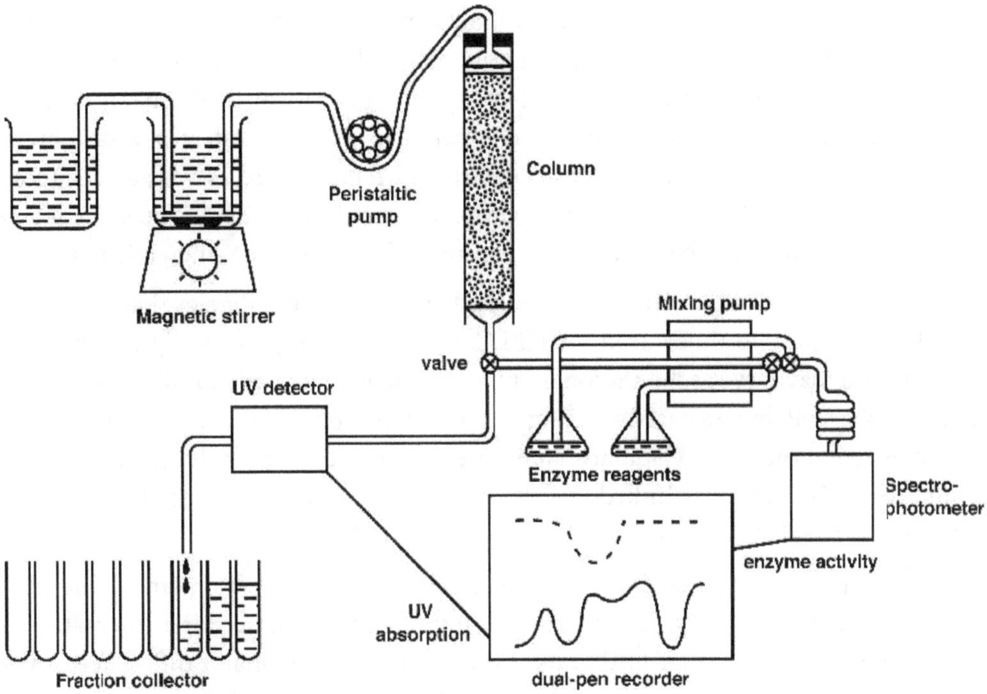

Fig. 1: System for low-pressure column liquid chromatography

A system for low-pressure column liquid chromatography. In this diagram, the column effluent is being split between two detectors in order to monitor both enzyme activity (at Right) and UV absorption (at Left). The two tracings can be recorded simultaneously by using a dual-pen recorder.

The detector response, in the form of an electrical signal, may be recorded (the chromatogram), using either a chart recorder or a computerized software, and used for qualitative or quantitative analysis. The fraction collector may be set to collect eluate at specified time intervals or after a certain volume or number of drops has been collected. Components of the sample that have been chromatographically separated and collected then can be further analyzed as needed.

2.3.5 Supercritical Fluid Chromatography

SFC refers to chromatography performed above the critical pressure (*Pc*) and critical temperature (*Tc*) of the mobile phase. A supercritical fluid (or compressed gas) is neither a liquid nor a typical gas. The combination of *Pc* and *Tc* is known as the critical point.

A supercritical fluid can be formed from a conventional gas by increasing the pressure, or from a conventional liquid by raising the temperature. Carbon dioxide frequently is used as a mobile phase for SFC; however, it is not a good solvent for polar and highmolecular- weight compounds. A small amount of a polar, organic solvent such as methanol can be added to a nonpolar supercritical fluid to enhance solute solubility, improve peak shape, and alter selectivity. Other supercritical fluids that have been used in food applications include nitrous oxide, trifluoromethane, sulphur hexafluoride, pentane, and ammonia. Supercritical fluids confer chromatographic properties intermediate to LC and GC. The high diffusivity and low viscosity of supercritical fluids mean decreased analysis times and improved resolution compared to LC. SFC offers a wide range of selectivity adjustment, by changes in pressure and temperature as well as changes in mobile phase composition and the stationary phase. In addition, SFC makes possible the separation of nonvolatile, thermally labile **compounds** that are not amenable to GC.

SFC can be performed using either **packed columns** or **capillaries**. Packed column materials are similar to those used for HPLC. Small particle, porous, high surface area, hydrated silica may serve as the stationary phase itself, or simply as a support for a bonded stationary phase. Polymer-based packing has been used, but is less satisfactory owing to long solute retention times. Capillaries are generally coated with a polysiloxane (–Si–O–Si) film, which is then cross-linked to form a polymeric stationary phase that cannot be washed off by the mobile phase.

Polysiloxanes containing different functional groups, such as methyl, phenyl, or cyano, may be used to vary the polarity of this stationary phase. Instrumentation for packed column SFC is similar to that used for HPLC with one major difference: A back pressure regulator is used to control the **outlet pressure** of the system. Without this device, the fluid would expand to a low-pressure, low-density gas. Besides the advantages of decreased analysis time and improved resolution, SFC offers the possibility to use a wide variety of detectors, including those designed for GC.

SFC has been used primarily for nonpolar compounds. Fats, oils, and other lipids are compounds to which SFC is increasingly applied. For example, the noncaloric fat substitute, Olestra°R, was characterized by SFC-MS (mass spectroscopy). Other researchers have used SFC to detect pesticide residues, study thermally labile compounds from members of the *Allium* genus, fractionate citrus essential oils, and characterize compounds extracted from microwave packaging.

2.4 PHYSICO-CHEMICAL PRINCIPLES OF CHROMATOGRAPHIC SEPARATION

Several physico-chemical principles are involved in chromatography mechanisms employed to separate or fractionate various compounds of interest, regardless of the specific techniques applied. The mechanisms described below apply mainly to liquid chromatography. Although it is more convenient to describe each of these phenomena separately, it must be emphasized that more than one mechanism may be involved in a given fractionation. For example, many cases of partition chromatography also involve adsorption.

2.4.1 Adsorption (Liquid–Solid) Chromatography

Adsorption chromatography is the oldest form of chromatography, originated with Tsvet in 1903 in the experiments that spawned modern chromatography. In this chromatographic mode, the stationary phase is a finely divided solid to maximize the surface area.

The stationary phase (adsorbent) is chosen to permit differential interaction with the components of the sample to be resolved. The intermolecular forces thought to be primarily responsible for chromatographic adsorption include the following:

- van der Waals forces
- Electrostatic forces
- Hydrogen bonds
- Hydrophobic interactions

Sites available for interaction with any given substance are heterogeneous. Binding sites with greater affinities, the most active sites, tend to be populated first, so that additional solutes are less firmly bound. The net result is that adsorption is a concentration-dependent process, and the adsorption coefficient is *not* a constant (in contrast to the partition coefficient). Sample loads exceeding the adsorptive capacity of the stationary phase will result in relatively poor separation. Classic adsorption chromatography utilizes silica (slightly acidic), alumina (slightly basic), charcoal (nonpolar), or a few other materials as the stationary phase. Both silica and alumina possess surface hydroxyl groups, and Lewis acid-type interactions determine their adsorption characteristics. The elution order of compounds from these adsorptive stationary phases can often be predicted on the basis of their relative polarities. Compounds with the most polar functional groups are retained most strongly on polar adsorbents and, therefore, are eluted last.

Nonpolar solutes are eluted first. One model proposed to explain the mechanism of liquid–solid chromatography is that solute and solvent molecules are competing for active sites on the adsorbent. Thus, as relative adsorption of the mobile phase increases, adsorption of the solute must decrease. Solvents can be rated in order of their strength of adsorption on a particular adsorbent, such as silica. Such a solvent strength (or polarity) scale is called a eluotropic series. A eluotropic series for alumina is listed. Silica has a similar rank ordering. Once an adsorbent has been chosen, solvents can be selected from the eluotropic series for that adsorbent. Mobile phase polarity can be increased (often by admixture of more polar solvents) until elution of the compound(s) of interest has been achieved.

Adsorption chromatography separates aromatic or aliphatic nonpolar compounds, based primarily on the type and number of functional groups present. The labile, fat-soluble chlorophyll and carotenoid pigments from plants have been studied extensively by adsorption column chromatography. Adsorption chromatography also has been used for the analysis of fat-soluble vitamins. Frequently, it is used as a batch procedure for removal of impurities from samples prior to other analyses. For example, disposable solid-phase extraction cartridges containing silica have been used for food analyses, such as lipids in soybean oil, carotenoids in citrus fruit, and vitamin E in grain.

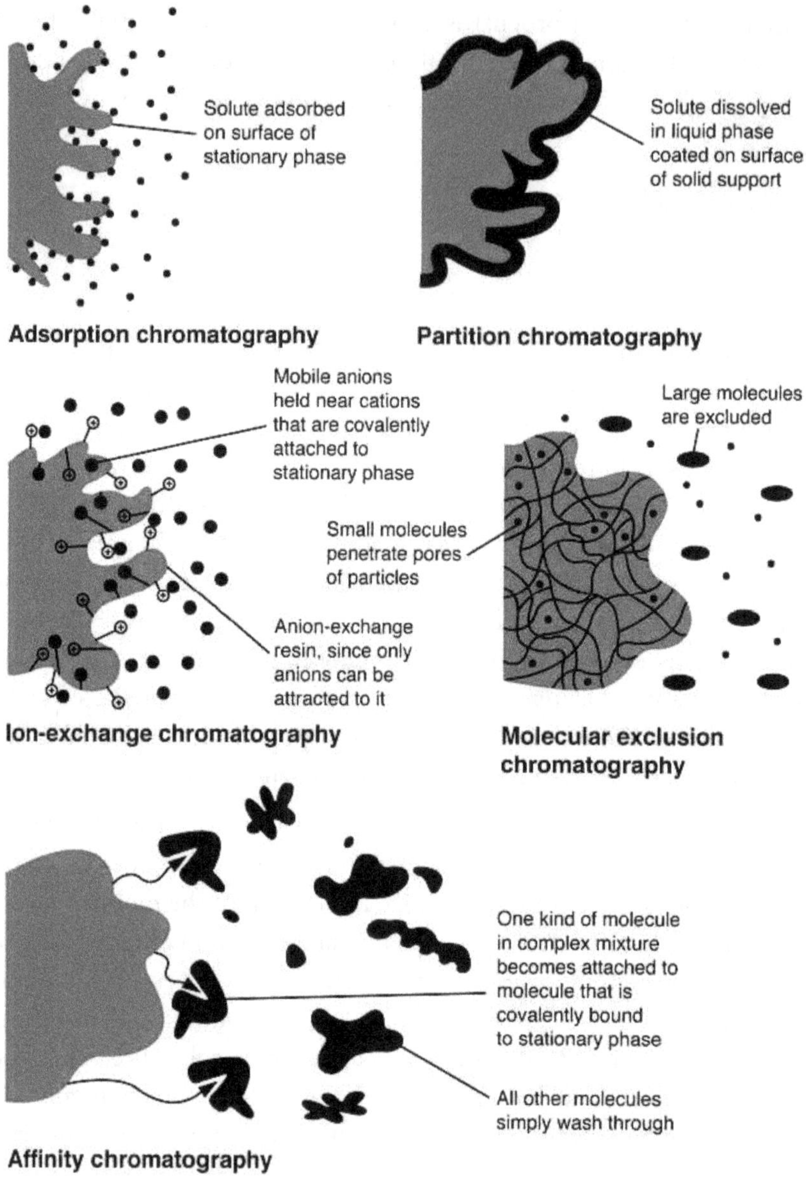

Fig. 2: Adsorption, Partition, Ion-exchange, Molecular exclusion and Affinity chromatographies

2.4.2 Partition (Liquid–Liquid) Chromatography

2.4.2.1 Introduction

In 1941, Martin and Synge undertook an investigation of the amino acid composition of wool, using a counter current extractor of 40 tubes with chloroform and water flowing in opposite directions.

The efficiency of the extraction process was improved enormously when a column of finely divided inert support material was used to hold one liquid phase (stationary phase) immobile, while the second liquid, an immiscible solvent (mobile phase), flowed over it, thus providing intimate contact between the two phases. Solutes partitioned between the two liquid phases according to their partition coefficients, hence the name partition chromatography.

A partition system is manipulated by changing the nature of the two liquid phases, usually by combination of solvents or pH adjustment of buffers. Often, the more polar of the two liquids is held stationary on the inert support and the less polar solvent is used to elute the sample components (normal-phase chromatogra phy). Reversal of this arrangement, using a nonpolar stationary phase and a polar mobile phase, has come to be known as reversed-phase chromatography.

Polar **hydrophilic** substances, such as amino acids, carbohydrates, and water-soluble plant pigments, are separable by **normal-phase** partition chromatography. **Lipophilic** compounds, such as lipids and fat-soluble pigments, and **polyphenols** may be resolved with **reversed-phase** systems. Liquid–liquid partition chromatography has been invaluable to carbohydrate chemistry. Column liquid chromatography on finely divided cellulose has been used extensively in preparative chromatography of sugars and their derivatives. Paper chromatography is a simple method for distinguishing between various forms of sugars (following normal-phase partition chromatography) or phenolic compounds (following reverse-phase partition chromatography) present in foods.

Compounds Class Polarity Scalea	
Fluorocarbons	Esters ≈ ketones ≈ aldehydes
Saturated hydrocarbons Olefins	Alcohols ≈ amines
Aromatics	Amides
Halogenated compounds	Carboxylic acids
Ethers	From (9), used with permission.
Nitro compounds	aListed in order of increasing polarity.

Table Eluotropic Series for Alumina	
Solvent	Methylene chloride
1-Pentane	Tetrahydrofuran
Isooctane	Acetone
Cyclohexane	Ethyl acetate
Carbon tetrachloride	Aniline
Xylene	Acetonitrile
Toluene	2-Propanol
Benzene	Ethanol
Ethyl ether	Methanol
Chloroform	Acetic acid

2.4.2.2 Coated Supports

In its simplest form, the stationary phase for partition chromatography consists of a liquid coating on a solid matrix. The solid support should be as inert as possible and have a large surface area in order to maximize the amount of liquid held. Some examples of solid supports that have been used are silica, starch, cellulose powder, and glass beads. All are capable of holding a thin film of water, which serves as the stationary phase. It is important to note that materials prepared for adsorption chromatography must be activated by drying them to remove surface water. Conversely, some of these materials, such as silica gel, may be used for partition chromatography if they are deactivated by impregnation with water or the desired stationary phase. One disadvantage of liquid–liquid chromatographic systems is that the liquid stationary phase is often stripped off. This problem can be overcome by chemically bonding the stationary phase to the support material, as described in the next section.

2.4.2.3 Bonded Supports

The liquid stationary phase may be covalently attached to a support by a chemical reaction. These bonded phases have become very popular for HPLC use, and a wide variety of both polar and nonpolar stationary phases is now available. Especially widely used is reversed-phase, with a nonpolar bonded stationary phase (e.g., silica covered with C8 or C18 groups) and a polar solvent (e.g., water– acetonitrile). It is important to note that mechanisms other than partition may be involved in the separation using bonded supports. Bonded-phase HPLC columns have greatly facilitated the analysis of vitamins in foods and feeds. Additionally, bonded-phase HPLC is widely used for the separation and identification of polyphenols such as phenolic acids (e.g., *p*-coumaric, caffeic, ferulic, and sinapic acids) and flavonoids (e.g., flavonols, flavones, isoflavones, anthocyanidins, catechins, and proanthocyanidins).

2.4.3 Ion-Exchange Chromatography

Ion exchange is a separation/purification process occurring naturally, for example, in soils and is utilized in water softeners and deionizers. Three types of separation may be achieved: (1) ionic from nonionic, (2) cationic from anionic, and (3) mixtures of similarly charged species. In the first two cases, one substance binds to the ion-exchange medium, whereas the other substance does not. Batch extraction methods can be used for these two separations; however, chromatography is needed for the third category.

The basis of ion-exchange chromatography.

a. Schematic of the ion-exchange process;
b. Ionic equilbria for cation- and anion-exchange processes.

a

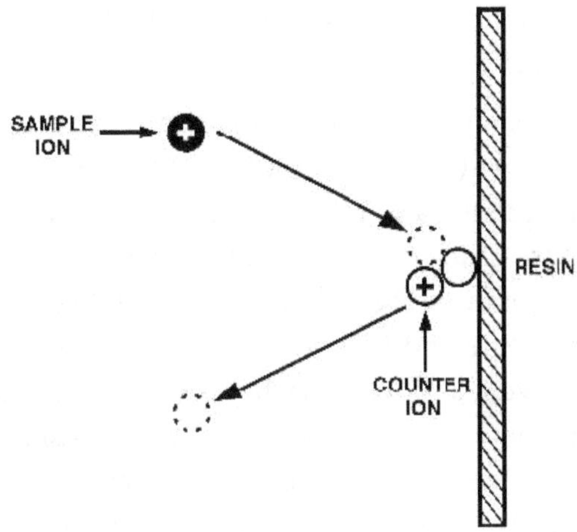

b

$$M^+ + (Na^{+\,-}O_3S-\text{Resin}) \rightleftharpoons (M^{+\,-}O_3S-\text{Resin}) + Na^+$$

M^+ = Cation

$$X^- + (Cl^{-\,+}R_4N-\text{Resin}) \rightleftharpoons (X^{-\,+}R_4N-\text{Resin}) + Cl^-$$

X^- = Anion

Type of adsorption chromatography in which interactions between solute and stationary phase are primarily electrostatic in nature. The stationary phase (ion exchanger) contains fixed functional groups that are either negatively or positively charged.

Exchangeable counterions preserve charge neutrality. A sample ion (or charged sites on large molecules) can exchange with the counterion to become the partner of the fixed charge. Ionic equilibrium is established. The functional group of the stationary phase determines whether cations or anions are exchanged. Cation exchangers contain covalently bound negatively charged functional groups, whereas anion exchangers contain bound positively charged groups. The chemical nature of these acidic or basic residues determines how stationary-phase ionization is affected by the mobile-phase pH.

The strongly acidic sulfonic acidmoieties (RSO3–) of "strong"-cation exchangers are completely ionized at all pH values above 2. Strongly basic quaternary amine groups (RNR'3 +) on "strong"-anion exchangers are ionized at all pH values below 10. Since maximum negative or positive charge is maintained over a broad pH range, the exchange or binding capacity of these stationary phases is essentially constant, regardless of mobile-phase pH. "Weak"-cation exchangers contain weakly acidic carboxylic acid functional groups, (RCO2–); consequently, their exchange capacity varies considerably between ca. pH 4 and 10.

Weakly basic anion exchangers possess primary, secondary, or tertiary amine residues (R-NHR'2 +), which are deprotonated in moderately basic solution, thereby losing their positive charge and the ability to bind anions. Thus, one way of eluting solutes bound to an ion-exchange medium is to change the mobile-phase pH. A second way to elute bound solutes is to increase the ionic strength (e.g., use NaCl) of the mobile phase, to weaken the electrostatic interactions.

Chromatographic separations by ion exchange are based upon differences in affinity of the exchangers for the ions (or charged species) to be separated. The factors that govern selectivity of an exchanger for a particular ion include the ionic valence, radius, and concentration; the nature of the exchanger (including its displaceable counterion); and the composition and pH of the mobile phase. To be useful as an ion exchanger, a material must be both ionic in nature and highly permeable. Synthetic ion exchangers are thus cross-linked polyelectrolytes, and they may be inorganic (e.g., aluminosilicates) or, more commonly, organic compounds. Polystyrene, made by crosslinking styrene with divinyl benzene (DVB), may be modified to produce either anion- or cation-exchange resins. Polymeric resins such as these are commercially available in a wide range of particle sizes and with different degrees of cross-linking (expressed as weight percent of DVB in the mixture). The extent of cross-linking controls the rigidity and porosity of the resin, which, in turn, determines its optimal use.

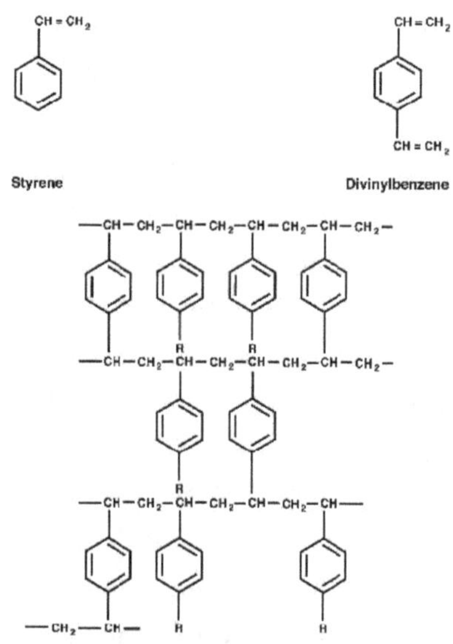

Crosslinked styrene-divinylbenzene copolymer

R = H, Plain polystyrene

R = $CH_2N^+(CH_3)_3Cl^-$, Anion-exchanger

R = $SO_3^-H^+$, Cation-exchanger

Lightly cross-linked resins permit rapid equilibration of solute, but particles swell in water, thereby decreasing charge density and selectivity (relative affinity) of the resin for different ions. More highly cross-linked resins exhibit less swelling, higher exchange capacity, and selectivity, but longer equilibration times. The small pore size, high charge density, and inherent hydrophobicity of the older ion-exchange resins have limited their use to small molecules [molecular weight (MW) <500]. Ion exchangers based on polysaccharides, such as cellulose, dextran, or agarose, have proven very useful for the separation and purification of large molecules, such as proteins and nucleic acids. These materials, called gels, are much softer than polystyrene resins, and thus may be derivatized with strong or with weak acidic or basic groups via OH moieties on the polysaccharide backbone. They have much larger pore sizes and lower charge densities than the older synthetic resins. Food-related applications of ion-exchange chromatography include the separation of amino acids, sugars, alkaloids, and proteins. Fractionation of amino acids in protein hydrolyzates was initially carried out

Chemical structure of polystyrene-based ion-exchange resins. by ion-exchange chromatography; automation of this process led to the development of commercially produced amino acid analyzers. Many drugs, fatty acids, and the acids of fruit, being ionisable compounds, may be chromatographed in the ion-exchange mode.

2.4.4 Size-Exclusion Chromatography

Size-exclusion chromatography (SEC), also known as molecular exclusion, gel permeation (GPC), and gel-filtration chromatography (GFC), is probably the easiest mode of chromatography to perform and to understand. It is widely used in the biological sciences for the resolution of macromolecules, such as proteins and carbohydrates, and also is used for the fractionation and characterization of synthetic polymers. Unfortunately, nomenclature associated with this separation mode developed independently in the literature of the life sciences and in the field of polymer chemistry, resulting in inconsistencies.

In the ideal SEC system, molecules are separated solely on the basis of their size; no interaction occurs between solutes and the stationary phase. In the event that solute/support interactions do occur, the separation mode is termed nonideal SEC. The stationary phase in SEC consists of a column packing material that contains pores comparable in size to the molecules to be fractionated. Solutes too large to enter the pores travel with the mobile phase in the interstitial space (between particles) outside the pores. Thus, the largest molecules are eluted first from an SEC column. The volume of the mobile phase in the column, termed the column void volume, Vo, can be measured by chromatographing a very large (totally excluded) species, such as Blue Dextran, a dye of MW = 2 × 106.

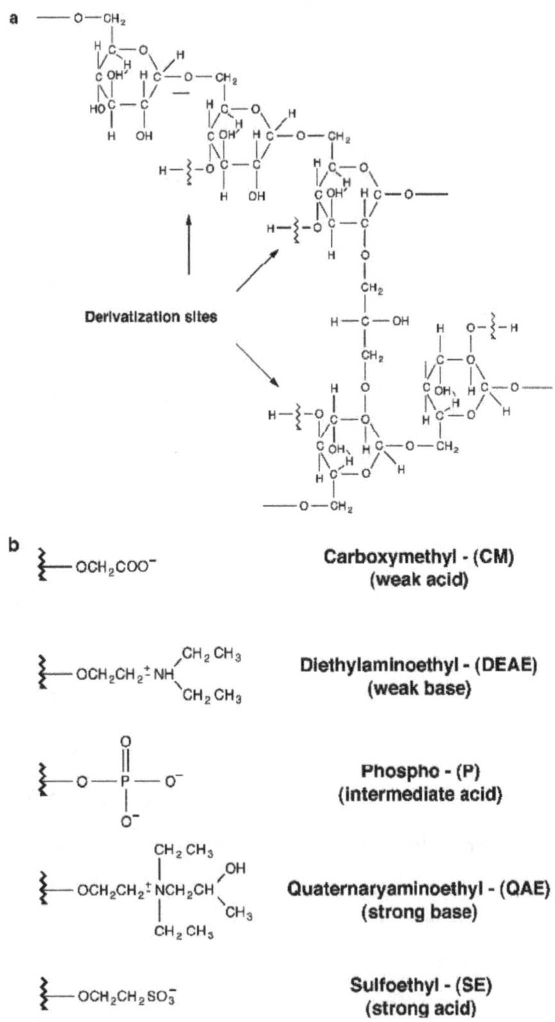

Chemical structure of one polysaccharide-based ion-exchange resin. (**a**): Matrix of cross-linked dextran (**b**): functional groups that may be used to impart ion-exchange properties to the matrix.

As solute dimensions decrease, approaching those of the packing pores, molecules begin to diffuse into the packing particles and, consequently, are slowed down. Solutes of low molecular weight (e.g., glycyltyrosine) that have free access to all the available pore volume are eluted in the volume referred to as *V*t. This value, *V*t, which is equal to the column void volume *V*o, plus the volume of liquid inside the sorbent pores, *V*i, is referred to as the total permeation volume of the packed column (*V*t = *V*o + *V*i). Solutes are ideally eluted between the void volume and the total liquid volume of the column. Because this volume is limited, only a relatively small number of solutes (ca. 10) can be completely resolved by SEC under ordinary conditions.

The behavior of a molecule in a size-exclusion column may be characterized in several different ways. Each solute exhibits an **elution volume**, *V*e. However, *V*e depends on column dimensions and the way in which the column was packed.

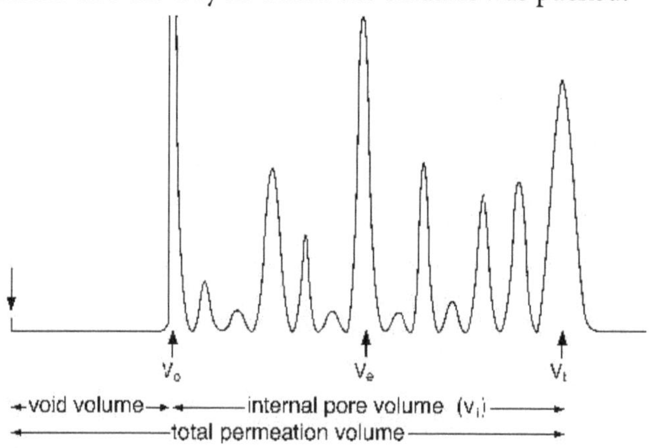

The available partition coefficient is used to define solute behavior independent of these variables:

$$Kav = (Ve - Vo)/(Vt - Vo) \quad [3]$$

where:

*K*av = available partition coefficient

*V*e = elution volume of solute

*V*o = column void volume

*V*t = total permeation volume of column

The value of *K*av calculated from experimental data for a solute chromatographed on a given SEC column defines the proportion of pores that can be occupied by that molecule. For a large, totally excluded species, such as Blue Dextran or DNA, *V*e = *V*o and *K*av = 0.

For a small molecule with complete access to the internal pore volume, such as glycyltyrosine, *V*e = *V*t and *K*av = 1. For each size-exclusion packing material, a plot of *K*av vs. the logarithm of the molecular weight for a series of solutes, similar in molecular shape and density, will give an S-shaped curve . In the case of proteins, *K*av is actually better related to the **Stokes radius**, the average radius of the protein in solution. The central, linear portion of this curve describes the **fractionation range** of the matrix, wherein maximum separation among solutes of similar molecular weight is achieved. This correlation between solute elution behavior and molecular weight (or size) forms the basis for a widely used method for characterizing large molecules such as proteins and polysaccharides. A size-exclusion column is calibrated with a series of solutes of known molecular weight (or Stokes radius) to obtain a curve similar. The value of *K*av for the unknown is then determined, and an estimate of molecular weight (or size) of the unknown is made by interpolation of the calibration curve.

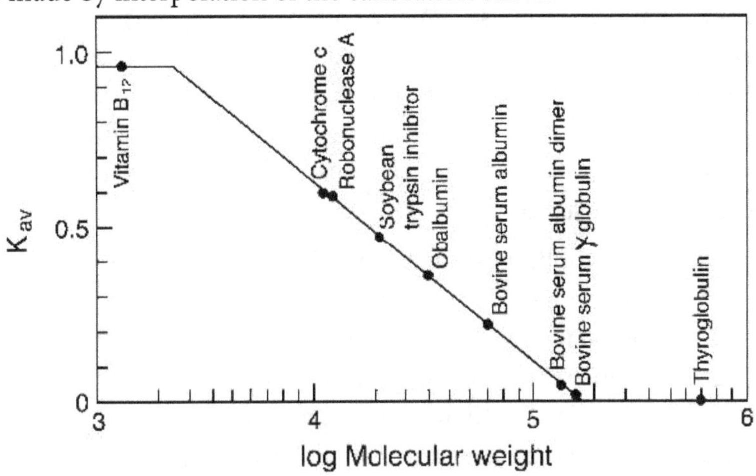

Relationship between *K*av and log (molecular weight) for globular proteins chromatographed on a column of Sephadex G-150 Superfine.

Two groups: semirigid, **hydrophobic media**, and soft, **hydrophilic gels**. The former are usually derived from **polystyrene** and are used with organic mobile phases (GPC or nonaqueous SEC) for the separation of polymers, such as rubbers and plastics. Soft gels, **polysaccharide**-based packings, are typified by Sephadex, a cross-linked dextran .

These materials are available in a wide range of pore sizes and are useful for the separation of water-soluble substances in the molecular weight range 1–2.5 × 107. In selecting an SEC column packing, both the purpose of the experiment and size of the molecules to be separated must be considered. If the purpose of the experiment is group separation, where molecules of widely different molecular sizes need to be separated, a matrix is chosen such that the larger molecules, e.g., proteins, are eluted in the void volume of the column, whereas small molecules are retained in the total volume. A common example of group separation is buffer exchange and desalting. When SEC is used for separation of macromolecules of different sizes, the molecular sizes of all the components must fall within the fractionation range of the gel. As discussed previously, SEC can be used, directly, to fractionate mixtures or, indirectly, to obtain information about a dissolved species. In addition to molecular weight estimations, SEC is used to determine the molecular weight distribution of natural and synthetic polymers, such as dextrans and gelatin preparations. Fractionation of biopolymer mixtures is probably the most widespread use of SEC, since the mild elution conditions employed rarely cause denaturation or loss of biological activity. It is also a fast, efficient alternative to dialysis for desalting solutions of macromolecules, such as proteins.

2.4.5 Affinity Chromatography

Affinity chromatography is unique in that separation is based on the specific, reversible interaction between a solute molecule and a ligand immobilized on the chromatographic stationary phase. While discussed here as a separate type of chromatography, affinity chromatography could be viewed as the ultimate extension of adsorption chromatography. Although the basic concepts of so-called biospecific adsorption were known as early as 1910, they were not perceived as potentially useful laboratory tools until ca. 1968. Affinity chromatography usually involves immobilized biological materials as the stationary phase.

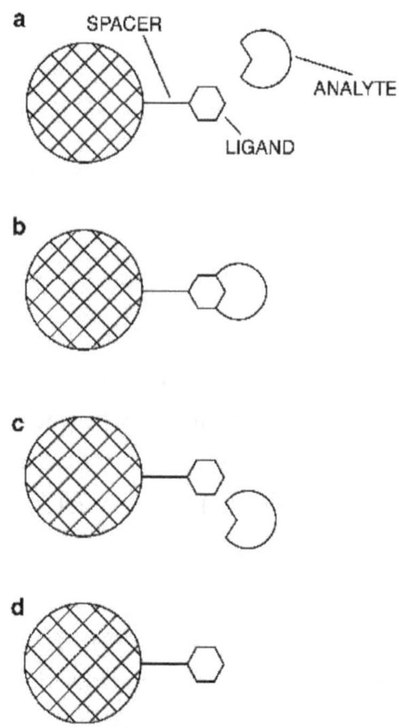

These **ligands** can be antibodies, enzyme inhibitors, lectins, or other molecules that selectively and reversibly bind to complementary analyte molecules in the sample. Separation exploits the lock and key binding of biological systems. Although both ligands and the species to be isolated are usually biological macromolecules, the term affinity chromatography also encompasses other systems, such as separation of small molecules containing *cis*-diol groups via phenylboronic acid moieties on the stationary phase.

A ligand, chosen based on its specificity and strength of interaction with the molecule to be isolated (analyte), is immobilized on a suitable support material. As the sample is passed through this column, molecules that are complementary to the bound ligand are adsorbed while other sample components are eluted. Bound analyte is subsequently eluted via a change in the mobile-phase composition as will be discussed below. After reequilibration with the initial mobile phase, the stationary phase is ready to be used again. The ideal support for Principles of bioselective affinity chromatography.

a. The support presents the immobilized ligand to the analyte to be isolated.
b. The analyte makes contact with the ligand and attaches itself.
c. The analyte is recovered by the introduction of an eluent, which dissociates thecomplex holding the analyte to the ligand.
d. The support is regenerated, ready for the next isolation.

Affinity chromatography should be a porous, stable, high-surface-area material that does not adsorb anything itself. Thus, polymers such as agarose, cellulose, dextran, and polyacrylamide are used, as well as controlled-pore glass. Affinity ligands are usually attached to the support or matrix by covalent bond formation, and optimum reaction conditions often must be found empirically. Immobilization generally consists of two steps: **activation** and **coupling**. During the activation step, a reagent reacts with functional groups on the support, such as hydroxyl moieties, to produce an activated matrix. After removal of excess reagent, the ligand is coupled to the activated matrix. (Preactivated supports are commercially available, and their availability has greatly increased the use of affinity chromatography.) The coupling reaction most often involves free amino groups on the ligand, although other functional groups can be used. When small molecules such as phenylboronic acid are immobilized, a **spacer arm** (containing at least four to six methylene groups) is used to hold the ligand away fromthe support surface, enabling it to reach into the binding site of the analyte. Ligands for affinity chromatographymay be either **specific** or **general** (i.e., group specific). Specific ligands, such as antibodies, bind only one particular solute. General ligands, such as nucleotide analogs and lectins, bind to certain classes of solutes. For example, the lectin concanavalin A binds to all molecules that contain terminal glucosyl and mannosyl residues.

Bound solutes then can be separated as a group or individually, depending upon the elution technique used.

Elution methods for affinity chromatography may be divided into nonspecific and (bio)specific methods. Nonspecific elution involves disrupting ligand analyte binding by changing the mobile-phase pH, ionic strength, dielectric constant, or temperature. If additional selectivity in elution is desired, for example, in the case of immobilized general ligands, a biospecific elution technique is used. Free ligand, either identical to or different from the matrix-bound ligand, is added to the mobile phase. This free ligand competes for binding sites on the analyte. For example, glycoproteins bound to a concanavalin A (lectin) column can be eluted by using buffer containing an excess of lectin. In general, the eluent ligand should display greater affinity for the analyte of interest than the immobilized ligand.

In addition to protein purification, affinity chromatography may be used to separate supramolecular structures such as cells, organelles, and viruses; concentrate dilute protein solutions; investigate binding mechanisms; and determine equilibrium constants Affinity chromatography has been useful especially in the separation and purification of enzymes and glycoproteins. In the case of the latter, carbohydratederivatized adsorbents are used to isolate specific lectins, such as concanavalin A, and lentil or wheatgerm lectin. The lectin then may be coupled to agarose, such as concanavalin A- or lentil lectin-agarose, to provide a stationary phase for the purification of specific glycoproteins, glycolipids, or polysaccharides.

2.5 ANALYSIS OF CHROMATOGRAPHIC PEAKS

Once the chromatographic technique and chromatographic mechanism have been chosen, the analyst has to ensure adequate separation of constituents of interests from a mixture, in a reasonable amount of time. After separation is achieved and chromatographic peaks are obtained, qualitative as well as quantitative analysis can be carried out. Basic principles of separation and resolution will be discussed in the subsequent sections. Understanding these principles allows the analyst to optimize separation and perform qualitative and quantitative analysis.

2.5.1 Separation and Resolution

2.5.1.1 Developing a Separation

There may be numerous ways to accomplish a chromatographic separation for a particular compound. In many cases, the analyst will follow a standard laboratory procedure or published methods. In the case of a sample that has not been previously analyzed, the analyst begins by evaluating what is known about the sample and defines the goals of the separation. How many components need to be resolved? What degree of resolution is needed? Is qualitative or quantitative information needed? Molecular weight (or molecular weight range), polarity, and ionic character of the sample will guide the choice of chromatographic separation mechanism (separation mode). For example, small ionic compounds may be separated by ion-exchange, ion-pair reversed-phase, or reversed-phase LC. In this case, the analyst's choice may be based on convenience, experience, and personal preference. Having chosen a separation mode for the sample at hand, one must select an appropriate stationary phase, elution conditions, and a detection method. Trial experimental conditions may be based on the results of a literature search, the analyst's previous experience with similar samples, or general recommendations from chromatography experts.

To achieve separation of sample components by all modes except SEC, one may utilize either isocratic or gradient elution. Isocratic elution is the most simple and widely used technique, in which solvent composition and flow rate are held constant. Gradient elution involves reproducibly varying mobile phase composition or flow rate (flow programming) during the LC analysis. Gradient elution is used when sample components possess a wide range of polarities, so that an isocratic mobile phase does not elute all components within a reasonable time. The change may be continuous or stepwise. Gradients of increasing ionic strength are extremely valuable in ion-exchange chromatography. Gradient elution is commonly used for desorbing large molecules, such as proteins, which can undergo multiple-site interaction with a stationary phase. Increasing the "strength" of the mobile phase, either gradually or in a stepwise fashion, shortens the analysis time.

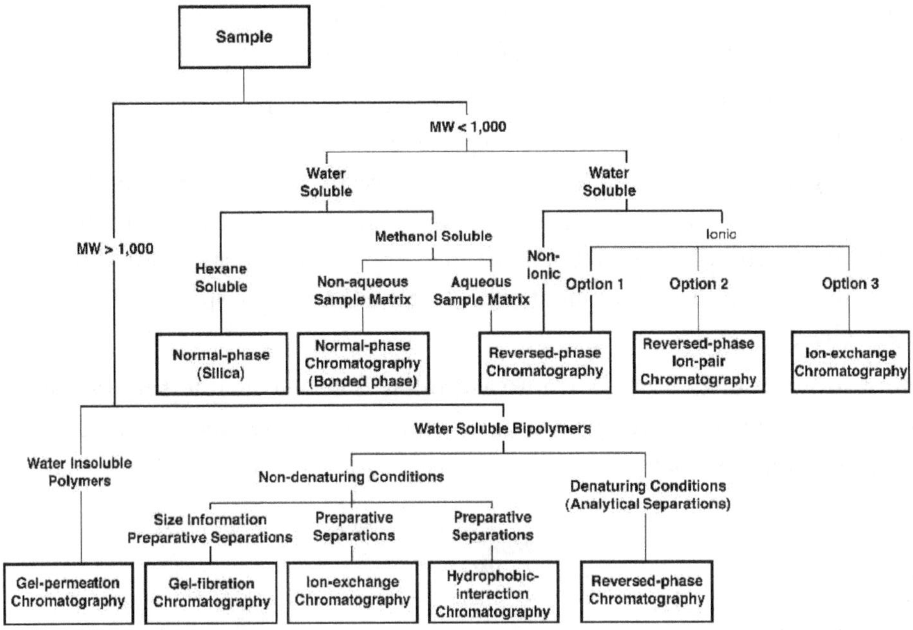

C

A schematic for choosing a chromatographic separation mode based on sample molecular weight and solubility.

Method development may begin with an is ocratic mobile phase, possibly of intermediate solvent strength; however, using gradient elution for the initial separation may ensure that some level of separation is achieved within a reasonable time period and nothing is likely to remain on the column. Data from this initial run allow one to determine if is ocratic or gradient elution is needed, and to estimate optimal is ocratic mobile phase composition or gradient range. The use of a gradient run does not presuppose that the final method will use gradient elution.

Once an initial separation has been achieved, the analyst can proceed to optimize resolution. This generally involves manipulation of mobile phase variables, including the nature and percentage of organic components, pH, ionic strength, nature and concentration of additives (such as ion-pairing agents), flow rate, and temperature. In the case of gradient elution, gradient steepness (slope) is another variable to be optimized. However, the analyst must be aware of the principles of chromatographic resolution as will be discussed in the following section.

2.5.1.2 Chromatographic Resolution

2.5.1.2.1 Introduction

The main goal of chromatography is to segregate components of a sample into separate bands or peaks as they migrate through the column. A **chromatographic peak** is defined by several parameters including **retention time, peak width**, and **peak height** . The volume of the mobile phase required to elute a compound from an LC column is called the **retention volume**, *VR*.

The associated time is the **retention time**, *t*R. Shifts in retention time and changes in peak width greatly influence **chromatographic resolution.**

Differences in column dimensions, loading, temperature, mobile phase flow rate, system deadvolume, and detector geometrymay lead to discrepancies in retention time. By subtracting the time required for the mobile phase or a nonretained solute (*t*m or *t*o) to travel through the column to the detector, one obtains an **adjusted retention time**, *t*'R (or volume). The adjusted retention time (or volume) corrects for differences in system deadvolume; it may be thought of as the time the sample spends adsorbed on the stationary phase.

The **resolution** of two peaks from each other is related to the **separation factor**, *a*. Values for *a* depend on temperature, the stationary phase, and mobile phase used. Resolution is defined as follows:

$$Rs \frac{2\Delta t}{W2+W1}$$

Where:

R_S = resolution

Dt = difference between retention times of peaks 1 and 2

w_2 = width of peak 2 at baseline

w_1 = width of peak 1 at baseline

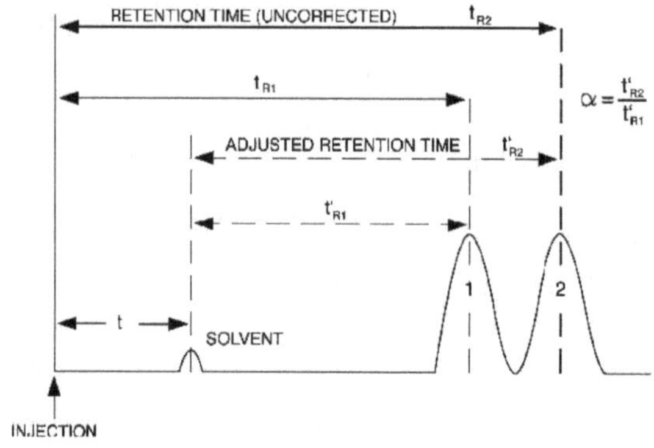

Fig. 12: illustrates the measurement of peak width [part (a)] and the values necessary for calculating resolution [part (b)].

Units, i.e., time or volume.)

Measurement of chromatographic retention.

Mathematically, thisrelationshipisexpressedasfollows:

$$R_s = \underbrace{1/4\sqrt{N}}_{a}\underbrace{\left(\frac{\alpha-1}{\alpha}\right)}_{b}\underbrace{\left(\frac{k'}{k'+1}\right)}_{c} \qquad [5]$$

Where:

a = column efficiency term
b = column selectivity term
c = capacity term

These terms, and factors that contribute to them, will be discussed in the following sections.

2.5.1.2.2 Column Efficiency

If faced with the problem of improving resolution, a chromatographer should first examine the efficiency of the column. An efficient column keeps the bands from spreading and gives narrow peaks. Column efficiency can be calculated by:

$$N\left(\frac{t_R}{\sigma}\right)^2 = 16\left(\frac{t_R}{w}\right)^2 = 5.5\left(\frac{t_R}{w_{1/2}}\right)^2 \qquad [6]$$

Where:

N = number of theoretical plates
tR = retention time
s = standard deviation for a Gaussian peak
w = peak width at baseline (w = 4s)
w1/2 = peak width at half height

The measurement of tR, w, and w1/2 is illustrated in

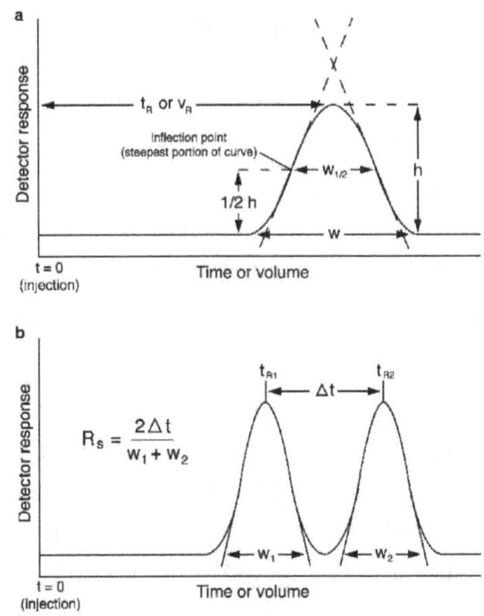

Fig.13: (Retention volume may be used instead of tR; in this case, band width is also measured in units of volume.) Although some peaks are not actually Gaussian in shape, normal practice is to treat them as if they were. In the case of peaks that are incompletely resolved or slightly asymmetric, peak width at half height is more accurate than peak width at baseline.

The value N calculated from the above equation is called the number of theoretical plates. The theoretical plate concept, borrowed from distillation theory, can best be understood by viewing chromatography as a series of equilibrations between mobile and stationary phases, analogous to counter current distribution.

Thus, a column would consist of N segments (theoretical plates) with one equilibration occurring in each. As a first approximation, N is independent of retention time and is therefore a useful measure of column performance. One method of monitoring column performance over time is to chromatograph a standard compound periodically, under constant conditions, and to compare the values of N obtained. It is important to note that columns often behave as if they have a different number of plates for different solutes in a mixture. Different solutes have different partition coefficient and thus have distinctive series of equilibrations between mobile and stationary phases. Band broadening due to column deterioration will result in a decrease of N for a particular solute. Band broadening is a result of an extended time for a solute to reach equilibrium between mobile and stationary phases.

The number of theoretical plates is generally proportional to column length. Since columns are available in various lengths, it is useful to have a measure of column efficiency that is independent of column length. This may be expressed as follows:

HETP =

here:

HETP = height equivalent to a theoretical plate

L = column length

N = number of theoretical plates

The so-called HETP is sometimes more simply described as plate height (H). If a column consisted of discrete segments, HETP would be the height of each imaginary segment. Small plate height values (a large number of plates) indicate good efficiency of separation. Conversely, reduced number of plates results in poor separation due to the extended equilibrium time in a deteriorating column.

In reality, columns are not divided into discrete segments and equilibration is not infinitely fast. The plate theory is used to simplify the equilibration concept. The movement of solutes through a chromatography column takes into account the finite rate at which a solute can equilibrate itself between stationary and mobile phases. Thus, band shape depends on the rate of elution and is affected by solute diffusion. Any mechanism that causes a band of solute to broaden will increase HETP and decrease column efficiency. The various factors that contribute to plate height are expressed by the Van Deemter equation:

$$HETP = \frac{A+B}{u+} \quad Cu \qquad [8]$$

Where:

HETP = height equivalent to a theoretical plate

A, B, C = constants

u = mobile phase rate

The constants A, B, and C are characteristic for a given column, mobile phase, and temperature. The A term represents the eddy diffusion or multiple flow paths. Eddy diffusion refers to the different microscopic flow streams that the mobile phase can take between particles in the column (analogous to eddy streams around rocks in a brook). Sample molecules can thus take different paths as well, depending on which flow streams they follow. As a result, solute molecules spread from an initially narrow band to a broader area within the column. Eddy diffusion may be minimized by good column packing techniques and the use of small diameter particles of narrow particle size distribution.

The B term of the Van Demeter equation, sometimes called the longitudinal diffusion term, exists because all solutes diffuse from an area of high concentration (the center of a chromatographic band) to one of low concentration (the leading or trailing edge of a chromatographic band). In LC, the contribution of this term to HETP is small except at low flow rate of the mobile phase. With slow flow rates there will be more time for a salute to spend on the column, thus its diffusion will be greater.

The C (mass transfer) term arises from the finite time required for solute to equilibrate between the mobile and stationary phases. Mass transfer is practically the partitioning of the solute into the stationary phase, which does not occur instantaneously and depends on the solute's partition and diffusion coefficients. If the stationary phase consists of porous particles, a sample molecule entering a pore ceases to be transported by the solvent flow and moves by diffusion only. Subsequently, this solute molecule may diffuse back to the mobile phase flow or it may interact with the stationary phase. In either case, solute molecules inside the pores are slowed down relative to those outside the pores and band broadening occurs. Contributions to HETP from the C term can be minimized by using porous particles of small diameter or pellicular packing materials.

As expressed by the Van Deemter equation, mobile phase flow rate, u, contributes to plate height in opposing ways – increasing the flow rate in crases the equilibration point ($Au1/3$ and Cu), but decreases longitudinal diffusion of the solute particles (B/u). A Van Deemter plot may be used to determine the mobile phase flow rate at which plate height is minimized and column efficiency is maximized. Flow rates above the optimum may be used to decrease analysis time if adequate resolution is still obtained. However, at very high flow rates, there will be less time to approach equilibrium, which will lead to broadening of the band. In addition to flow rate, temperature can affect the longitudinal diffusion and the mass transfer. Increasing the temperature causes enhanced movement of the solute between the mobile phase and the stationary phase, and within the column, thus leading to faster elution and narrower peaks.

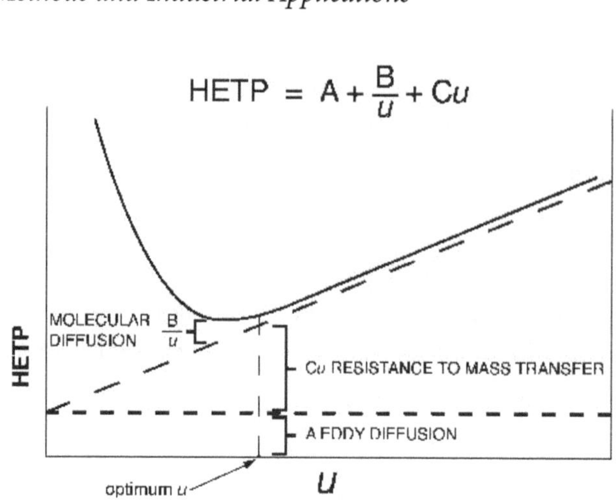

Fig. 13: Van Demeter plot of column efficiency (HETP) vs. mobile phase rate (u). Optimum u is noted.

27.5.1.2.3 Column Selectivity

Chromatographic resolution depends on column selectivity as well as efficiency. Column selectivity refers to the distance, or relative separation, between two peaks and is given by:

$$\alpha = \frac{t_{R2} - t_o}{t_{R1} - t_o} = \frac{t'_{R2}}{t'_{R1}} = \frac{K_2}{K_1} \qquad [9]$$

Where:

a = separation factor

t_{R1} and t_{R2} = retention times of components l and 2, respectively

to (or tm) = retention time of unretained components (solvent front)

t'_{R1} and t'_{R2} = adjusted retention times of components l and 2, respectively

K_1 and K_2 = distribution coefficients of components l and 2, respectively

Retention times (or volumes) are measured as shown in

The time, to, can be measured by chromatographing a solute that is not retained under the separation conditions (i.e., travels with the solvent front). When this parameter is expressed in units of volume, Vo or Vm, it is known as the dead-volume of the system. Selectivity is a function of the stationary and/or mobile phase. For example, selectivity in ion-exchange chromatography is influenced by the 14 nature and number of ionic groups on the matrix but also can be manipulated via pH and ionic strength of the mobile phase. Good selectivity is probably more important to a given separation than high efficiency (Fig. 14), since resolution is directly related to selectivity but is quadratically related to efficiency; thus a fourfold increase in N is needed to double RS (Equation [5]).

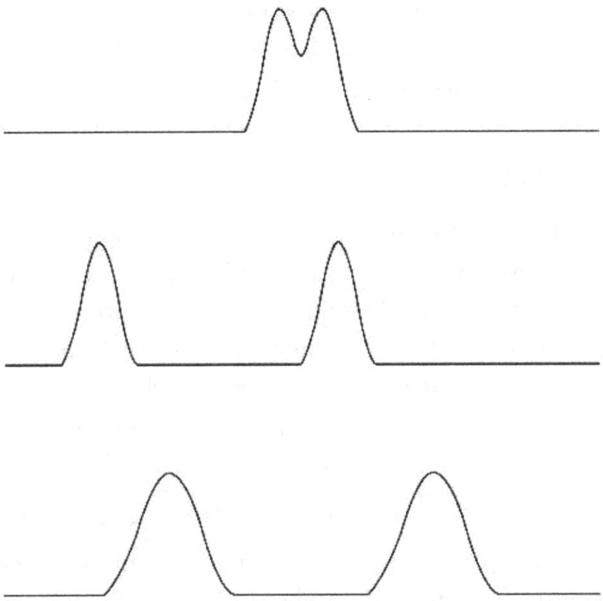

Chromatographic resolution: efficiency vs. selectivity. (a) Poor resolution; (b) good resolution due to high column efficiency; (c) good resolution due to column selectivity.

2.5.1.2.4 Column Capacity

Factor The capacity or retention factor, k′, is a measure of the amount of time a chromatographed species (solute) spends in/on the stationary phase relative to the mobile phase. The relationship between capacity factors and chromatographic retention (which may be expressed in units of either volume or time) is shown below:

$$k' = \frac{KV_s}{V_m} = \frac{V_R - V_m}{V_m} = \frac{t_R - t_o}{t_o} \qquad [10]$$

where:

k'= capacity factor

K = distribution coefficient of the solute

Vs = volume of stationary phase in column

Vm = volume of mobile phase

VR = retention volume of solute

tR = retention time of solute

to = retention time of unretained components (solvent front)

Small values of k′ indicate little retention, and components will be eluted close to the solvent front, resulting in poor separations. Overuse or misuse of the column may lead to the loss of some functional groups, thus resulting in small k′ values. Large values of k′ result in improved separation but also can lead to broad peaks and long analysis times. On a practical basis, k′ values within the range of 1–15 are generally used. (In the equation for Rs, k′ is actually the average of k1′ and k2′ for the two components separated.)

2.5.2 Qualitative Analysis

Once separation and resolution have been optimized, identification of the detected compounds can be achieved. Comparing *V*R or *t*R to that of standards chromatographed under identical conditions often enables one to identify an unknown compound. When it is necessary to compare chromatograms obtained from two different systems or columns, it is better to compare **adjusted retention time,** *t′*R . Different compounds may have identical retention times. In other words, even if the retention time of an unknown and a standard are equivalent, the two compounds might not be identical.

Therefore, other techniques are needed to confirm peak identity. For example:

1. Spike the unknown sample with a known compound and compare chromatograms of the original and spiked samples to see which peak has increased. Only the height of the peak of interest should increase, with no change in retention time, peak width, or shape.
2. A diode array detector can provide absorption spectra of designated peaks (see Sects. 23.2.6 and 28.2.4.1). Although identical spectra do not prove identity, a spectral difference confirms that sample and standard peaks are different compounds.
3. In the absence of spectral scanning capability, other detectors, such as absorption or fluorescence, may be used in a ratioing procedure. Chromatograms of sample and standard are monitored at each of two different wavelengths. The ratio of peak areas at these wavelengths should be the same if sample and standard are identical.
4. Peaks of interest can be collected and subjected to additional chromatographic separation using a different separation mode.
5. Collect the peak(s) of interest and establish their identity by another analytical method (e.g., mass spectrometry, which can give a mass spectrum that is characteristic of a particular compound; see Chap. 26).

27.5.3 Quantitative Analysis

Assuming that good chromatographic resolution and identification of sample components have been achieved, quantification involves measuring peak height, area, or mass and comparing these data with those for standards of known concentration. When strip chart recorders were commonly used, measurement of the peak height, area (usually using: peak area = width at half height × height) or mass was done manually. This was followed by standalone integrators that produced a chromatogram on a strip of paper and digitally integrated the peaks by area for a tabular report. Nearly all chromatography systems (especially GC and HPLC systems) now use data analysis software, which recognizes the start, maximum, and end of each chromatographic peak, even when not fully resolved from other peaks. These values then are used to determine retention times and peak areas. At the end of each run, a report is generated that lists these data and postrun calculations, such as relative peak areas, areas as percentages of the total area, and relative retention times. If the system has been standardized, data fromexternal or internal standards can be used to calculate analyte concentrations.

Data analysis software to quantify peaks is not as common in low-pressure, preparative chromatography, in which postchromatography analysis of collected fractions is used to identify samples eluted. Examples of postchromatography analysis include the BCA (bicinchoninic acid) protein assay and the phenol-sulfuric acid assay for carbohydrate . After obtaining the absorbance reading on a spectrophotometer for such assays, the results are plotted as fraction number on the x-zxis and absorbance on the y-axis, to determine which fractions contain protein and/or carbohydrate.

Having quantified sample peaks, one must compare these data with appropriate standards of known concentration to determine sample concentrations. Comparisons may be by means of external or internal standards. Comparison of peak height, area, or mass of unknown samples with standards injected separately (i.e., external standards) is common practice.

Standard solutions covering the desired concentration range (preferably diluted from one stock solution) are chromatographed, and the appropriate data (peak height, area, or mass) plotted vs. concentration to obtain a standard curve. An identical volume of sample is then chromatographed, and height, area, ormass of the sample peak is used to determine sample concentration via the standard curve (Fig. 15a). This absolute calibration method requires precise analytical technique and requires that detector sensitivity be constant from day to day if the calibration curve is to remain valid. Calibration curves for quantification of a sample component, x. (a): External standard technique; (b): internal standard technique.

Use of the internal standard (relative or indirect) method can minimize errors due to sample preparation, apparatus, and operator technique. In this technique, a compound is utilized that is structurally related to, but is eluted independently of, compounds of interest in the sample to be analyzed. Basically, the amount of each component in the sample is determined by comparing the height, area, or mass of that component peak to the height, area, or mass of the internal standard peak. However, variation in detector response between compounds of different chemical structure must be taken into account. One way to do this is by first preparing a set of standard solutions containing varying concentrations of the compound(s) of interest. Each of these solutions is made to contain a known and constant amount of the internal standard. These standard solutions are chromatographed, and peak height, area, or mass is measured. Ratios of peak height, area, or mass (compound of interest/internal standard) are calculated and plotted against concentration to obtain calibration curves such as those shown in Fig. 15b. A separate response curve must be plotted for each sample component to be quantified. Next, a known amount of internal standard is added to the unknown sample, and the sample is chromatographed. Peak height, area, or mass ratios (compound of interest/internal standard) are calculated and used to read the concentration of each relevant component from the appropriate calibration curve. The advantages of using internal standards are that injection volumes need not be accurately measured and the detector response need not remain constant since any change will not alter ratios.

The main disadvantage of the internal standard technique is the difficulty of finding a standard that does not interfere chromatographically with components of interest in the sample.

Standards should be included during each analytical session, since detector response may vary from day to day. Analyte recovery should be checked periodically.

This involves addition of a known quantity of standard to a sample (usually before extraction) and determination of how much is recovered during subsequent analysis. During routine analyses, it is highly desirable to include a control or check sample, a material of known composition. This material is analyzed parallel to unknown samples. When the concentration of analyte measured in the control falls outside an acceptable range, data from other samples analyzed during the same period should be considered suspect. Carefully analyzed food samples and other substances are available from the National Institute of Standards and Technology (formerly the National Bureau of Standards) for use in this manner.

3

SEPARATION TECHNIQUES CHROMATOGRAPHY

3.1 INTRODUCTION

Chromatography is an important biophysical technique that enables the separation, identification, and purification of the components of a mixture for qualitative and quantitative analysis. Proteins can be purified based on characteristics such as size and shape, total charge, hydrophobic groups present on the surface, and binding capacity with the stationary phase. Four separation techniques based on molecular characteristics and interaction type use mechanisms of ion exchange, surface adsorption, partition, and size exclusion. Other chromatography techniques are based on the stationary bed, including column, thin layer, and paper chromatography. Column chromatography is one of the most common methods of protein purification.

Initially chromatographic techniques were used to separate substances based on their color as was the case with herbal pigments. With time its application area was extended considerably. Nowadays, chromatography is accepted as an extremely sensitive, and effective separation method. Column chromatography is one of the useful separation, and determination methods. Column chromatography is a protein purification method realized especially based on one of the characteristic features of proteins. Besides, these methods are used to control purity of a protein. HPLC technique which has many superior features including especially its higher sensitivity, rapid turnover rate, its use as a quantitative method, can purify amino acids, proteins, nucleic acids, hydrocarbons, carbohydrates, drugs, antibiotics, and steroids.

Chromatography is based on the principle where molecules in mixture applied onto the surface or into the solid, and fluid stationary phase (stable phase) is separating from each other while moving with the aid of a mobile phase. The factors effective on this separation process include molecular characteristics related to adsorption (liquid-solid), partition (liquid-solid), and affinity or differences among their molecular weights. Because of these differences, some components of the mixture stay longer in the stationary phase, and they move slowly in the chromatography system, while others pass rapidly into mobile phase, and leave the system faster. Based on this approach three components form the basis of the chromatography technique. Stationary phase: This phase is always composed of a "solid" phase or "a layer of a liquid adsorbed on the surface a solid support". Mobile phase: This phase is always composed of "liquid" or a "gaseous component."

3.2 SEPARATED MOLECULES

The type of interaction between stationary phase, mobile phase, and substances contained in the mixture is the basic component effective on separation of molecules from each other. Chromatography methods based on partition are very effective on separation, and identification of small molecules as amino acids, carbohydrates, and fatty acids. However, affinity chromatographies (ie. ion-exchange chromatography) are more effective in the separation of macromolecules as nucleic acids, and proteins. Paper chromatography is used in the separation of proteins, and in studies related to protein synthesis; gas-liquid chromatography is utilized in the separation of alcohol, esther, lipid, and amino groups, and observation of enzymatic interactions, while molecular-sieve chromatography is employed especially for the determination of molecular weights of proteins. Agarose-gel chromatography is used for the purification of RNA, DNA particles, and viruses.

Stationary phase in chromatography, is a solid phase or a liquid phase coated on the surface of a solid phase. Mobile phase flowing over the stationary phase is a gaseous or liquid phase. If mobile phase is liquid it is termed as liquid chromatography (LC), and if it is gas then it is called gas chromatography (GC). Gas chromatography is applied for gases, and mixtures of volatile liquids, and solid material. Liquid chromatography is used especially for thermal unstable, and non-volatile samples.

The purpose of applying chromatography which is used as a method of quantitative analysis apart from its separation, is to achive a satisfactory separation within a suitable timeinterval. Various chromatography methods have been developed to that end. Some of them include column chromatography, thin-layer chromatography (TLC), paper chromatography, gas chromatography, ion exchange chromatography, gel permeation chromatography, high-pressure liquid chromatography, and affinity chromatography.

Types of chromatography

- Column chromatography
- Ion-exchange chromatography
- Gel-permeation (molecular sieve) chromatography
- Affinity chromatography
- Paper chromatography
- Thin-layer chromatography

- Gas chromatography
- Dye-ligand chromatography
- Hydrophobic interaction chromatography
- Pseudoaffinity chromatography
- High-pressure liquid chromatography (HPLC)

3.2.1 Column Chromatography

Since proteins have difference characteristic features as size, shape, net charge, stationary phase used, and binding capacity, each one of these characteristic components can be purified using chromatographic methods. Among these methods, most frequently column chromatography is applied. This technique is used for the purification of biomolecules. On a column (stationary phase) firstly the sample to be separated, then wash buffer (mobile phase) are applied (Figure 1). Their flow through inside column material placed on a fiberglass support is ensured. The samples are accumulated at the bottom of the device in a time and volume-dependent manner.

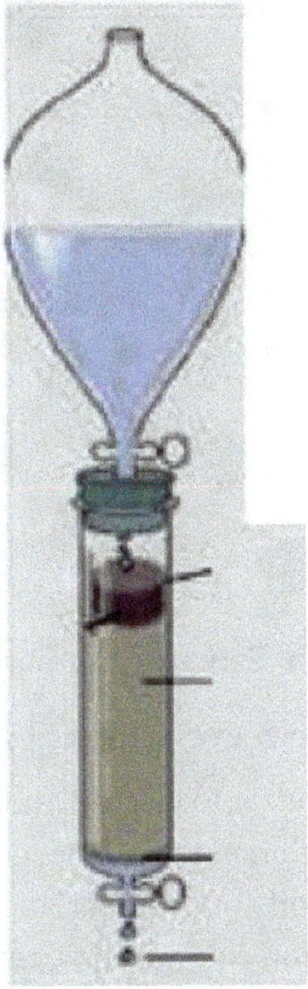

Fig. 1: Column chromatography.

3.2.2 Ion- Exchange Chromatography

Ion- exchange chromatography is based on electrostatic interactions between charged protein groups, and solid support material (matrix). Matrix has an ion load opposite to that of the protein to be separated, and the affinity of the protein to the column is achieved with ionic ties. Proteins are separated from the column either by changing pH, concentration of ion salts or ionic strength of the buffer solution. Positively charged ion-exchange matrices are called anion-exchange matrices, and adsorb negatively charged proteins. While matrices bound with negatively charged groups are known as cation-exchange matrices, and adsorb positively charged proteins (Figure 2).

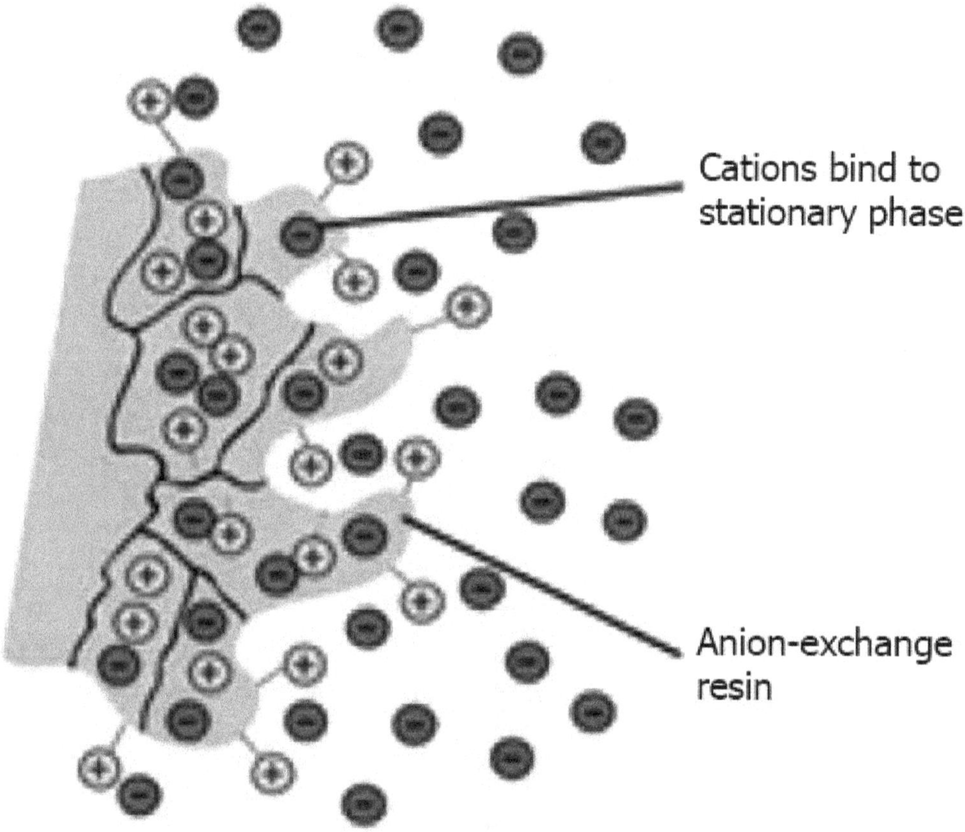

Fig. 2: Ion- Exchange Chromatography.

3.2.3 Gel- Permeation (Molecular Sieve) Chromatography

The basic principle of this method is to use dextran containing materials to separate macromolecules based on their differences in molecular sizes. This procedure is basically used to determine molecular weights of proteins, and to decrease salt concentrations of protein solutions. In a gel permeation column stationary phase consists of inert molecules with small pores. The solution containing molecules of different dimensions are passed continuously with a constant flow rate through the column. Molecules larger than pores cannot permeate into gel particles, and they are retained between particles within a restricted area. Larger molecules pass through spaces between porous particles, and move rapidly through inside the column. Molecules smaller than the pores are diffused into pores, and as molecules get smaller, they leave the column with proportionally longer retention times (Figure 3). Sephadeks G type is the most frequently used column material. Besides, dextran, agorose, polyacrylamide are also used as column materials.

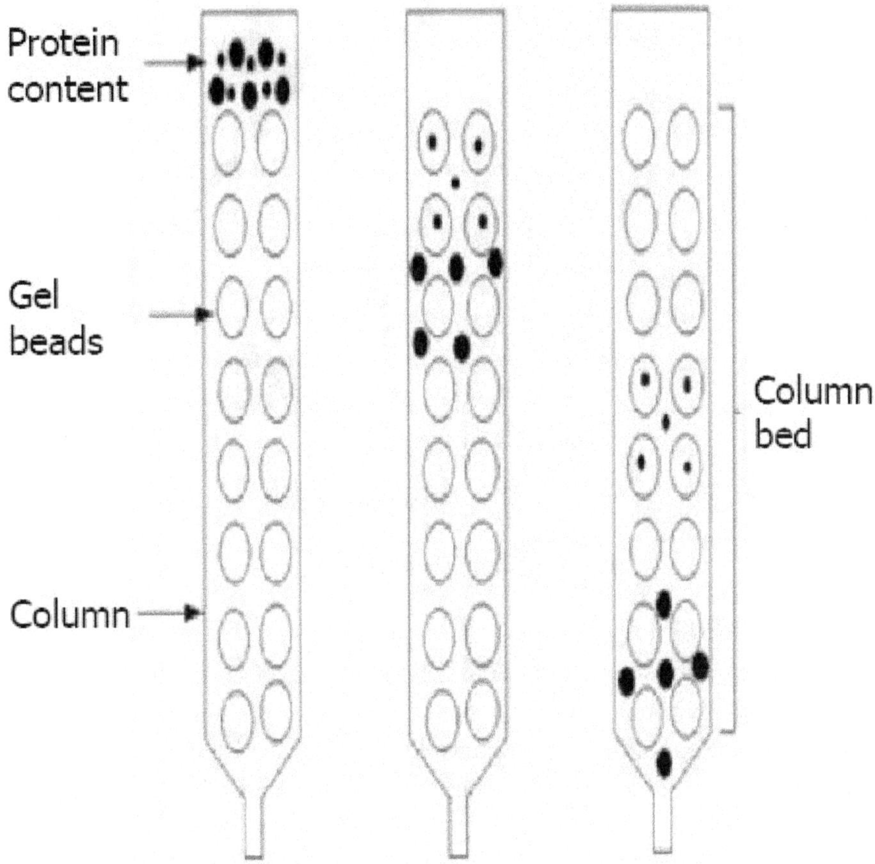

Fig. 3: Gel-permeation (molecular sieve) chromatography.

3.2.4 Affinity Chromatography

This chromatography technique is used for the purification of enzymes, hormones, antibodies, nucleic acids, and specific proteins. A ligand which can make a complex with specific protein (dextran, polyacrylamide, cellulose etc) binds the filling material of the column. The specific protein which makes a complex with the ligand is attached to the solid support (matrix), and retained in the column, while free proteins leave the column. Then the bound protein leaves the column by means of changing its ionic strength through alteration of pH or addition of a salt solution (Figure 4).

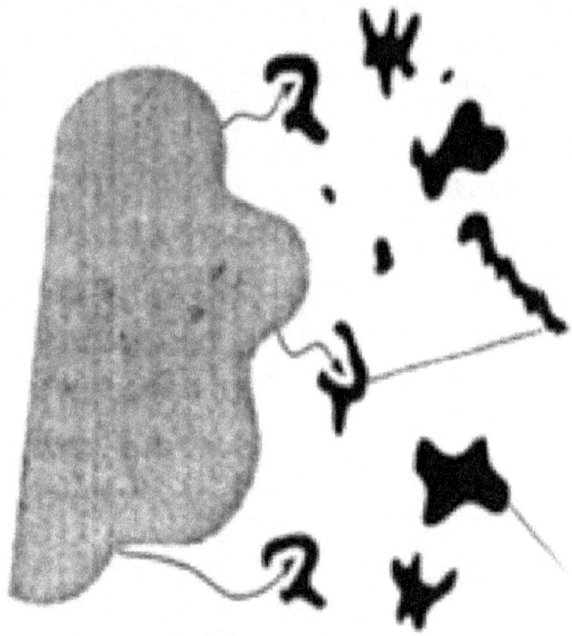

Fig. 4: Affinity chromatography.

3.2.5 Paper chromatography

In paper chromatography support material consists of a layer of cellulose highly saturated with water. In this method a thick filter paper comprised the support, and water drops settled in its pores made up the stationary "liquid phase." Mobile phase consists of an appropriate fluid placed in a developing tank. Paper chromatography is a "liquid-liquid" chromatography.

3.2.6 Thin-layer chromatography

Thin-layer chromatography is a "solid-liquid adsorption" chromatography. In this method stationary phase is a solid adsorbent substance coated on glass plates. As adsorbent material all solid substances used. in column chromatography (alumina, silica gel, cellulose) can be utilized. In this method, the mobile phase travels upward through the stationary phase The solvent travels up the thin plate soaked with the solvent by means of capillary action. During this procedure, it also drives the mixture priorly dropped on the lower parts of the plate with a pipette upwards with different flow rates. Thus the separation of analytes is achieved. This upward travelling rate depends on the polarity of the material, solid phase, and of the solvent.

In cases where molecules of the sample are colorless, florescence, radioactivity or a specific chemical substance can be used to produce a visible coloured reactive product so as to identify their positions on the chromatogram. Formation of a visible colour can be observed under room light or UV light. The position of each molecule in the mixture can be measured by calculating the ratio between the distances travelled by the molecule and the solvent. This measurement value is called relative mobility, and expressed with a symbol R_f. R_f value is used for qualitative description of the molecules.

3.2.7 Gas Chromatography

In this method stationary phase is a column which is placed in the device, and contains a liquid stationary phase which is adsorbed onto the surface of an inert solid. Gas chromatography is a "gas-liquid" chromatography. Its carrier phase consists of gases as He or N_2. Mobile phase which is an inert gas is passed through a column under high pressure. The sample to be analyzed is vaporized, and enters into a gaseous mobile phase phase. The components contained in the sample are dispersed between mobile phase, and stationary phase on the solid support. Gas chromatography is a simple, multifaceted, highly sensitive, and rapidly applied technique for the extremely excellent separation of very minute molecules. It is used in the separation of very little amounts of analytes.

3.2.8 Dye- Ligand Chromatography

Development of this technique was based on the demonstration of the ability of many enzymes to bind purine nucleotides for Cibacron Blue F3GA dye. The planar ring structure with negatively charged groups is analogous to the structure of NAD. This analogy has been evidenced by demonstration of the binding of Cibacron Blue F3GA dye to adenine, ribose binding sites of NAD. The dye behaves as an analogue of ADP-ribose. The binding capacity of this type adsorbents is 10–20-fold stronger rhat that of the affinity of other adsorbents. Under appropriate pH conditions, elution with high-ionic strength solutions, and using ion-exchange property of adsorbent, the adsorbed proteins are separated from the column.

3.2.9 Hydrophobic Interaction Chromatography (HIC)

In this method the adsorbents prepared as column material for the ligand binding in affinity chromatography are used. HIC technique is based on hydrophobic interactions between side chains bound to chromatography matrix.

3.2.10 Pseudoaffinity Chromatography

Some compounds as anthraquinone dyes, and azo-dyes can be used as ligands because of their affinity especially for dehydrogenases, kinases, transferases, and reductases The mostly known type of this kind of chromatography is immobilized metal affinity chromatography (IMAC).

3.2.11 High-Pressure Liquid Chromatography (HPLC)

Using this chromatography technique it is possible to perform structural, and functional analysis, and purification of many molecules within a short time, This technique yields perfect results in the separation, and identification of amino acids, carbohydrates, lipids, nucleic acids, proteins, steroids, and other biologically active molecules, In HPLC, mobile phase passes throuıgh columns under 10–400 atmospheric pressure, and with a high (0.1–5 cm//sec) flow rate. In this technique, use of small particles, and application of high presure on the rate of solvent flow increases separation power, of HPLC and the analysis is completed within a short time.

Essential components of a HPLC device are solvent depot, high- pressure pump, commercially prepared column, detector, and recorder. Duration of separation is controlled with the aid of a computerized system, and material is accrued.

4

FUNDAMENTALS OF GAS CHROMATOGRAPHY

4.1 OVERVIEW

Even though there are many manufacturers of gas chromatographs used in different applications, they all share the same fundamental components and theory of operation. The sample must be cleaned, dried, and the pressure controlled through an effective sample handling system. The analytical components of the gas chromatograph are contained in a temperature controlled oven. Various chromatographic columns separate the individual components of the sample. Low volume multi-port analytical valves control the sample injection and manipulate the analytical flow path. Detectors measure the concentration of each component as it elutes from the columns. The detector signal is processed by a controller that produces a chromatogram and calculates individual concentrations and other physical properties of the sample. Finally, the controller communicates to the outside world through serial or discrete outputs.

Gas chromatography is one of the most widely used techniques for analyzing hydrocarbon mixtures. Some of the advantages of chromatography are the range of measurement (from ppm levels up to 100 percent), the detection of a wide range of components, and the repeatability of the measurements. Chromatography is used in the laboratory, in permanently installed online systems, and in the field with portable systems. No matter the location, style, or brand, all gas chromatographs are composed of the same functional components: the sample handling system, the chromatograph oven, and the controller electronics (Figure 1).

This application note will cover the principles of sample handling, how chromatograph columns separate the components, why and how multi-port analysis valves are used, the common detector type used in the hydrocarbon applications, and the analysis processing that provides the component concentrations and the other calculated values (such as heating value and specific gravity) through physical reports or interfaces to other devices.

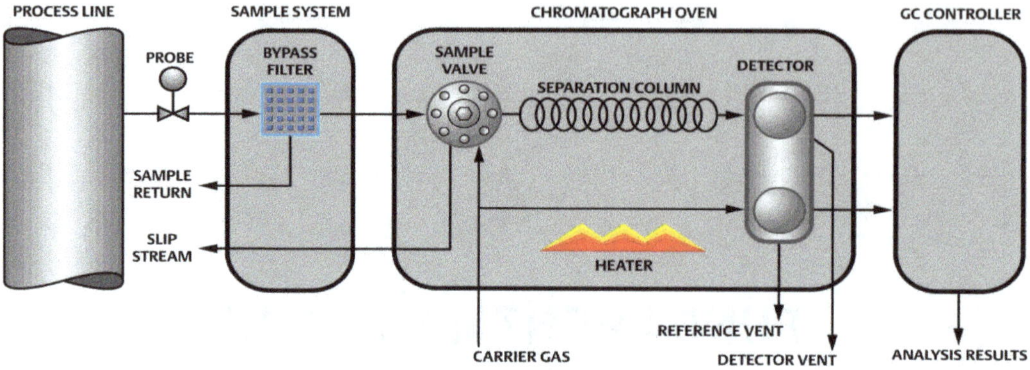

Fig. 1: Functional components: the sample handling system, the chromatograph oven, and the controller electronics

4.2 SAMPLE HANDLING SYSTEM

The gas chromatograph uses very small diameter gas paths and significant restrictions to flow; therefore, the sample injected to the analyzer must be dry and free of solids. Additionally, the gas chromatograph requires samples to be injected at a pressure significantly different than the source of the sample. The sample handling system should control the sample pressure and remove solid and liquids to ensure the sample presented to the gas chromatograph is clean and dry. The stream selection on multistream applications is also a function of the sample handling systems and is usually performed by simple solenoid valves. Most gas chromatographs typically require the sample to be controlled at a pressure between 15 and 30 PSIG (100 and 200 kPaG). In the majority of applications, the sample will be at considerably higher pressures, so the sample pressure must be regulated down to within the acceptable range. An important consideration is the Joules-Thompson effect in which the temperature of the gas will decrease as the pressure is reduced.

Whereas the amount of cooling is dependent on the composition of the gas, a typical value used for estimating the cooling of natural gas is 7 °F per 100 PSI (American Gas Association, n.d.) or 5.6 °C per 1000 kPa. As the temperature of a hydrocarbon gas mixture is reduced, eventually some of the larger hydrocarbon components will begin to drop out of the gas phase into the liquid phase. The temperature at which this begins to occur is the hydrocarbon dew point. If the sample is near the hydrocarbon dew point when it is sampled, the temperature of the gas may drop below this point when the pressure is regulated down, and the composition of the gas will change as the heavy components drop out as liquids.

The goal of the sample handling system is to preserve the composition of the gas, so it is important to consider the detrimental effects that the Joules-Thompson effect may have on the sample.

The use of heated regulators, insertion regulating probes (which use the heat from the flowing gas in the pipeline to maintain the temperature of the regulated sample), and heated sample handling components, such as heat-traced tubing is highly recommended to ensure that the temperature of the sample is maintained to at least 30 °F (16.7 °C) above the hydrocarbon dew point.

Inline filters are typically used to remove solid particles down to two microns in size. The larger the volume inside the sample handling system, the longer the lag time between when the sample enters the probe and when it enters the gas chromatograph. To avoid introducing large sample lag times into the system, particulate filters should not have large volumes, so coalescing filters are generally not recommended. Many sample probes in use today include a particulate filter at the filter tip. These are highly recommended as they stop contamination entering the system; however, users must routinely remove the probe and replace the filters to ensure the sample flow through the filter/probe is sufficient.

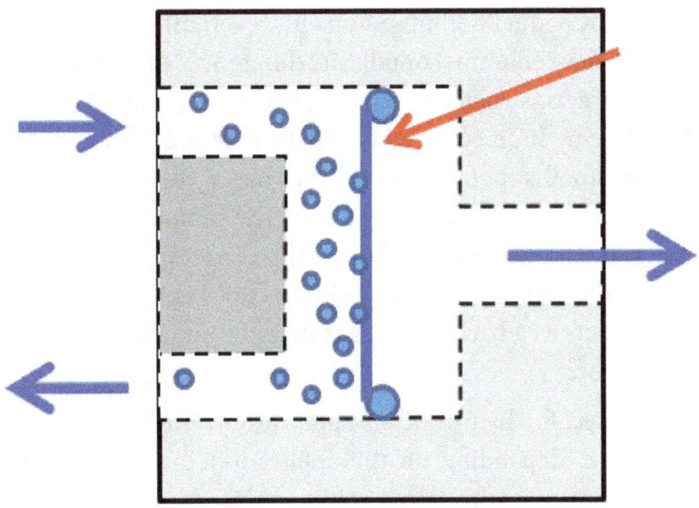

Fig. 2: - Moisture Membrane Filter

Liquids can be removed using membrane filters that use a specially designed membrane that only permits gas to pass through. The membrane can be located on the probe tip or in a membrane filter/ separator located on the gas chromatograph sample handling system. The typical liquid membrane filter passes the sample gas through the membrane and sweeps the removed liquids away through a bypass flow path (Figure 2). The bypass flow also allows the sample lag time to be reduced as it provides for a faster flow rate from the probe to the analyzer sample handling system (up to 500 cc/min) than the flow rate into the gas chromatograph, which is typically 30 cc/min to 50 cc/min.

4.3 CHROMATOGRAPH OVEN

The analytical components of the gas chromatograph (the columns, valves, and detectors) are enclosed in a heated oven compartment. The performance and response of the chromatograph columns and the detectors are very susceptible to changes in temperature, so the oven is designed to insulate these components from the affects of ambient temperature changes and maintain a very stable temperature internally. A gas chromatograph's performance is directly correlated to the temperature stability of the columns and detector, so the temperature control is typically better than ± - 0.5 °F (± - 0.3 °C) The temperature at which the oven is controlled is dependent on the application: the heavier the expected hydrocarbon mixture, the hotter the oven temperature. Natural gas applications have a typical oven temperature setting of around 176 °F (80 °C). The temperature of the oven is typically set in the factory and is rarely adjusted in the field.

4.4 CHROMATOGRAPH COLUMNS

The heart of the gas chromatograph is the chromatograph columns. The columns separate the gas mixture into its individual components using some physical characteristic. In the case of most hydrocarbon applications, "boiling point" columns are used and separate the components by their individual boiling points; however, other applications may use molecular size (molecular sieve columns) or polarity differences to achieve the separation. Chromatograph columns are constructed by packing a tube with column packing material. The material is held in place by sintered metal filters at either end of the tube. The packing material consists of very small support material that has a very thin coating of liquid solvent. This is called the stationary phase (Figure 3). The sample gas is carried through the columns by the carrier gas. The combination of the carrier gas and the sample gas is called the moving phase. The carrier gas is a gas which is not a component of interest (it is not being measured) and acts as a background gas that permits the easy detection of the components being measured.

Typically, helium is used for hydrocarbon applications; however, hydrogen, argon, and nitrogen are also used, depending on the application. As the gas sample moves through the column, components with lower boiling points move more slowly than the components with higher boiling points. The speed at which this separation occurs is dependent on the temperature of the column. The length of the column determines the amount of separation of the components.

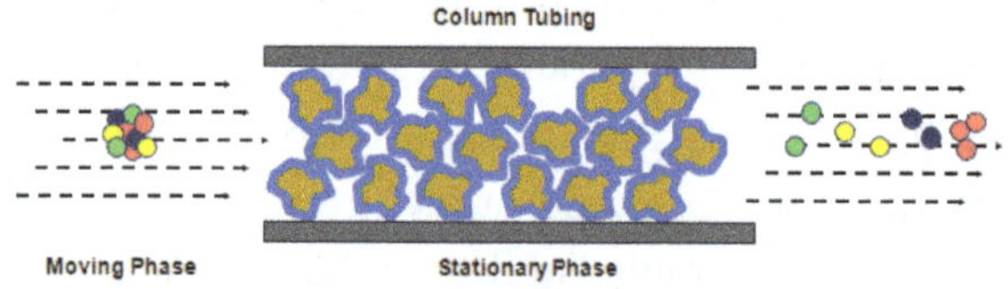

Fig. 3:- The Stationary and Moving Phases

Theoretically, a single column could separate all of the components of interest in a sample; however, the column would be very long, and the analysis time would be longer than 30 minutes (Gas Processors Association, 2000). To accelerate the analysis time to a more practical time, multiple columns are used to split the analysis into smaller and faster column applications. In a typical four-minute C6+ three-column arrangement (Figure 4), column one separates the hexanes and heavier components (the C6+ components) from the pentanes to methane hydrocarbons, nitrogen, and carbon dioxide. Column two then separates normal-pentane, iso-pentane, neo-pentane, normal-butane, and iso-pentane, while allowing the nitrogen, methane, carbon dioxide, and ethane to pass on to and be separated by column three.

4.5 ANALYSIS VALVES

The injection of sample into the analyzer (the sample valve) and the switching of the analytical flow paths to enable the use of multiple columns are achieved using specialized analysis valves. Analysis valves come in either six-port or ten-port configurations with very low internal volumes to avoid any dead volumes and to allow precise switching between components as they leave the columns. As the analysis valves are located inside the oven compartment, they are pneumatically operated through three-way solenoids mounted outside of the heat-controlled zone of the oven. In some applications, the carrier gas is used as the actuation gas; however, clean and dry air or nitrogen also can be used to reduce expensive carrier gas usage.

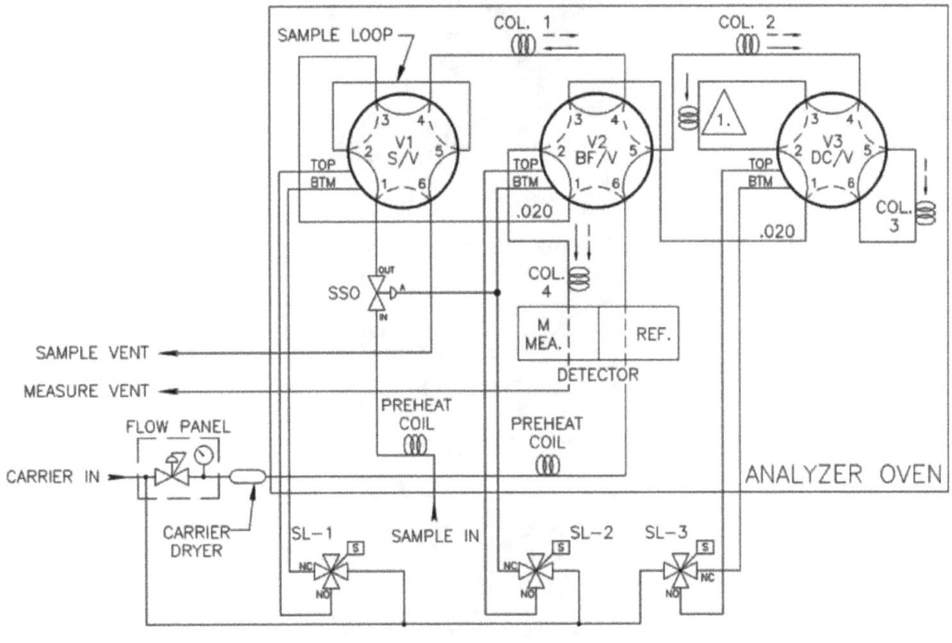

Fig. 4: - A Three-Valve, Three-Column C6+ Analysis Flow Diagram

The sample valve is used to inject a repeatable volume of sample into the analytical flow path at the beginning of the analysis cycle. Before the sample injection, the sample loop is flushed with new sample (Figure 5). To ensure the physical amount of sample is the repeatable from one analysis cycle to the next and is not affected by sample pressure, the sample flow is stopped for a short time before the sample injection using a sample shut-off valve, and the sample loop is allowed to equalize to atmospheric pressure via the sample vent. To inject the sample into the analytical flow path, the sample injection valve is actuated and the carrier gas is switched so as to push (or "carry") the sample out of the sample loop and into the first column (Figure 6).

Two very common column switching techniques are used with multiple column gas chromatographs: the back-flush and the dualcolumn arrangement. The back-flush arrangement (refer to Figure 7) is used to separate the individually measured components from the heavy components, which are sent to the detector as a single peak (such as the C6+ peak). The sample is pushed through the back-flush column until the last of the lighter components has left the backflush valve, but before any of the back-flushed components have left the column. The back-flush valve is then actuated and the carrier gas direction through the column is reversed so that the heavy components (which have been partially separated in the column) are recombined into a single peak (Figure 8).

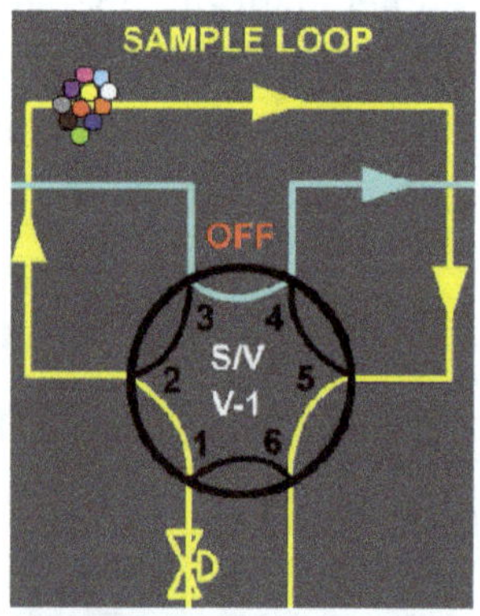

Fig. 5: Sample Loop Being Purged with New Sample (Valve off)

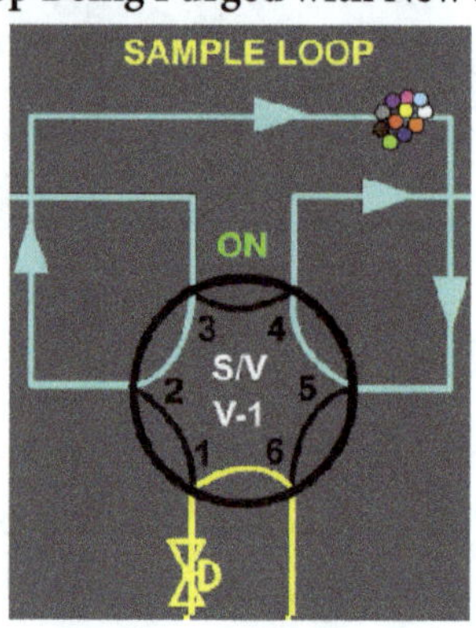

Fig. 6: Sample Loop Being Injected into the Analytical Flow Path (Valve on)

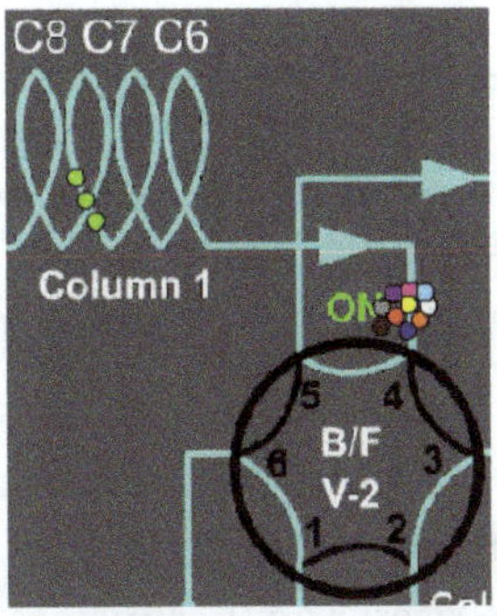

Fig. 7: The Back-Flush Valve in Forward-Flow Mode. Carrier Gas is Flowing Through the Column from Left to Right

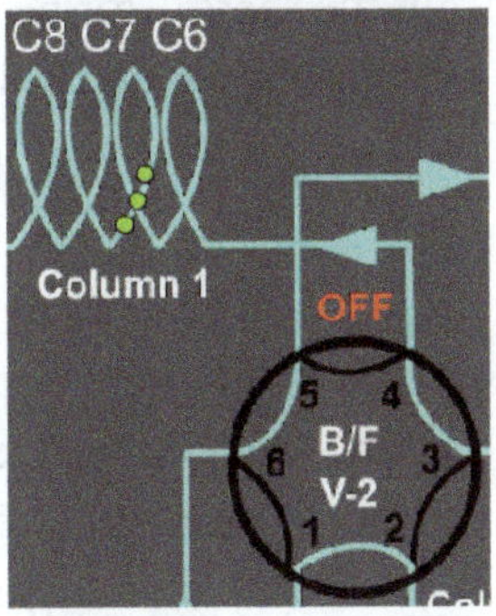

Fig. 9: The Back-Flush Valve in Back-flush Mode. Carrier Gas is Flowing Through the Column from Right to Left

The second column switching technique is the dual-column technique in which the analysis valve switches the flow path between two columns. Initially, lighter components are partially separated by a previous column and are directed into a column (column three in Figure 4) designed to fully separate the individual components.

When all of the lighter components have entered the column, but before the first of the fully-separated components have entered the valve, the dual-column valve is switched (Figure 9) and the lighter components are trapped. The flow of carrier gas and sample components is now switched to the second column (the restrictor column marked by a 1 in a triangle in Figure 4), which could either conduct further separation of the components, or only act as a restriction to flow (as shown). The flow rate through the analytical flow path is dependent on the supply pressure of the carrier gas (which is constant) and the restriction to flow through the entire analytical flow path. Therefore, the second column on the dualcolumn arrangement must have the same restriction to flow as the column that has been switched out of the analytical flow path.

Once all of the components being separated by the previous column have passed through the restrictor column and gone through to the detector, the dual-column valve is then actuated and the carrier gas flow is routed back through the original column to separate the lighter components and then carry them to the detector.

4.6 DETECTORS

After the components have been separated by the chromatograph columns, they then pass over the detector. Several types of detectors are available for gas chromatographs, including flame ionization detectors (for ppm-level hydrocarbons) and flame photometric detectors (for ppb- to ppm-level sulphur detection), but the most common detector used for most hydrocarbon gas measurements is the thermal conductivity detector (TCD) (Figure 10). The TCD can measure components from the low ppm levels up to 100 percent concentration. The TCD uses two thermistors that will reduce in resistance as their temperature rises. The thermistors are connected on either side of a Wheatstone bridge with a constant current power supply. As the carrier gas passes over the thermistors, it removes heat from the thermistor bead, dependent on the carrier gas's thermal conductivity. Helium is a commonly used carrier gas because it has a very large thermal conductivity, and, therefore, will reduce the temperature of the thermistor bead considerably.

On the reference side of the detector, only pure carrier gas will pass over the thermistor bead, so the temperature and resistance will remain relatively constant. On the measure side of the detector, the carrier gas and each component in series of elution from the columns passes over the thermistor bead, removing heat from the bead dependent on the thermal conductivity. When there is only carrier gas passing over the detector bead, the temperature of the bead will be similar to the reference detector (any difference is compensated for using the bridge balance).

However, the gas components will have different thermal conductivities than the carrier gas. As the component flows across the thermistor bead, less heat is removed from the bead, so the temperature of the thermistor increases, reducing the resistance. This change in resistance imbalances the electrical bridge and results in a millivoltage output. The amount of difference and, therefore, the output signal is dependent on the thermal conductivity and the concentration of the component. The detector output will then be amplified and passed to the gas chromatograph controller for processing.

4.7 GAS CHROMATOGRAPH CONTROLLER

The controller may be remote from the oven or integral, depending on the design and application, and performs several functions, including:

- Control the valve timing and oven temperature
- Control stream selection
- Store and analyze the detector output
- Calculate composition from the detector output
- Calculate physical properties from the composition (e.g. BTU, Specific Gravity)
- Communicate to supervisory systems such as flow computers or SCADA systems

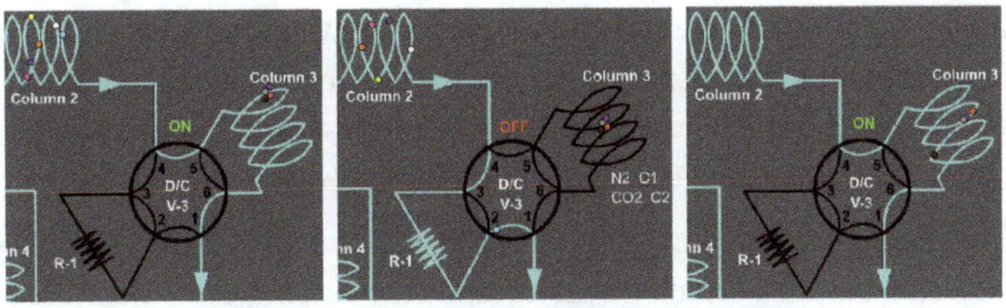

Fig. 9: The Dual-Column Operation Valve

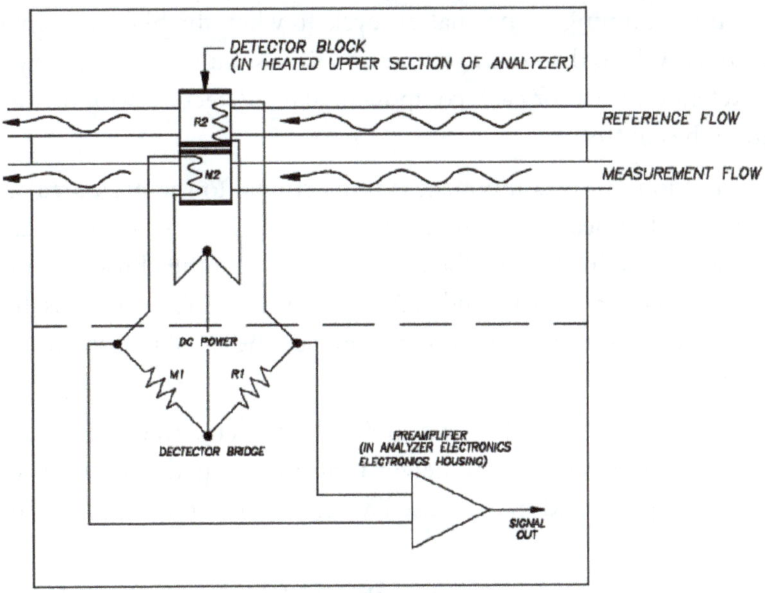

Fig. 10: Thermal Conductivity Detector

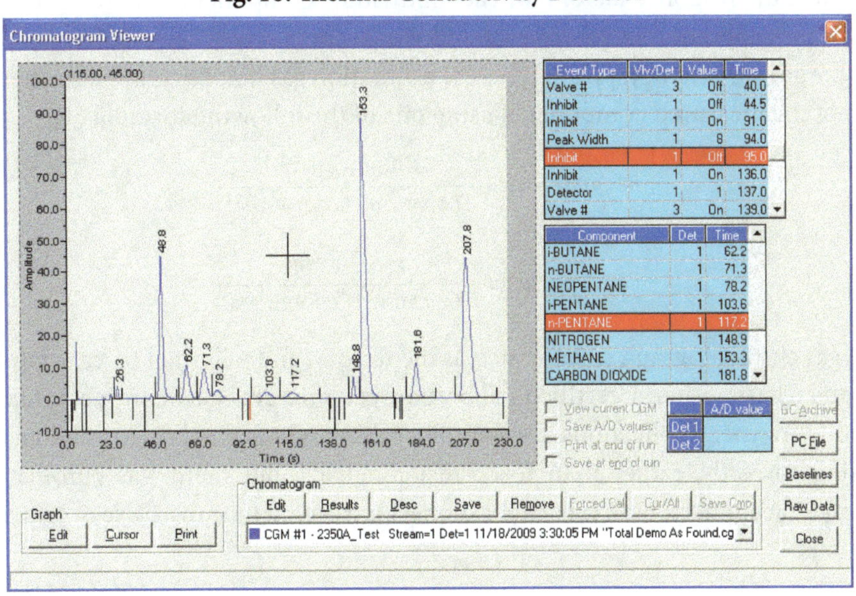

Fig. 11: A Typical C6+ Chromatogram

The unique function of the controller is the analysis of the detector output. The detector output is graphically displayed on a chromatogram (Figure 11) that will typically also show the valve timing, the expected retention times of each of the components to be measured, the composition and calculation results, and raw detector data.

Each measured component in the sample gas will have a peak (Figure 12), which is the change in detector output because of the component passing over the measure detector. The controller determines which component the peak represents by the retention time, the time from the beginning of the analysis cycle to when the highest point of the peak. From one analysis cycle to the next, the retention time should not change significantly, so the expected retention time for each component and a detection window (in ± seconds) is configured in the controller.

The concentration of the component is proportional to the peak area, and the peak height. To calculate the peak area, the controller detects the start and end of the peak by analyzing the change in slope of the detector trace. On the chromatogram shown in Figure 12, the detected peak start is indicated by a large line above the baseline at around 46 seconds, and the detected peak end is indicated by the small line above the baseline at around 56 seconds.

The controller will integrate the area under the detector trace and use this value to calculate a concentration value. Peak height is also calculated from the baseline to the highest point of the peak, and may be used as an alternative method to calculating the concentration.

Before a concentration can be calculated, the gas chromatograph needs to be calibrated to determine the response factors for each individual component. The response factor is effectively the calibration factor for each individual component. To calculate the response factors, the gas chromatograph analyzes the calibration gas that has known concentrations for each of the measured components using one of the following formula (1):

$$\text{Response Factor} = \frac{\text{Peak Area}}{\text{Calibration Concentration}}$$

or

$$\text{Response Factor} = \frac{\text{Peak Height}}{\text{Calibration Concentration}}$$

The gas chromatograph is calibrated at the factory, and will then be calibrated onsite on a regular basis using a certified calibration gas mixture. Gas chromatographs used for custody transfer applications are typically calibrated once a day (to validate the unit is working correctly, rather than to correct for drift), but many gas chromatographs can operate accurately between calibrations for several months in less critical plant applications.

$$Response\ Factor = \frac{Calibration\ Concentration}{Peak\ Area}$$

1. Some gas chromatographs may use the formula: The resultant number is different, but the concept is the same.

When performing an analysis, one of the following formulas is used to calculate the concentration of each component, dependent on the peak detection method (area or height):

$$Concentration = \frac{Peak\ Area}{Response\ Factor}$$

or

$$Concentration = \frac{Peak\ Height}{Response\ Factor}$$

Many gas chromatographs perform a complete analysis for all of the major components expected in the sample, such as the natural gas "BTU" measurement. Theoretically, the sum of the concentrations of all of the measured components should equal 100 percent. However, in reality the result rarely does equal 100 percent. The cause of this "error" can be the result of 1) the uncertainty of the detector measurement, 2) slight changes in the sample loop pressure (because of atmospheric pressure changes) resulting in changes in the amount of gas sample injected, or 3) measurement error. The total of the components as measured is referred to as the un-normalized total and is commonly expected to be between 98 percent and 100 percent for a correctly functioning gas chromatograph. As the analysis from the gas chromatograph is often used for calculating physical properties using calculation methods that presume the concentration total is 100 percent, the analysis results are mathematically adjusted, or normalized, so that the sum of the components is exactly 100 percent. Once the controller has calculated the full analysis, other physical property calculations may be performed. This may include:

- BTU/scf (according to GPA 2172-09 and GPA 2145-09)
- Standard Compressibility and Specific Gravity (AGA report No. 8–1994)
- Heating Value (ISO 6976)
- Hydrocarbon Dew Point

The analysis results, analyzer status, and calculation results are typically communicated to a supervisory system (such as a flow computer, RTU, or SCADA system) using a Modbus® serial link. Most chromatographs are apable of providing a 4–20 Ma output; however, because of the large number of values that a chromatograph produces (up to 16 component concentrations, several physical property calculations, and analyzer status signals) using discrete signals is impractical. The serial links are commonly RS-232, RS-485, RS-422 or TCP/IP signals. Recent developments have seen the use of FOUNDATION™ Fieldbus communications to DCS-style control system]

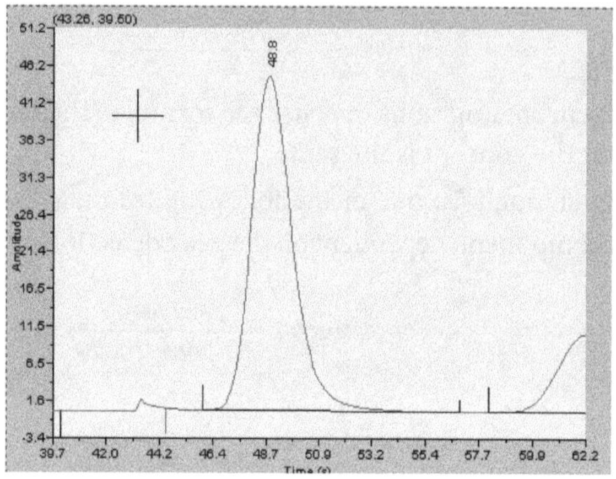

Fig. 12: A Component Peak Magnified on the Chromatogram Viewer

5

FUNDAMENTALS OF HIGH PERFORMANCE LIQUID CHROMATOGRAPHY

5.1 INTRODUCTION AND HISTORY

Michael Tswett (1872-1920) was credited as the father of chromatography due to his 1903 separation of the green-leaf pigments into bands of colors, a demonstration of liquid chromatography. While similar work was being conducted in the petroleum industry, Tswett is credited with coining the term "chromatography". Despite Tswett's results, chromatography did not develop quickly. The next major developments were the use of thin-layer chromatography (TLC) in 1937-38 and the use of paper chromatography in the mid-1940s, but thin layer chromatography quickly won popularity. Thin layer chromatography was originally developed by Nikolai Izmailov (1907-1961) and his graduate student Maria Shraiber (1904-1992) for pharmaceutical preparations. Early TLC was conducted with microscope slides that were coated with suspensions of calcium, magnesium, and aluminum oxides. As used today, a small spot of solution was placed on one end of the slide, the slide was dipped into a solvent, and the analytes migrated at different rates through the oxide coatings where they were later detected (today by a UV lamp or chemical stain).

TLC advanced slowly during the next few years but a major advancement was made in 1956 by Egon Stahl (1924-1986) when he attempted to standardize the preparation of the sorbents used to make the plates. While these advances and others such as forced flow TLC, significantly matured TLC into an accepted (and reproducible) practice; it was still only a qualitative technique, at best.

However, Izmailov and Shraiber's spot chromatography, commonly known today as TLC, is the workhorse of undergraduate organic synthesis labs where synthesis reactions are conducted and the resulting products are selected for using common glass open columns filled with silica. Eluent from these columns is collected in fractions that are then run by TLC to identify which column fraction contains the desired product.

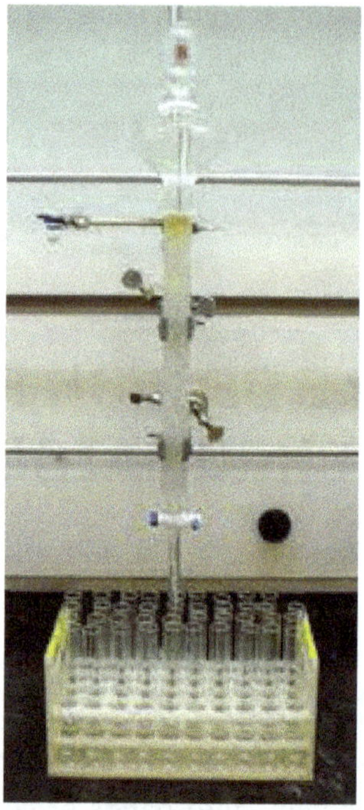

Fig. 1: An Atmospheric Pressure Open Column Chromatographic Column. Fractions are collected in the test tubes and later ran by TLC to determine the purity of the fraction and the presence of synthesis products.

Since the development of TLC, liquid chromatography needed a technique comparable to gas chromatography where a complex mixture of analytes could be quantitatively separated and identified. Several attempts were made to pressurize the relatively large glass preparatory column (shown above) with little success due to the fragile nature of the column. High-pressure liquid chromatography (HPLC) was later developed to meet this goal in the 1970s. The pressure was first delivered by a large syringe, but this approach limited the volume of solvent that would be passed through a column and therefore limited the analysis time. Syringes were later replaced by a single reciprocating pump but these delivery systems experienced flow surges, between strokes of the single piston, interfering with stable detector baselines. The placement of two reciprocating pump, operating opposite to each other with respect to flow, greatly minimized the flow fluctuations which were later removed completely with a pulse damper. This form of chromatography is referred to as HPLC, where the HP stands for high performance or high pressure. Some jokingly refer to the HP as meaning high priced since it replaced TLC plates, that only cost pennies, with $20 000 to $30 000 instruments. The inflation of the cost of a LC analysis is even greater when an HPLC-MS is considered, a minimum of $150 000. But regardless, HPLC-MS is considered the technique of choice for isolating a synthetic product and is widely utilized in most synthesis laboratories.

Chromatography, as noted in Chapter 1, is divided into gas, liquid and supercritical fluid techniques. Liquid chromatography can be divided up into a relatively large collection of techniques. Those mentioned above include atmospheric or low-pressure open column chromatography and thin layer chromatography. Pressurized liquid chromatography can be divided into ion exchange, exclusion, partition, and liquid-solid chromatography as summarized in Animation 1.

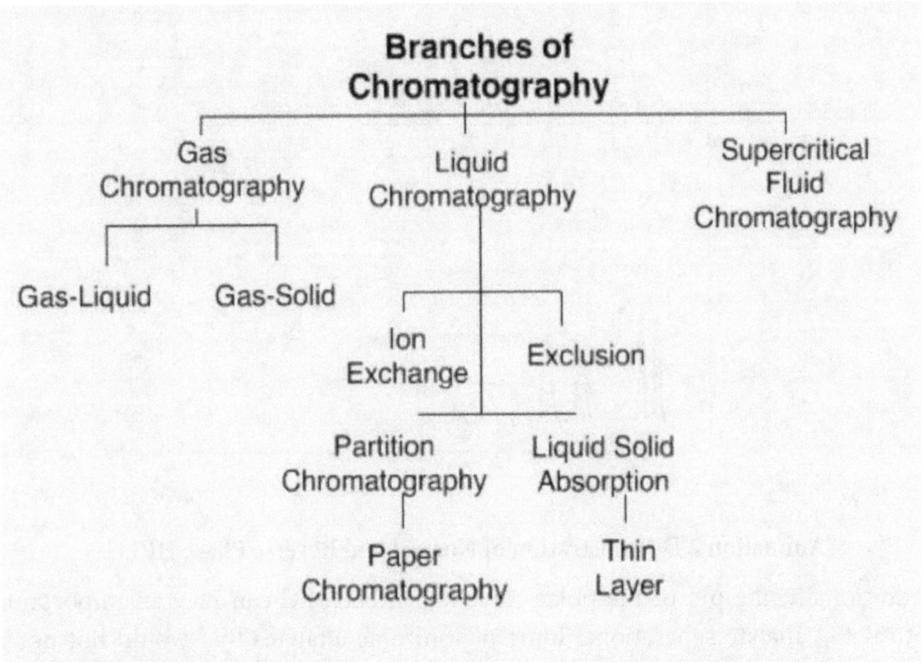

Animation 1. A Discussion of the Various Types of Chromatography.

As noted in Animation 1, a variety of separation techniques are available with high pressure liquid chromatography. Most relevant to this chapter is partition chromatography, although a few others will be discussed in later sections. The most important point here is to distinguish between normal and reverse phase chromatography. HPLC was first developed using normal phase conditions (NP-HPLC), that followed the logic of atmospheric open-column chromatography, where the stationary phase acted as the polar phase and the mobile phase was non-polar, specifically an organic solvent. Normal phase HPLC focused on the separation of analytes that were readily soluble in nonpolar solvents but had slight affinities for the polar stationary phase. However, NP-HPLC required the use of relatively expensive and large volumes of organic solvents that also led to high waste disposal costs. NP-HPLC was effectively replaced by reverse phase HPLC (RP-HPLC) that operates with a non-polar stationary phase and an aqueous, moderately polar mobile phase. Gradient programming, an additional development, changes the composition of the mobile phase during a chromatographic run which greatly enhanced the utility of RPHPLC. In RP-HPLC, the gradient is initially more polar (i.e. water) and as the chromatographic run progresses, more and more less-polar solvent is added to the mixture (i.e. methanol or acetonitrile) to end the gradient program with pure organic solvent. As noted in Animation.2, the retention order of analytes in RPHPLC is opposite NP-HPLC. For example, the first analyte that eludes on a RP system would elude last is separated by NP-HPLC.

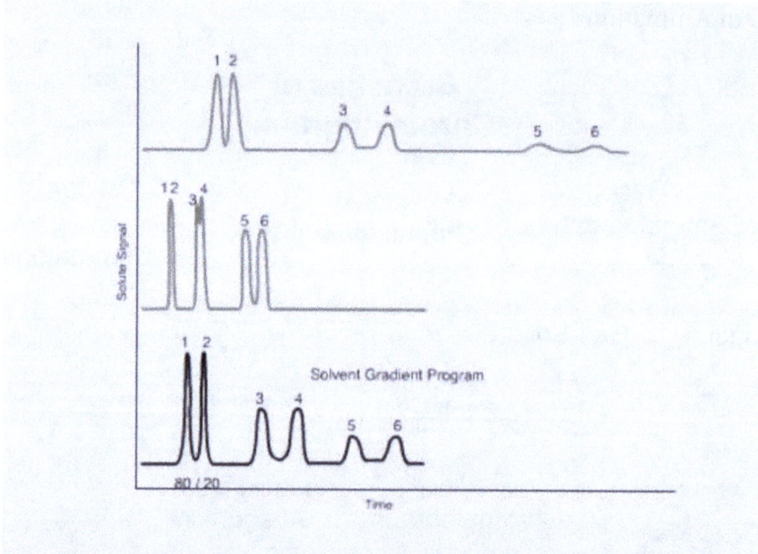

Animation 2 Demonstration of Normal and Reverse Phase HPLC.

Additionally, the pH of the polar solvent in RP-HPLC can play an important role in optimizing analyte separations. Ionic or ionizable analytes that would not normally be separated on a RP (non-polar) analytical column can also be analyzed by ion-pair chromatography. In this type of chromatography, the ionic analyte is bound to another ion (usually a large organic counter-ion such as quaternary ammonium or alkyl sulfonate) to form a neutral pair that has selective affinity for the non-polar stationary phase. Even many chiral compounds can be separated on special chiral stationary phases or by the additional of chiral resolving agents that selectively bind to one of the enantiomers. One final point should be made with respect to HPLC. Chemists use HPLC for two completely different purposes. Organic chemists, especially in the pharmaceutical industry, use large-scale systems, referred to as preparatory HPLC or flash chromatography, to recover relatively large-scale milligram quantities of their products. The main differences of a preparatory HPLC, as opposed to an analytical HPLC, are the pump flow rates and the size of the columns. In contrast, analytical chemists use HPLC to separate and identify nanogram or smaller quantities of analytes. Whichever practice is needed, the overall chromatography is the same.

5.2 TYPES OF ANALYTES, SAMPLES AND SAMPLE INTRODUCTION

Gas chromatography is somewhat limited in that the analyte has to be volatile below 300 C and thermally stable with molecular weights less than 1000 Daltons. HPLC greatly expands the range of possible analytes and with reverse phase HPLC and can include many of the same analytes as GC (but with less resolution as compared to capillary column GC). Analysis of compounds commonly used in HPLC increases the analyte range of molecular weights to just less than one million Daltons. Many chemicals, such as petroleum hydrocarbons, solvents, illicit drugs and environmental chemicals, can be analyzed by both GC and HPLC. But with the ability of HPLC to analyze nonvolatilechemicals many bio-molecules can be analyzed, including sugars, amino acids, proteins, and a large variety of other non-volatile compounds.

Samples analyzed by HPLC are always in liquid form, in either aqueous or organic solvents. Little to no sample preparation is needed, except that it is good practice to filter all samples through a 0.2 µm low-volume cartridge filter prior or during injection. Filtration of samples and solvents avoids the buildup and eventual clogging of the in-line filter or guard column. Samples containing high concentrations of analytes may need to be diluted in order to avoid overwhelming the capacity of the stationary phase and remaining in the linear range of the detector (governed by Beer's law in UV-Vis applications). Solvent exchange of the samples may be necessary depending on the solvent gradient conditions needed for separation. Samples containing compounds with significantly different chemical structures may interfere with the detector of the analyte(s) and may require removal with a micro-column clean up (such as silica gel or alumina) or may require the use of a specialty clean up cartridge (such as those used in GC for concentration of the analytes from aqueous samples described in Section 2.2). Samples containing relatively low concentrations of analytes may need to be concentrated using one of these same micro-resin columns.

5.3 THE LIQUID CHROMATOGRAPH

5.3.1 Overview

In recent decades the modern HPLC system has become increasing complex. Automatic injection systems have all but replaced manual injections, manual six-port values are now pressure actuated, the instrument is usually computer controlled, and data are collected and processed via a specified computer method. It should come as no surprise that the cost of a basic HPLC system (without mass spectrometry detection) has risen from $10 000 a few decades ago to nearly $40 000 today. In this section, each component of an HPLC will be presented. Figure 3.2 illustrates an overview of a modern HPLC-MS system.

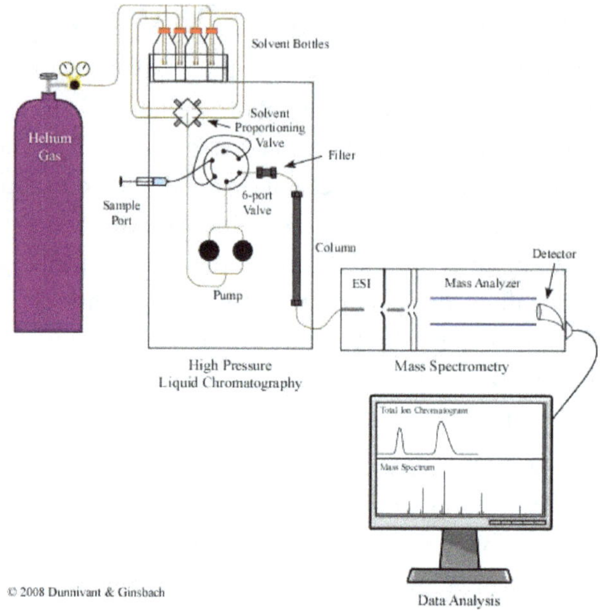

Fig. 2: An Overview of a Modern HPLC-MS System. More modern changes to this system eliminates the He purge gas with a vacuum degasser.

5.3.2 Actuator Gas

Pressurized gas is required to degas solvents and to drive many of the mechanical functions of modern HPLCs. Solvents used as the mobile phase contain trace concentrations (ppm) of atmospheric gases (N2 and O2) which create bubbles in the pump (during the low pressure intake of solvents) causing pumping problems. Dissolved gases can also evolve in the low pressure detectors in ultrahigh pressure LC, ion exchange, and supercritical fluid chromatography giving the appearance of a chromatographic peak depending on the specific detector being used. Pressurized gas is also needed to turn the sixport injection valve during automatic injection mode.

5.3.3 Solvents and Solvent Preparation

Solvent purity requirements depend on the type of samples being analyzed. All solvents used as the mobile phase must be filtered through 0.2 μm filters prior to entry to an HPLC system in order to avoid the scratching of the pump pistons by particles and to avoid clogging the in-line filter or guard column. Organic chemists are usually not concerned with the purity of solvents since their analytes are present in high concentrations and their samples usually contains many side-products and solvents. Hence, relatively low purity, and therefore inexpensive, solvents can be used. In contrast, the analytical chemist must purchase HPLC grade solvents that are considerable more expensive. For example a 20-L metal can of ACS grade methanol costs less than $70, while a 4-L glass bottle of HPLC UV-Vis grade bottle costs $150. Many solvent gradients call for acetonitrile that in recent years (2009-2009) has risen considerable in costs due to a decrease in production, from \$100 in 2005 to \$250 in 2009 (if supplies allow you to even order it). The increase in cost has tempted many chromatographers to switch their acetonitrile solvent gradients to one based on methanol while maintaining the same polarity index (a measure of the polarity of the solvent mixture that is used to optimize analyte separation). However, as noted by William Campbell in a recent Supelco Analytical Note (volume 27.1, page 13), while analyte retention orders may (or may not) remain the same, the relative retention times and resolution can significantly change, in some cases where analyte separations are no longer achievable. Solvents used for elution are contained in glass reservoirs (usually one liter in capacity) and connected to HPLC with Teflon tubing. The solvent bottle is purged with He gas to remove dissolved gases, mentioned above, and kept under a He atmosphere during HPLC operation.

Tubing and Connections: Different components of an HPLC system operate under different pressures ranging from atmospheric to 4 500 psi (3.10 x 107 Pa) in standard HPLC and to 20 000 psi (1.38 x 108 Pa) in ultra high pressure liquid chromatography. Systems are currently under development that will operate at 100 000 psi. While GC fittings are usually standardized to Swagelok fittings, many HPLC manufactures have established their own type of high pressure fittings. All operate on the same ferrule-nut configuration as in GC but the shape of the ferrule and receiving system is different. A few of the more common fitting are shown in Figure 3. Tubing sizes are much smaller in HPLC as compared to GC. Common tubing materials are stainless steel and PEEK (polyaryletheretherketone), a very strong organic polymer.

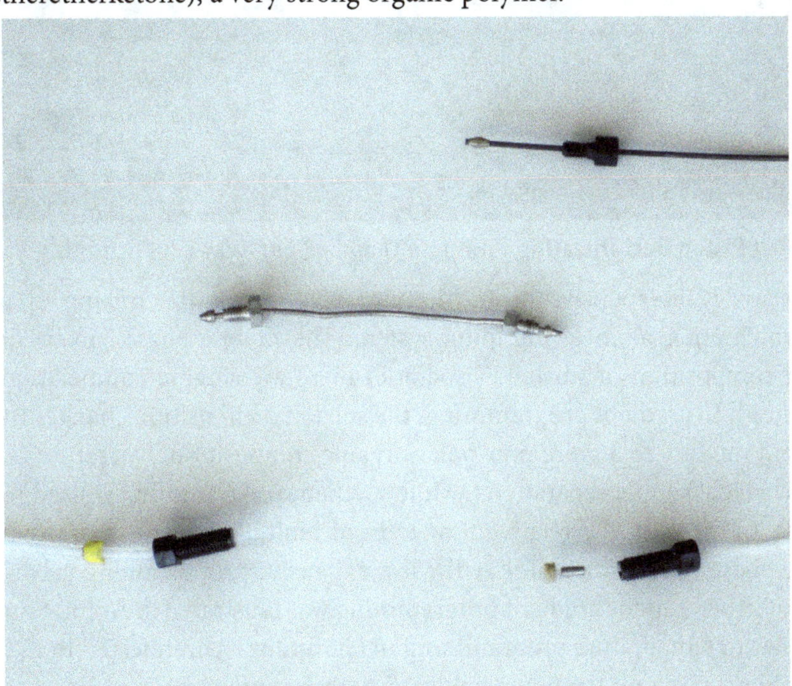

Fig. 3: A Selection of HPLC Tubing and Connectors.

5.3.4 PROPORTIONING VALVE

Up to four separate elution solvents enter the HPLC first through a proportioning valve that adjusts th e flow of each solvent to a predetermined amount. A proportioning valve is shown in Figure 4. Prior to the development of the proportioning valve, a separate pumping system was needed for each solvent. Since the pumping system (Figure 4) is typically the most expensive component of an HPLC system, the proportioning valves help reduce the cost of an HPLC system.

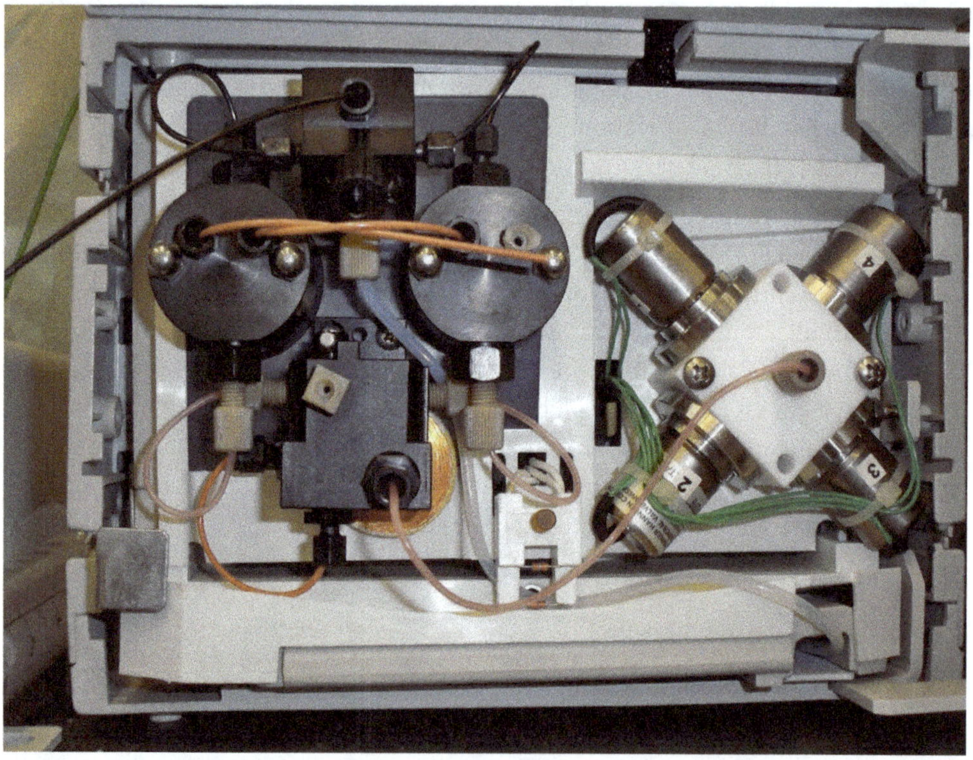

Fig. 4: A Dual Piston Reciprocating Pump (left) and a Four-Way Proportioning Valve (right).

The relatively inexpensive proportioning valve allows the mixing of up to four solvents which enter a single pumping system. The mobile phase can be delivered as a constant composition of solvents (isocratic) or as a changing composition (gradient programming). In gradient programming, the solvent composition changes from a more polar solvent mixture to a more non-polar organic composition. Isocratic operations are used for relatively simple separations, while gradient programming is used for complex separations. In the past, the relative flow rates of multiple pumps were used to control the mobile phase composition. Today the less expensive proportioning valves adjust the composition. The need for gradient programming was illustrated in Animation 2 (earlier) and is similar to temperature programming in GC and is again referred to as the general elution problem. Click on the figure below to start the animation.

5.3.5 Pump

As noted earlier, HPLC pumping systems have evolved more than any other component of LC. Gravity-fed open column systems were later pressurized with gas, which was replaced by syringe pumps in early HPLC, then by single piston pumps, and today with specially engineered dual reciprocating piston pumps. Today's pumps provide constant pressure and mobile phase flow by alternating the pumping actions between two pumping systems. When either pump is in full stroke, constant flow is easily provided but as each pump reaches the end of a stroke flow rates may be altered. This is overcome in some systems by an oval-shaped caming device that speeds the pumping rate at the end of each piston stroke. At the end of the stroke, one pump speeds up the flow of mobile phase as the other pump head decreases the pumping rate. This combined action provides constant flow rates in modern HPLC systems. While the simultaneous operation of two pumps and their associated pistons, cams, and check valves are difficult to describe in words, an animation makes the process easy to understand. Such an animation is shown in Animation 3. The animation can be viewed by clicking on the figure below.

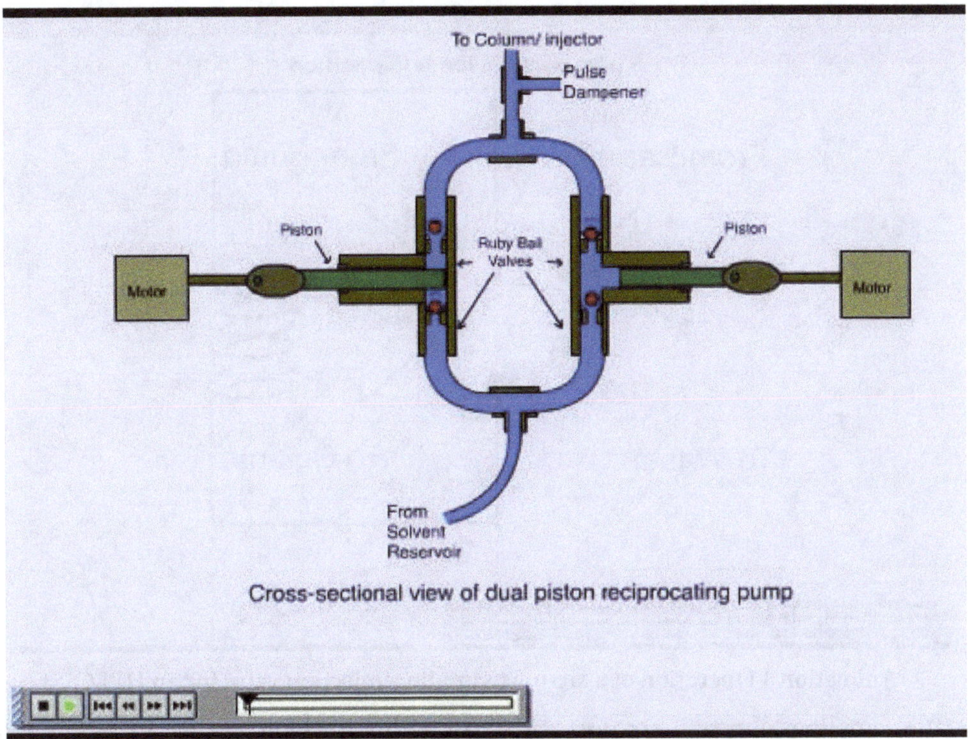

Animation 3 Illustration of a dual reciprocating piston HPLC pump.

Pulse Damper:

Early HPLC pumps produced pressure, and therefore flow pulses, this was especially true with single piston pumps which required a pulse damper to be installed down-line from the pump. Today's systems, with advanced pump design, produce little pressure fluctuations and either do not require a pulse damper or electronically compensate for the very small pressure fluctuations.

5.3.6 Six-Port Injector

High pressure systems require a special type of sample injection since injections with standard syringes are not possible. Four-, six-, and eight-port valves are used for high pressure systems. These systems are equipped with a fixed volume loop of tubing that serves as the sample loop that is loaded with a standard blunt-tipped syringe. These multiple port systems allow un-interrupted flow to the column during the loading of a sample on a sample loop and during injection of a sample under high pressure. As with a dual reciprocating HPLC pump, it is easier to show how a six-port valves works through an animation. Animation 3.4 shows the alignment of the valves during the loading of a sample onto a sample loop, switching of the values, and injection of the contents of the sample loop onto an HPLC column. View the animation by clicking on the figure below.

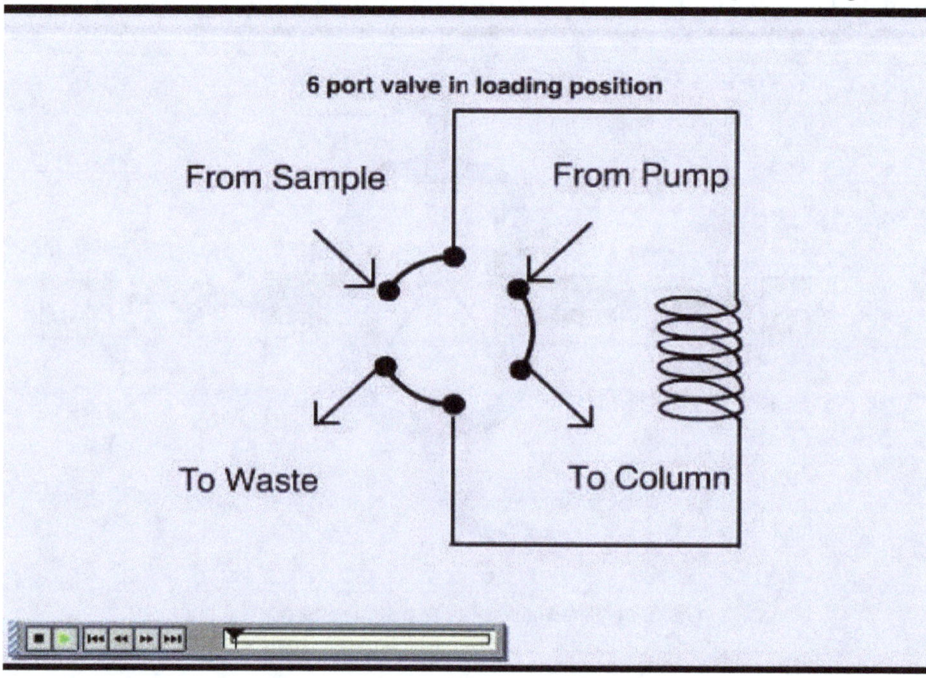

Animation 4 Operation of a six-port sampling/injection valve for an HPLC.

The sampling loop of a six-port valve can be loaded by a syringe or by pumping liquid sample into the valve by an automatic sampler. Sample loop sizes range from 1 - 100 μL for analytical-scale HPLC to milliliter volumes in preparatory-scale HPLC. The advantage of a syringe (manual) injection is that less sample volume is required (since filling of the tubing in an automatic sampler is not required). The obvious advantage of an automatic sampler is that numerous samples can be automatically analyzed by a computer-controlled system.

In-Line Filter:

Although samples and solvents are filtered prior to injection or use with an HPLC, an in-line filter is usually present immediately after the injection valve. This filter removes any remaining particles in the mobile phase that may clog the guard or analytical column. In-line filters are one of the most commonly maintained items for HPLC systems.

5.3.7 Columns

Guard Column

The next component of an HPLC is the guard column. Guard columns are miniature versions of the analytical (separation) column and they contain the same stationary phase. The purpose of the guard column is to adsorb any permanently adsorbing chemicals that could destroy the more expensive analytical column. Guard columns typically cost one-fourth or less of the cost of an analytical column.

Analytical Column

The heart of an HPLC system is the analytical column. This is where the mixture of chemicals injected in the system is separated into individual analytes that appear as peaks in the chromatogram. Columns are available in a variety of diameters and lengths ranging from large preparatory columns (20 - 50 mm in diameter by 50 - 250 mm in length), to analytical columns (typically 4.5 mm in diameter by 12 - 25 mm in length), to narrow bore analytical columns for improved performance and MS applications (1-2 mm in diameter by 10 cm in length), to capillary columns for MS detectors (from 0.075 - 0.1 mm in diameter). Larger columns require a larger mobile phase volume to push the analytes through the system. A collection of HPLC columns is shown in Figure 5.

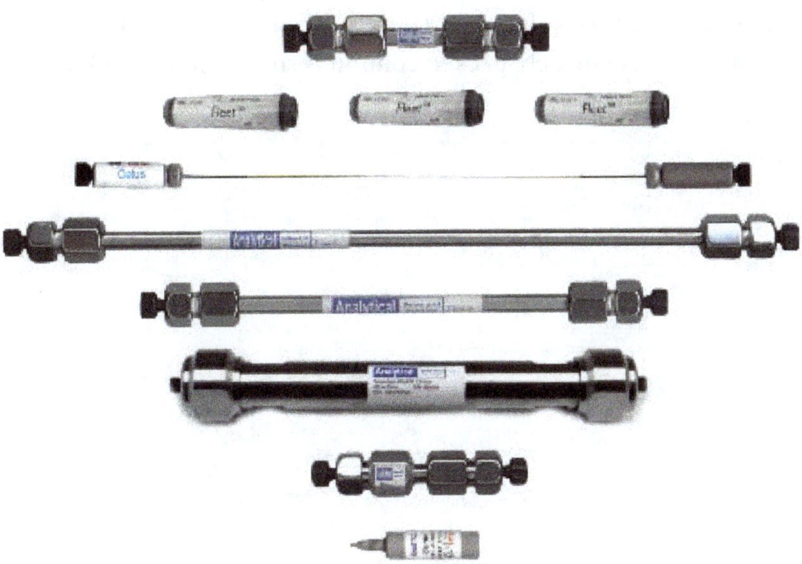

Fig. 5: Various Sizes of HPLC Analytical and Quard Columns..

With the exception of relatively new capillary columns, all HPLC columns are packed with small resin beads (typically 2-5 μm in diameter or less) that contain a coating of stationary phase. As in GC, stationary phases are selected based on the expected intermolecular forces available to the analyte for bonding. The adage "like dissolves like" is also appropriate for liquid chromatography applications. Separations in HPLC are considerable more complicated then in GC. In GC, the chromatographer is only concerned with the stationary phase and the temperature program. In HPLC, one must also be concerned with the polarity of the mobile phase . Adequate analyte separations are only achieved with an appropriate match of gradient polarity and stationary phase interactions.

In more advanced applications, the pH and ionic composition of the mobile phase must also be controlled. The most common stationary phases used in reversephase HPLC include alkylamine, octodecyl (C18), octyl (C8), butyl (C4), cyanopropyl (CN), and methyl functional groups. Additional and custom-made stationary phases can be ordered from many suppliers.

Early chromatography columns that were open to the atmosphere were easy to pack. The chromatographer filled a glass column (1 to 5 cm in diameter) with solvent, slowly poured in silica gel with stirring to remove air pockets, and applied the solution to be separated to the top of the column. These types of columns yielded adequate separation for organic chemists attempting to isolate their synthesis products but have extremely limited use for analytical chemists who require more theoretical plates in the column and on-line detection of column effluents. Today, analytical chromatographers almost exclusively purchase their separation columns from manufactures because of the need for perfectly packed columns with no void spaces. The emphasis here is on perfect column packing since the presence of only a few void areas in a column will significantly affect the theoretical plate height in the column and decrease resolution. A comparison of a poorly packed column and an adequately packed column is shown in Animation5.

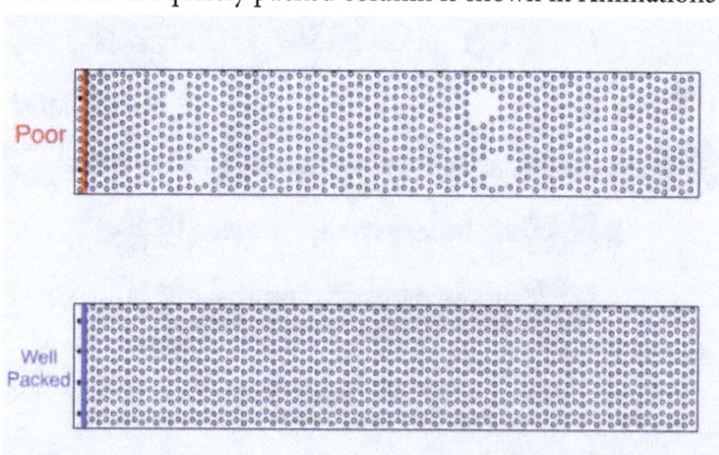

Animation 5 The Effect of Adequate Column Packing on Peak Broadening in HPLC Columns

The particle size of the stationary phase also significantly affects the operation of an HPLC by determining the shape of the van Deemter curve. Figure 3.6 shows the van Deemter curves for a variety of particle-sized stationary phases. Recall from our discussions in Chapter 1, better separations occur at the minimum point (linear velocity) in the van Deemter curve and a more stable system operation will occur with a large range of linear velocities. In HPLC, smaller sized stationary phase particles will yield a van Deemter curve that is essentially flat (refer to Figure.6). Thus, a range of flow rates will produce optimum analyte separation.

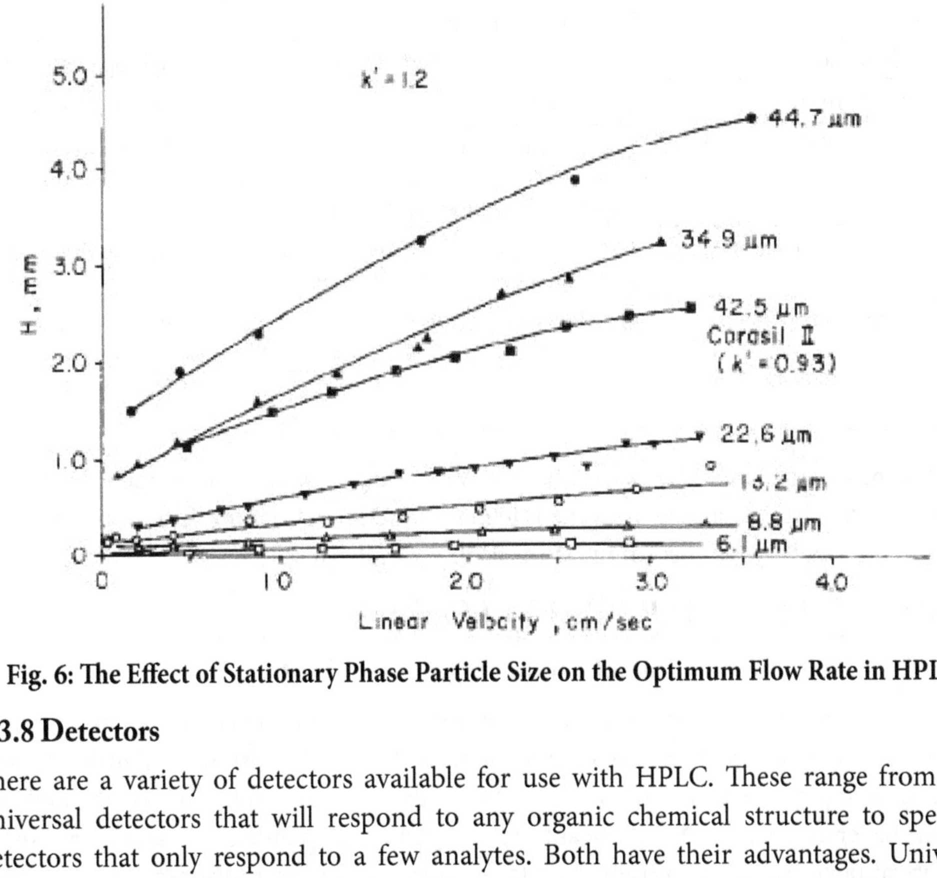

Fig. 6: The Effect of Stationary Phase Particle Size on the Optimum Flow Rate in HPLC.

5.3.8 Detectors

There are a variety of detectors available for use with HPLC. These range from near universal detectors that will respond to any organic chemical structure to specialty detectors that only respond to a few analytes. Both have their advantages. Universal detectors provide economic use for a large diversity of chemicals but specific analyte detection can be hampered if two or more analytes elute from the analytical column at the same retention time. Specialty detectors provide near-interference free detection of only a few analytes. Common detectors, along with their use and approximately detection limits are given in Table 1.

Table 1. Common Commercially Available HPLC detectors, Applications, and Detection Limits.

Detector	Application(s)	Detection Limits
Uv-visible absorbance	For compounds that absorb in the uv or visible range	Pg quantities
Fluorescence	For compounds capable of fluorescence (especially polyaromatic hydrocarbons)	Fg quantities
Refractive index (ri)	For alcohol, sugar, saccharide, fatty acid, and polymer analysis with refractive indices different from the solvent	Ng quantities
Electrochemical	For analyzing a wide range of compounds	High pg quantities
Conductivity for ic	Mainly for inorganic ions	~Ng quantities
Evaporative light scattering (els)	For a wide variety of compounds that lack uv/vis chromophores including triglycerides, sugars and natural products	Mg quantities
Fourier transfer infrared	For compounds with vibrational functional groups	Mg quantities
Mass spectrometry	A universal detector	Mg to pg quantities depending on the type of mass spectrometer

The most common detector is the UV-Vis detector that comes standard on basic HPLC systems. This is a near-universal detector since most organic compounds have a chromophore capable of adsorbing UV or visible wavelengths.

5.4 ADVANCED AND SPECIALTY LC SYSTEMS

5.4.1 Ultra-High Pressure Liquid Chromatography, U-HPLC:

HPLC technology made a significant advance with the development of ultra high pressures systems. These systems, currently operate at 15 000 to 20 000 psi, offer significant time and cost saving per analysis. The increased pressure allows samples to pass through the column faster and results in significant increases in column efficiency (and decreases in theoretical plate height) by minimizing longitudinal diffusion processes. These systems offer higher column efficiencies (resolution per column length) and allow the systems to operate at a much wider range of linear velocities, flow rates, and backpressures. Nanoboresized columns offer up to 95 percent decreases in solvent use and interface well with mass spectrometers. Higher pressure systems are under development that will possibly operate at up to 100 000 psi and yield even greater column efficiencies.

5.4.2 Ion Chromatography, IC:

Ion chromatography is a form of HPLC where the system is modified to analyze for mainly inorganic ions such as nutrient ions (i.e. nitrate, sulfate, phosphate, etc.) as well as many organic anions and metal cations. The systems require a special ion exchange analytical column that is specific to the analytes of interest, a relatively simple conductivity detector, and a unique ion suppressor column for removing ions other than the analytes that would generate electrical conductivity in the detector. Columns and tubing are usually composed of PEEK to avoid reactions on metal surfaces.

Analytical columns are mainly divided into columns for cations and anions and are similar in diameter to standard HPLC columns but slightly longer in length. Cationic exchange columns have active strong acid sites such as sulfonic acid (- SO3 -H+) or weak acid sites such as carboxylic acid groups (-COO-H+) that preferentially exchange with analyte cations. Common anionic exchange sites include the strongly basic tertiary amines (-N(CH3)3 +OH-) and the weakly basic primary amine group (-NH3+OH-). Both exchange groups are usually placed on porous microparticles of silica. As the injected sample passes through the analytical column, shown below for the sulfonic acid exchange surface, the following reaction occurs

$$\text{-R-SO}_3\text{ –H}^+{}_{\text{solid}} + \text{Metal Cation}^+{}_{\text{solution}} < \text{---} >$$

$$(\text{-R-SO}_3\text{ -})_n\text{ Metal Cation}^+{}_{\text{solid}} + \text{nH}^+{}_{\text{Solution}}$$

Similarly for anion analysis, the reaction would be

$$\text{-NH}_3\text{ +OH}_{\text{solid}} + \text{Anionsolution} < \text{---} >$$

$$(\text{-NH}_3\text{ +})_n\text{ Anion}^-{}_{\text{solid}} + \text{nOH}_{\text{solution}}$$

At appropriate flow rates, an equilibrium migration front is set up for each analyte passing through the column. Since each metal cation will have a different affinity for the stationary ion exchange resin, each analyte will elute from the column at a different time. As in all types of chromatograph (in the absence of mass spectrometry detection), identification is largely based on retention time.

The detector in IC is a simple conductivity detector where a potential is placed across two electrodes. In the absence of ions in the solution passing through the detector, minimal current is transmitted and no signal is generated by the electronics. As each packet of analytes, cations or anions, pass through the detector a current is generated that is proportional to the concentration of analytes. In order to detect low concentrations of analytes, the background signal (current) of the mobile phase must be maintained as low as possible.

Since all samples have counterions (an equal concentration of cations or anions) half of the ionic components must be removed prior to entering the detector. For example, detecting the presence of cations require the removal of anions before the column effluent reaches the detector. The removal of these ions is accomplished with an ion suppressor device shown in Figure 3.7. This device is positioned within the ion chromatography system as shown in Figure 3.8.

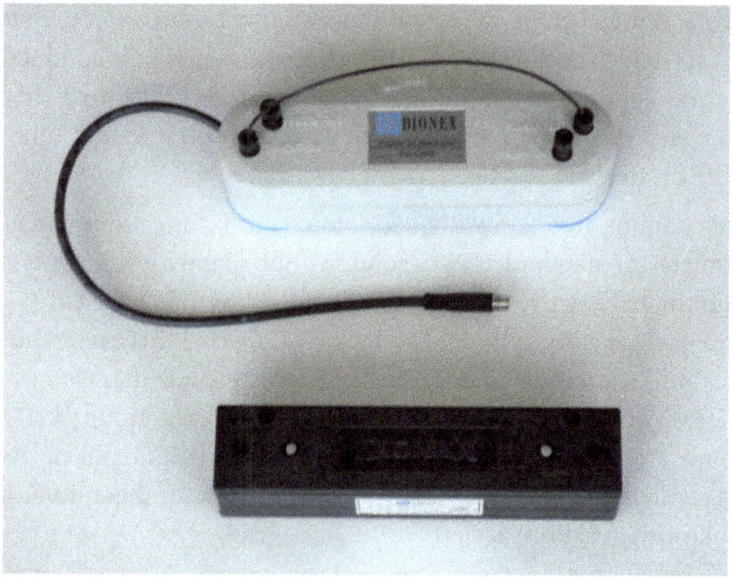

Fig. 7: Photograph of Two Ion Suppressor Devices.

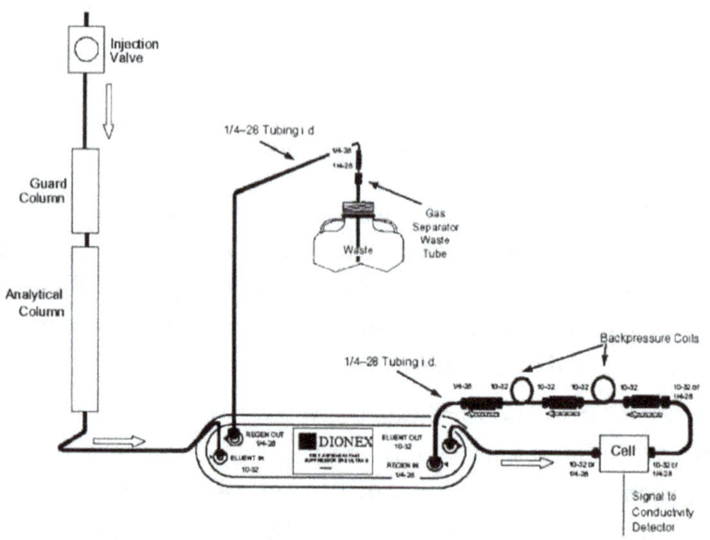

Fig. 8: Overview of the Components used to Separate Analytes in an Ion Chromatography System.

As presented in Figure 8, the sample first passes through the analytical column where the cations or anions are separated relative to their affinity for the solid-phase exchange resin. The effluent from the analytical column then enters the ion suppressor device, along with acid or base to keep the ion suppressor column active. For anionic analysis, strong acid (H+) is used in the suppressor device; for cationic analysis, strong base (OH- or COO-) is used in the device. In both cases the analytes of interest are unchanged and unretained. As a consequence, the following suppression reactions occur depending on the mode of operation (cation or anion analysis).

In anion analysis, cations in the sample are exchanged for H+ in the suppressor column and neutralized in the center flow through portion of the suppressor column by the following reaction

$$H^+ \text{ solution} + \text{Clsolution} + \text{resin}{+}OH_{solid} \text{ --- > resin+Cl solid} + H_2O$$

Thus, only anions, such as Cl-, contribute to the conductivity in the detector. In cation analysis, anions in the sample are exchanged with HCO3 - (or OH-) in the suppressor column and neutralized in the center flow through portion of the suppressor column by the following reaction cation+ solution + HCO_3 - solution + resin-H+ solid --- > resin-cation+ solid + H_2CO_3solution

Thus, only cations, such as Ca^{2+}, contribute to the conductivity in the detector. Again, note that these reactions convert all ionic species, except for the analytes, to non-conductive chemicals. These reactions in relation to the flow through the ion suppressor are illustrated in Figure 3.9. In both cases, non-ionic species are produced and the interfering ions are removed (suppressed) from the solution.

This significantly decreases the background ions in solution and allows for low detection limits (parts per trillion to parts per billion in the injected sample). Ion chromatography systems also operate well with MS detectors.

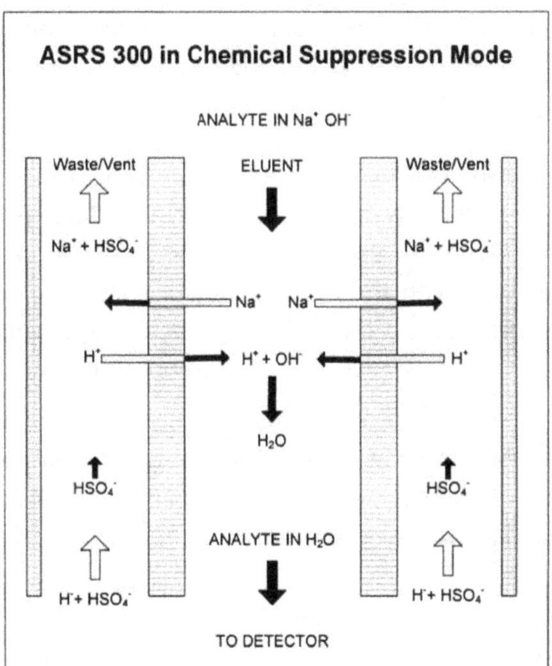

Fig. 9: Cation suppression in an ion suppression device.

5.4.3 Super Critical Fluid Chromatography, SCF

A liquid turns into a super critical fluid (SCF) when its temperature rises above the critical temperature—the temperature where it can no longer exist as a liquid no matter how much pressure is applied. Super critical fluids exist in a state between a liquid and a gas and have the penetration abilities of a gas and the dissolving power of a liquid. As a result, SCF is a mixture of GC and HPLC and in some cases is superior to GC or HPLC. Super critical CO_2, the same matrix used tonselectively extract caffeine from coffee beads and nicotine from tobacco products, can be used as a mobile phase in chromatography. Super critical fluid chromatography is a special form of HPLC where a near-identical system is used but the mobile phase, as noted above, is super critical CO_2,. The system is therefore pressured and temperature controlled to maintain the super critical fluid. SCF is a form of normal phase chromatography that is used for the analysis of thermally labile molecules. The same types of packed and capillary columns that are used in HPLC are utilized in SCF. Due to the nature of the super critical fluid, packed columns can actually contain more theoretical plates than capillary column. Also, the shape of the van Deemter curve is different from those observed in GC and HPLC in that a minimum plate height exists over a very broad range of linear velocities. SCF can be used for a variety of separations but it is most commonly used in the separation of chiral compounds in the pharmaceutical industry. In place of a solvent gradient the chromatographer uses pressure programming and the affinity of the stationary phases to separate complex mixtures of analytes. Pressure programming in SCF is analogous to temperature programming in GC and gradient programming in HPLC. Methods of detection include UV/Vis, mass spectrometry, FID (as in GC but unlike in HPLC) and evaporative light scattering.

5.5 APPLICATIONS

HPLC extends the capabilities of chromatographic separations past the abilities of capillary GC by allowing the analysis of thermally unstable analytes but at the expense of poorer resolving power due to the packed column nature of the system. These limitations are being slowly overcome with the use of capillary columns and ultra high pressure systems. This section highlights a few HPLC applications. As with GC, major column manufacturers and distributors provide very useful websites that contain example chromatograms for common analytes.

These sources can be extremely helpful in selecting the appropriate column for your needs. In addition, technical help is available from professional chromatographers at these companies with more complex separations.

The first example we will illustrate is for polyaromatic hydrocarbons that can be analyzed very efficiently by GC. While HPLC also offers adequate separation (as shown in Figure 3.10 below) as compared to GC, HPLC allows improved detection limits with the use of fluorescence detection.

PAHs on 3 mm Columns

1. Napthalene
2. Acenaphtalene
3. Fluorene
4. Phenanthrene
5. Anthracene
6. Fluoranthene
7. Pyrene
8. Benzo(a)anthracene
9. Chrysene
10. Benzo(a)fluoranthene
11. Benzo(k)fluoranthene
12. Benzo(a)pyrene
13. Dibenzo(a,h)anthracene
14. Benzo(g,h,i)perylene
15. Indeno(1,2,3-cd)pyrene

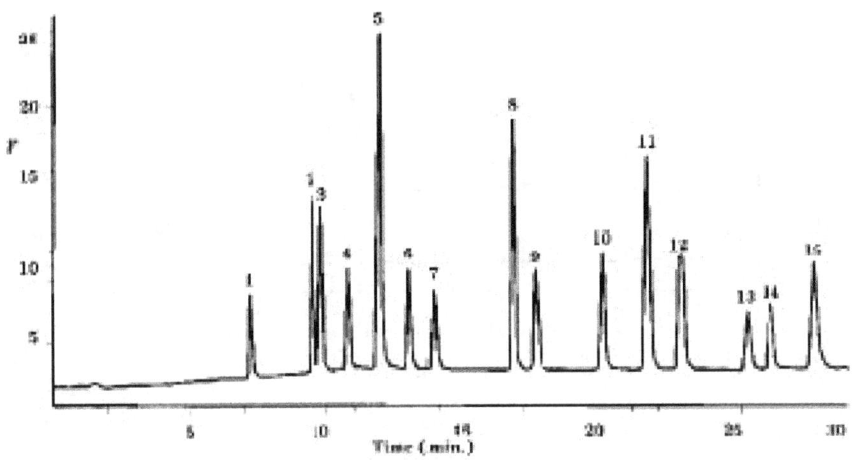

Sample:	PAH Standard
Column:	LiChrospher PAH, 3.0 X 250 mm, 5 μm, (P/N: 79925PA-583)
Mobile phase:	A = Water, B = Acetonitrile
Gradient:	0 min 50% B, 3 min 60% B, 15.4 min 100% B, 23.5 min 50% B
Flow rate:	0.8 mL/min
Temperature:	27°C
Detection:	254 nm

Fig. 10: Separations of PAHs by HPLC.

Vitamin A and Beta Carotene

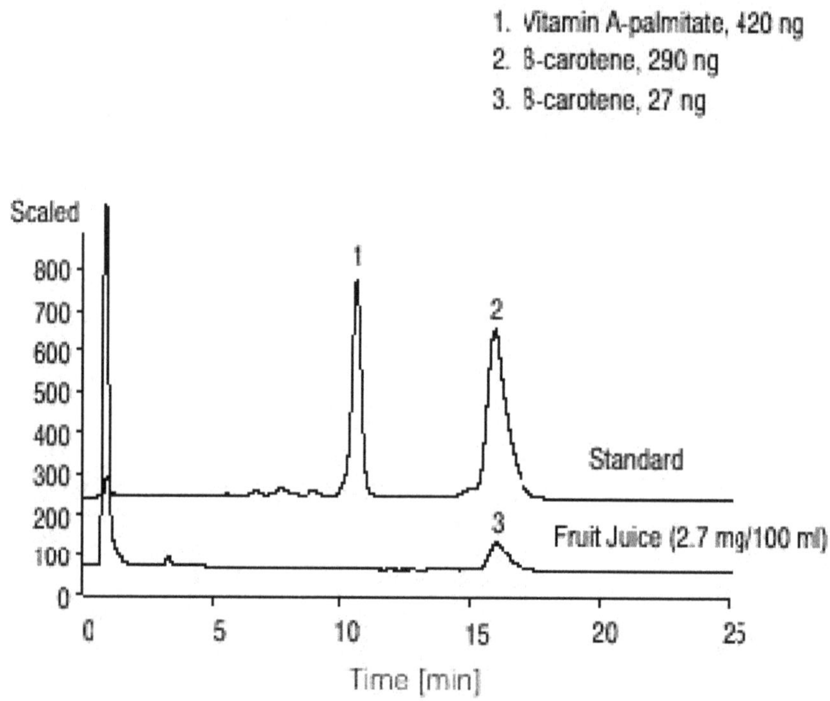

Sample:	ß -carotene in fruit juice
Column:	Lichrospher RP 18, 4 x 125 mm, 5 µm (P/N: 799250D-564)
Mobile Phase:	Methanol + 5 g/L Lithiumperchlorate + 1 g/L Acetic Acid
Flow Rate:	1.5 mL/mn
Temperature:	30°C
Detection:	Amperometric, 1 V (range 0.5 µA, ref. AgCl/KCl)

Pub No.: 5091-3194E

Fig. 11: Isolation of Vitamin A by HPLC. Source: Copyright 2006 Agilent Technologies, Inc. Reproduced with Permission

Proteins: Separation by SEC

1. Mouse iGM	900,000 D
2. Bovine thyroglobin	670,000 D
3. Sweet potato b-amylase	200,00 D
4. Bovine serum albumin	67,000 D
5. Chicken albumin	45,000 D
6. Bovine RNAase	13,700 D
7. Azide	65 D

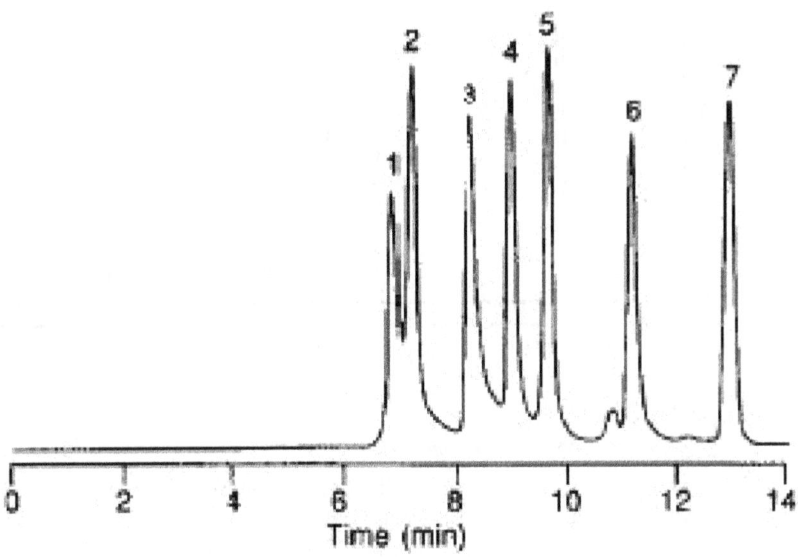

Sample:	Proteins
Column:	ZORBAX GF-250, 9.4 x 250 mm, 4 μm (P/N: 884973-901)
Mobile Phase:	130 mM NaCl, 20 mM KCl, 50 mM Na_2PO_4, pH 7.0
Flow rate:	1 ml /min
Temperature:	Ambient
Detection:	UV 210 nm

Fig. 12: The separation of proteins by HPLC. Source: Copyright 2006 Agilent Technologies, Inc. Reproduced with permission.

Food Colors, FD&C

1. Yellow # 5	C16H9N4Na3O9S2	MW= 534
2. Red # 40	C18H14N2Na2O8S2	MW= 496
3. Blue # 1	C37H34N2Na2O9S3	MW= 760
4. Propylparaben	C10H12O3	MW= 180
5. Red # 3	C20H4I4Na2O5	MW= 878

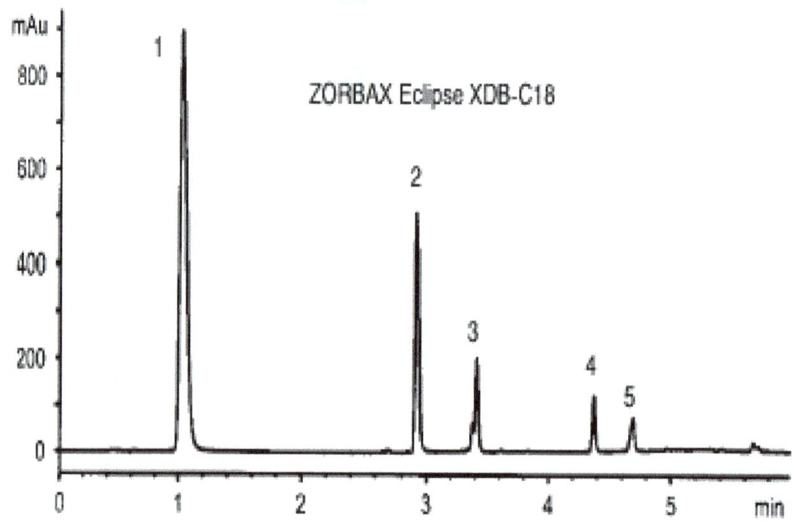

Operating Conditions:

HPLC System:	Agilent 1100 with quaternary pump
Column:	ZORBAX Eclipse XDB-C18 Rapid-Resolution (3.5 μm), 4.6 x 50 mm PN 935967-902
Mobile Phase:	A: 0.1% TFA, pH to 4.4 with TEA, B: MeOH
Gradient:	17 to 100% B / 4 min
Detector:	UV 254 nm
Flow:	1 mL/ min.
Temperature:	ambient
Pub:	LI FD16

Fig. 13: The separation of food coloring agents by HPLC. Source: Copyright 2006 Agilent Technologies, Inc. Reproduced with permission. A separation of anions in drinking water by IC is shown in Figure 3.14.

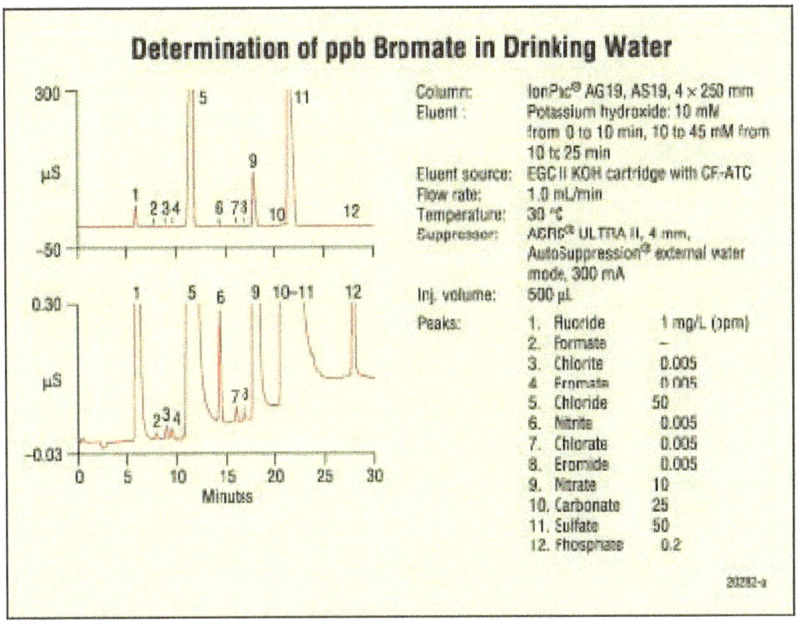

Fig. 14: Separation of anions by IC. Source: Dionex Corporation, Inc. Reprinted with permission, courtesy of Dionex Corporation, Sunnyvale, California.

5.6 SUMMARY

This chapter described the basic applications and operation of an HPLC system. Modern HPLC systems significantly expand the capabilities of chromatography to include the relatively delicate biomolecules that are so abundant in nature. Current research in ultra high pressure systems may result in HPLC performances that match those of capillary column GC in the near future. Basic non-confirmatory detectors systems were described in this chapter and confirmatory identification by mass spectrometer techniques that are applicable to all forms of chromatography will be covered in Chapter 4.

6

FUNDAMENTALS OF PAPER CHROMATOGRAPHY

6.1 INTRODUCTION

The technique of paper chromatography was first discovered by Synge and Martin in 1943. Paper chromatography is specific type of technique that operates on a specific piece of paper. It is a type of planar chromatography, in which separation of compounds is performed using a filter paper made up of cellulose which acts as a stationary phase. The method is comparatively cheap and helps to separate dissolved chemical substance by their different rates of migration through paper sheets. The method requires very minute quantity of sample for analysis.

Chromatography is a well known technique used for separation of compounds. Among all, chromatographic techniques, paper chromatography is a type of analytical tool which is used for separation of colored components. The principle involved may be separation and partition of components based on their affinity towards stationary phase. Literature survey reveals that paper chromatography has been used from ancient times, especially in countries like United State for separation of organic and inorganic components. Further, investigation should be made to make new advancement in the field of chromatography separation involving identification of types of paper used and gel permeation process.

6.2 PRINCIPLE

The basic principle involved in paper chromatography is partition in which the various components get distributed or partitioned between liquid phases. It involves use of aqueous solvent held in pores of filter paper which acts as stationary phase whereas mobile phase travels over the paper. Due to differences in their affinity towards water (in stationary phase) and mobile phase solvents, the compounds in the mixture get separated through capillary action of the pores in the paper. The components may also be separated on the basis of principle of adsorption between solid and liquid phases, where solid surface of paper serves as stationary phase and mobile phase is a liquid solvent. Although the main working principle of paper chromatography is partitioning this is employed in many pharmaceutical applications.

6.3 MIGRATION PARAMETERS

The positions of migrated spots on the chromatography are indicated by different terms such as RF, RI, RM and RC. These parameters are also qualitative and quantitative parameters of substance.

RF= The term RF is associated with the migration of the solute relative to the solvent front as: $Rf = Distance\ traveled\ by\ t'e\ solute\ from\ origin\ lineDistance\ traveled\ by\ t'e\ solvent\ from\ origin\ line$. The RF is the function of the partition coefficient, it is constant for the given substance, provided the condition of chromatographic systems are kept constant with respect to temperature, type of paper, duration, direction of development, nature and the shape and size of the wick. The RF values of a substance depends on no of factors such as:

- Solvent system
- Medium of separation i.e. quality of paper
- Nature of mixture
- Temperature
- Dimensions of vessel in which operation is to be carried out.

6.4 EXPERIMENTAL DETAILS FOR QUALITATIVE ANALYSIS

- **Choice of proper chromatographic technique:** The first and prime job is to identify the mode of paper chromatographic technique required which mainly depends upon nature of the sample to be analyzed.
- **Choice of the filter paper:** The success of paper chromatography depends upon appropriate use of filter paper. The choice of paper is dependent upon the type of problem i.e. sample under investigation. The prime factors, that governs the choice are as follows:
- Type of analysis is qualitative or quantitative in nature or it is analytical or preparative technique
- Nature of the substance i.e. hydrophilic or lipophilic neutral or charged species.

6.5 TYPES OF PAPERS

A wide variety of papers, which are very uniform from lot to let are commercially available in different sizes, porosities, shapes, thickness and chemical treatments. In general, cellulose fiber is the main component of filter paper. The cellulose fiber is a linear polymeric carbohydrate chains owning hydrophilic character and is further cross linked with stable hydrogen bonded system. Water or other very popular types solvents are tightly held within the hydrophilic cellulose system and can be considered to be different from bulk water or polar system.

The cellulose papers can be altered in different ways to modify its chromatographic comportment. For example, paper can be impregnated with diatomaceous earth, alumina, silica gel, ion exchange resins. These kinds of papers will exhibit properties of these adsorbents and consequently, influence the retention of the stationary liquid and the adsorption or partition sequence of a mixture. The ion exchange resin impregnated paper will have either cation or anion exchange properties. If the paper is acetylated, the paper takes on a hydrophobic property. That is, it tends to retain a hydrophobic type solvent rather than a hydrophilic type solvent as a stationary phase. This type of application is referred to as reverse phase chromatography. This paper can be made hydrophobic by silicone treatment or by impregnating it with inert nonpolar-type organic polymers.

A glad fiber type paper can be used if very corrosive eluting conditions are needed. Adsorption symptoms due to the glass can be reduced by special treatment. Different types of Whatman chromatography papers are available and the choice of paper relies on type of separation. The most commonly used Whatman filter papers have a content of 90% of α-cellulose. Cellulose paper is also used as a support for various adsorbents like alumina, silica, zirconium oxide etc. which get precipitated in the pores of the filter paper to produce a thin sheet of adsorbent with the flexibility of the paper but having adsorbing characteristics of the precipitation.

6.6 STATIONARY AND MOBILE PHASE

Paper chromatography is essentially partition chromatography and there is wide variety of useful combinations of stationary and mobile phases. It is not necessary that the two systems be immiscible. The types of stationary phase that are used can be classified as aqueous, hydrophilic and hydrophobic systems.

6.6.1 Stationary phases

1. **Aqueous stationary phase:** Water is readily held by paper. Therefore, water-equilibrated paper is attached by suspending paper in a closed chamber whose atmosphere is saturated with water. If an aqueous buffer or salt phase is required, the paper is drawn through the respective solution and then exposed to water saturated atmosphere in a chamber. This type of system is particularly suited for separation of moderate polar to extremely polar mixtures.
2. **Hydrophilic stationary phase:** An organic solvent can be used for hydrophilic stationary phase. If the solvent is volatile enough, the paper can be equilibrated in a chamber whose atmosphere is saturated with solvent. Alternatively, the stationary phase solvent is dissolved in a very volatile diluent evaporates leaving the stationary phase liquid uniformly distributed throughout the paper. Commonly used hydrophilic solvents include, form-amide, methanol, glycerol and glycols.

3. **Hydrophobic stationary phase:** The paper must be modified previously, before it will exhibit a tendency to retain hydrophobic stationary phase. Equilibration in the vapours of the solvent is the dipping technique in a solution of the solvent and a volatile diluent are used for introducing the hydrophobic solvent into the modified paper. Solvents such as dimethylformamide, aromatic and aliphatic hydrocarbons and kerosene are commonly used.

6.6.2 Mobile Phase

Mobile phase in various combinations can be used in paper chromatography. Choosing optimum eluting condition is a trial and error process. However, certain guidelines can be used to predict eluting conditions. For example, the characteristics of the components in the mixture and the type of stationary phase being employed should be considered. The solvent system used in paper chromatography is usually a mixture of organic solvent with water. The ionization of analytes can be controlled by addition of acids (HCl, HNO_3, acetic acid) and bases (NH_3). Different combination of solvent system can be used for identification of compounds based on their chemical nature for example; for amino acids the solvent system used is acetic acid: water: n-butanol in ratio of 1:5: 4. For separation of sugar; solvent system composed of ethylacetate: pyridine: water; water: ethyl acetate is suitable. For inorganic ions, solvents like pyridine: water or HCl: water more popular.

6.7 MODES OF PAPER CHROMATOGRAPHY

Paper chromatography can be performed on the basis of method of development of chromatogram. Broadly; it can be classified as:

6.7.1. Ascending Chromatography

The technique of ascending chromatography was introduced by Consden, Gordan and Martin and was later modified by Williams and Kirby. As the name indicates, the chromatogram is allowed to ascend and development of chromatogram takes place due to movement of solvent in upward direction on the paper. In this technique, solvent reservoir is kept at the bottom of the beaker and piece of paper with loaded sample is dipped in solvent. It is always recommended to take care that the spot should remain above the solvent system. Furthermore, it is also important to take care about size of paper to avoid bending and crumpling.

6.7.2. Descending Chromatography

In this chromatography, the development of chromatogram is done by allowing the solvent to travel dZown the paper. The solvent reservoir is kept at the top and process of movement of solvent is assisted against the gravity. This method is preferred over simple ascending chromatography due to (i) constant flow rate if solvent, (ii) Less time consuming, (iii) the ease of separation of solutes with low Rf value. It only drawback of this technique compared with ascending chromatography is requirement of extensive apparatus.

6.7.3. Ascending Descending Chromatography

This describes modified form of paper chromatography which involves ascending and descending flow of solvent on the same piece of paper. The advantages of this method over other methods are (i) The run time is reduced, i.e. needs short span of time, (ii) Components with RF value > 0.50 can be detected individually as they will have own individual channel, (iii) Longer flow distance available which gives better resolution. The RF value obtained by ascending descending chromatography is not significantly different those from ordinary techniques.

6.7.4. Radial Chromatography

The term radial chromatography was described by Rutter and involves use of circular filter paper in which components get separated in the form of concent rings rather than a single spot. A list of advantages was also given which includes (i) Sharpness and resolution of separation, (ii) speedy separation, (iii) Simplicity and compactness of the employed apparatus, (iv) Control on rate of flow of solvent, (v) Reproducibility, (vi) Ease of removal of test samples during and after development. The technique employs use of filter paper immersed in elutant and placed horizontally in a petri dish. The elutant flows from the center towards periphery of the paper and kept in covered petri plate for development of the chromatogram. The wick of the paper is dipped in mobile phase and solvent flows over the paper and the spots appear as concentric rings.

6.7.5. Two-dimensional Chromatography

Two-dimensional chromatography (Circular chromatography); is one of best method for separation of organic and inorganic compounds. Development of chromatogram is carried out in two different directions at right angle to each other. The sample is spotted on one of the corner of Circular paper and is allowed to develop in one direction followed by immersing of paper again in same solvent at right angle to the first one. The diagrammatic represents of various types of paper chromatography is depicted in **Figure 1.**

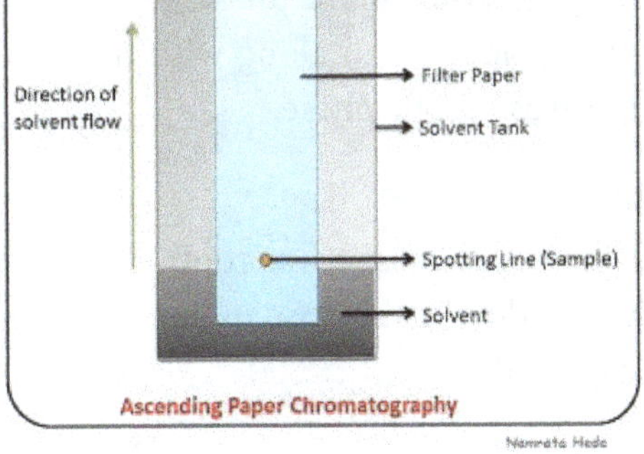

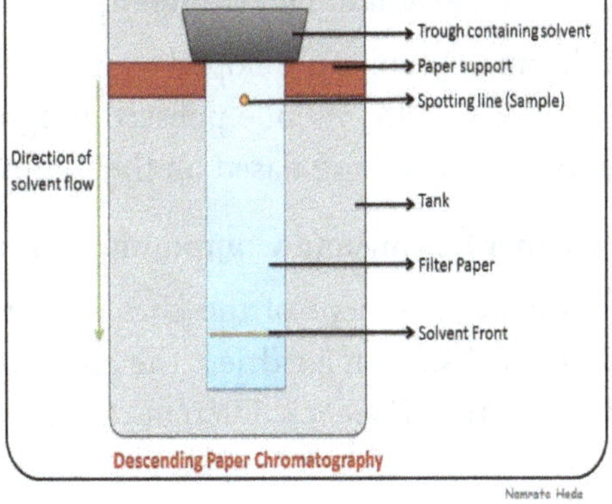

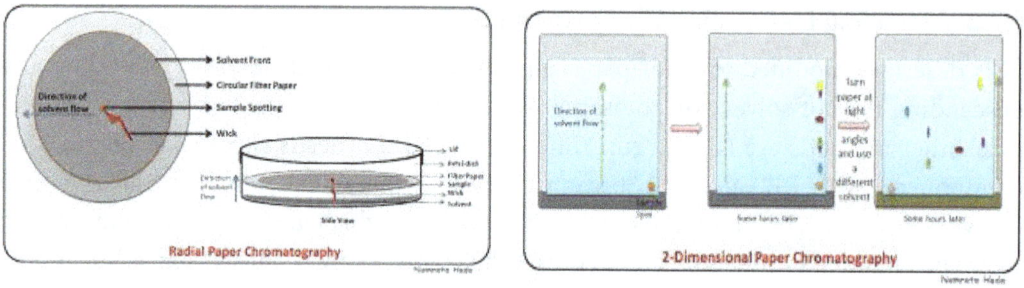

Fig. 1: Different types of paper chromatography techniques

6.8 PROCEDURE

The basic procedure for performing paper chromatography is explained below:

6.8.1 Selecting a Suitable Type of Development

The type of development is selected on the basis of complexity of the solvent, type of paper used and nature of sample etc. Most of the time, radial chromatography is preferred due to its ease of operation and high resolution. Moreover, it is easy to perform, less time consuming and results are reproducible.

6.8.2 Selecting a Suitable Filter Paper

Sample quality and pore size of filter paper helps to identify the appropriate type of filter paper to be used. Usually, a thin layer of filter paper (Whatman No. 1) is commonly used.

6.8.3 Prepare the Sample

Sample is prepared by dissolving it in a suitable solvent and should be inert with sample under analysis and most of the times mobile phase is used for the purpose.

6.8.4 Spot the Sample on the Paper

Using a capillary tube, the sample should be appropriately spotted in the center on the paper and should be at a proper position.

6.8.5 Chromatogram Development

The chromatogram is developed by immersing paper in the mobile phase. As soon as the filter paper gets the mobile phase through capillary action and along with the samples components also move based on their affinity towards mobile phase.

6.8.6 Paper Drying and Compound Detection

After the development of the chromatogram, the paper is dried at room temperature followed by using an air drier. The components are identified by use of detecting agents and are characteristic for different chemical compounds.

6.9 APPLICATIONS OF PAPER CHROMATOGRAPHY

- Paper chromatography is a useful method to identify number of constituents present in a sample, with a correctly chosen mobile phase.
- This method requires small scale setup, involves very minute quantity of sample and is also cost effective.
- Paper chromatography is an effective tool for separation of free amino acids present in human serum.
- It also offers a rapid method of separating and estimating sugars quantitatively; however the identification depends upon determination of their physical constants and formation of characteristic derivatives.
- Paper chromatographic technique is also used for carrying out assay of pharmaceutical compounds such as mixture of phenylephrine hydrochloride, chlorpheniramine hydrochloride and dextromethorphan hydrochloride.
- The technique is also useful in isolation of pair of components having sample RF values using two-dimensional paper chromatography.
- Paper chromatography also involves inorganic applications such as separation of cations like cadmium, zinc, mercury, beryllium and calcium.
- It technique is also helpful in identification of accelerator and anti-oxidant in rubber and is useful for determining its quality.
- Paper chromatography is widely used in detection of various plant constituents such as opium, quinine alkaloids.
- Paper chromatography is also used to determine rate of ongoing reaction. Therefore, is a valuable tool in synthetic chemistry.

7

CHROMATOGRAPHY AND MASS SPECTROMETER

7.1 INTRODUCTION

We know that the biochemistry or biological chemistry deals with the study of molecules present in organisms. These molecules are called as biomolecules and they form the basic unit of every cell. These include carbohydrates, proteins, lipids and nucleic acids. To study the biomolecules and to know their function, they have to be obtained in purified form. Purification of the biomolecules includes many physical and chemical methods. This topic gives about two of the commonly used methods namely, chromatography and mass spectrometry. These methods deals with purification and separation of biomolecules namely, protein and nucleic acids.

7.2 CHROMATOGRAPHY

When we have a mixture of colored small beads, it is easily separated by visual examination. The same holds true for many chemical molecules. In 1903, **Chromatography and Mass Spectrometer** Mikhail, a botanist (person studies plants) described the separation of leaf pigments (different colors) in solution by using solid adsorbents. He named this method of separation called chromatography. It comes from two Greek words:

Chroma – colour

graphein – to write/detect

Modern separation methods are based on different types of chromatographic methods. The basic principle of any chromatography is due to presence of two phases:

- **Mobile phase** – substances to be separated are mixed with this fluid; it may be gas or liquid; it continues moves through the chromatographic instrument
- **Stationary phase** – it does not move; it is packed inside a column; it is a porous matrix that helps in separation of substances present in sample. It works on simple physical process called **adsorption**.

7.2.1 Components of Chromatography

The equipment used for separation of mixture in a sample is called a chromatograph. The major components of a chromatograph:

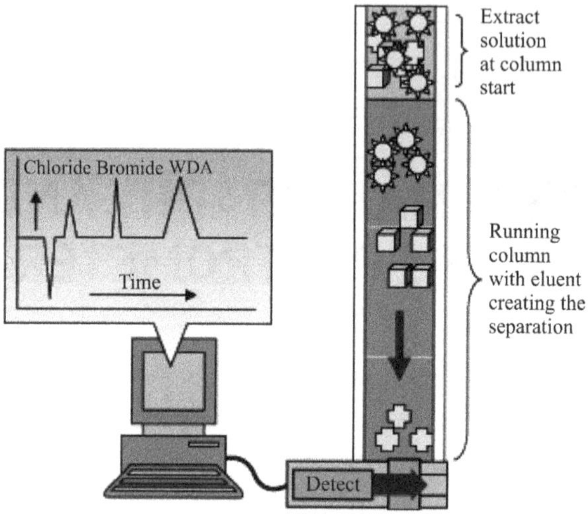

Fig. 1:

- Mobile phase source – separate chambers or containers having either gas or liquid
- Analyte – sample to be separated
- Sample injection – chamber to inject the sample
- Separating column – having the stationary phase
- Detector – identify the change in separation of a molecule; this calculates the time between injection and detection of a sample, which is called as **retention time.**
- Recorder and analyser – usually a computer attached to the chromatograph.

This records the response of detector as a graph called chromatogram

7.2.2 Types

There are many types of chromatographic methods present today. Some of the important methods are described as follows:

- Gas chromatography
- Liquid chromatography
- Gel filtration chromatography
- Ion exchange chromatography
- Affinity chromatography
- Others- hydrophobic interaction chromatography (HIC), Isochromatic focusing (ICF), etc.

7.2.2.1 Gas Chromatography

The gas chromatography principle involves separation of the components of the sample due to separation in between the gaseous mobile phase and stationary phase (usually silica). The components separated into gas comes out first while other comes later. It detects compounds like fatty acids, essential organic solvents, flavored oils, etc.

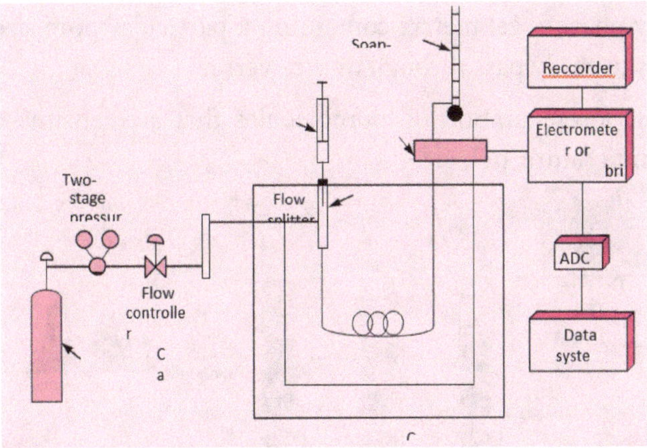

Fig. 2: Gas chromatography

In GC, a liquid sample is injected into the column. The GC column is usually coated with stationary phase and placed inside an oven chamber. The sample is vaporized as it passes the column which is more than its boiling point. The sample compounds are carried to column by gas (usually helium or nitrogen) and then to a detector. The detector signals a chart recorder for chromatogram (Fig.22.2)

7.2.2.2 Liquid Chromatography

It is separation technique where the mobile phase is a liquid. It can be performed in a column or a plane surface. This has been advanced to many types :

- HPLC – high performance liquid chromatography (Fig.3)
- FPLC – fast protein liquid chromatography

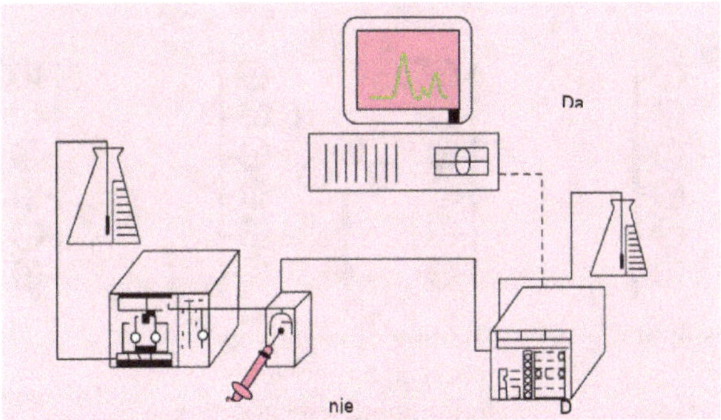

Fig. 3: HPLC

7.2.2.3 Gel Filtration Chromatography

This chromatographic technique is also known as size exclusion chromatography or molecular sieve chromatography. It involves the separation of molecules based on their molecular weight and size. (Fig.4) It involves:

- Mobile phase – liquid mixed with sample mixture
- Stationary phase – gel matrix containing a particular pore size to allow smaller molecules to easily pass through or vice versa.

It well suitable for separation of biomolecules that are sensitive to environmental conditions like temperature, pH, etc.

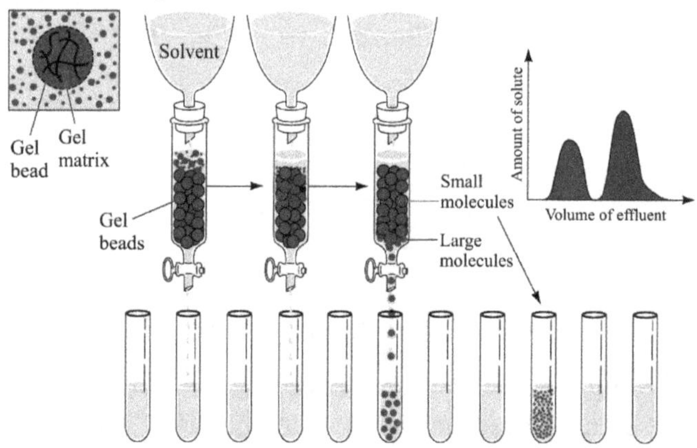

Fig. 4: Gel Filtration Chromatography

7.2.2.4 Ion Exchange Chromatography

Every chemical molecule that is capable of forming into ions (different charges – positive, **cations** and negative, **anions**) is separated by using this chromatography (Fig. 22.5). In this:

- Mobile phase – liquid
- Stationary phase – matrix of beads with either positive charge (cationic) to separate anionic sample or negative charge (anionic) to separate cationic sample.

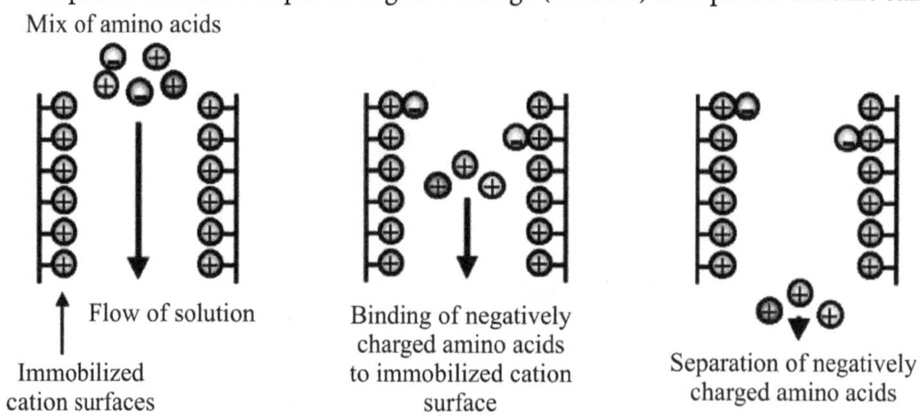

Fig. 5: Ion Exchange Chromatography

It is used to separate charged proteins and peptides in a given sample

7.2.2.5 Affinity Chromatography

Affinity chromatography separates proteins on the basis of a specific interaction between the molecules in the sample with a compound called ligand in the column (Fig.22.6). Some of the ligands used in matrix are:

- enzymes
- antibody
- metal ions like Nickel,etc.

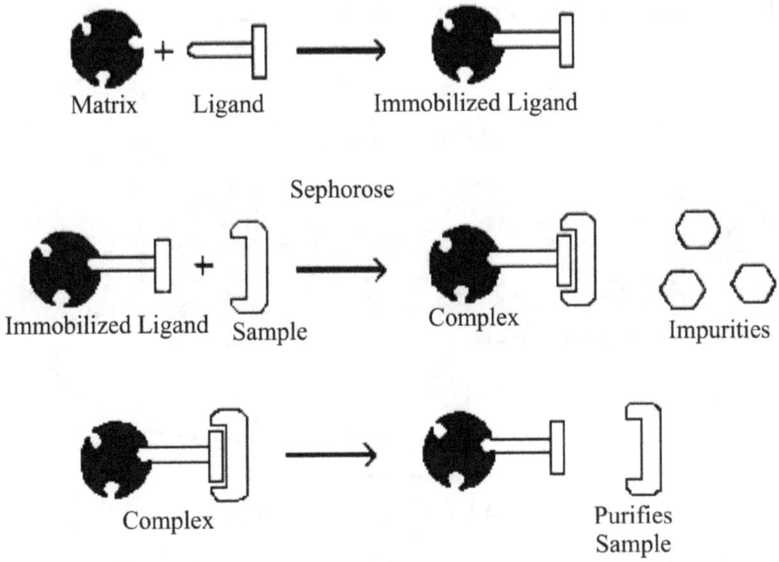

Fig. 6: Affinity Chromatography

This technique is highly specific and is used mostly for purification of proteins and peptides.

7.3 MASS SPECTROMETRY

The mass spectrometer is an instrument that measures the masses and relative concentrations of atoms and molecules. It makes use of the basic magnetic force or velocity on a moving charged particle.

7.3.1 Basic Principle

Suppose you had a cannonball travelling past you and you wanted to deflect (change of direction) it. All you've got is a jet of water from a hose-pipe. But, it's not going to make a lot of difference! Because the cannonball is so heavy, it will hardly be deflected at all from its original course. If suppose instead of water, you try to deflect the table tennis ball travelling at same speed using jet of water. Because the ball is so light, it will be deflected easily.

Thus from the above examples, we can understand that amount of deflection depends on the weight (mass) of the object. The same principle holds true for atom-sized molecules.

7.3.1.1 Working Principle

"The technique for studying the masses of atoms or molecules or fragments of molecules separated by their mass-to-charge ratio (m/z)".

7.3.2 Components

There are many different kinds of mass spectrometers, but all use magnetic and/ or electric fields to exert forces on the charged particles produced from the chemicals to be analyzed (Fig. 7). A basic mass spectrometer consists of three parts:

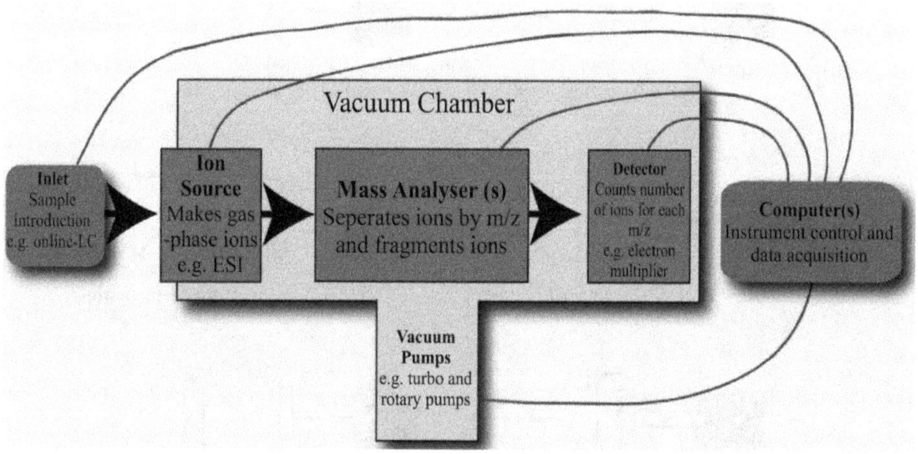

Fig. 7: A basic mass spectrometer

1. 1. **Source** - in which ions are produced from the chemical substances to be analyzed.
2. 2. **Analyzer** - in which ions are separated according to mass.
3. 3. **Detector** - which produces a signal from the separated ions.

A magnetic field or electric field separates ions according to their momentum (mass x velocity)

7.3.3 Types of Mass Spectrometry

There are many different kinds of mass spectrometers based on the type of ion source, mass analyzer and detector used. Let us see the various types of ion source and mass analyzers:

7.3.3.1 ESI – TQ Mass Spectrometer

The basic workflow of ESI tandem MS is as follows:

- The proteins pass through the source become ions;
- Then the ions pass through the first analyzer and some specific ions are selected;

- Then these selected ions are broken up by a procedure called collision induced dissociation (CID);
- Finally the second analyzer is used to catch the ions produced after CID.

ESI – Electro Spray Ionization – helps in ionizing all molecules of a given sample (Fig. 8) Small dry

Capillary

Sample Droplets ions Mass flow with ions analyzer → → →

Fig. 8: Work flow of ESI ion source Mass Analyzer – it may be of:

- Triple Quadrapole – consists of three single quadrupole mass analyzers Q1, Q2 and Q3.
- Ion Trap – traps specific molecules/ions with a particular m/z ratio and leaves all other to pass through; voltage is then suddenly increased leading to colliding of molecules/ions and then detection (Fig. 9)

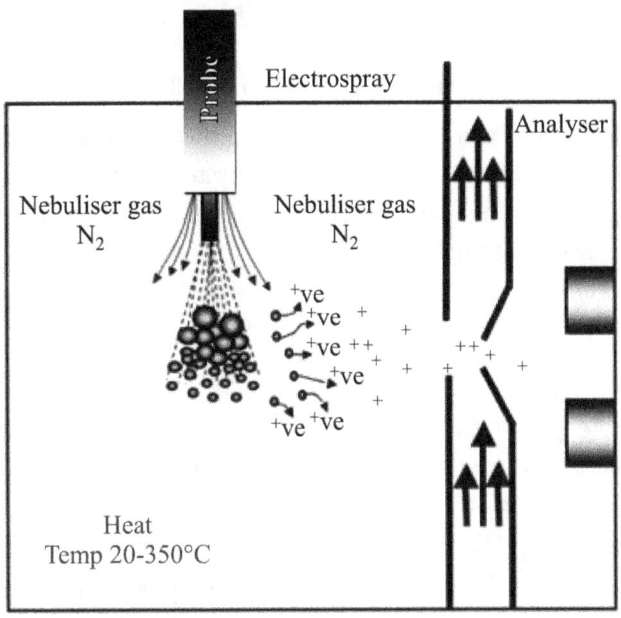

Fig. 9: ESI Tandem Mass Spectrometer

7.3.3.2 MALDI TOF Mass Spectrometer

MALDI – Matrix Assisted Laser Desorption and Ionisation, is a soft ionization technique used for analysis of biomolecules (Fig.10)

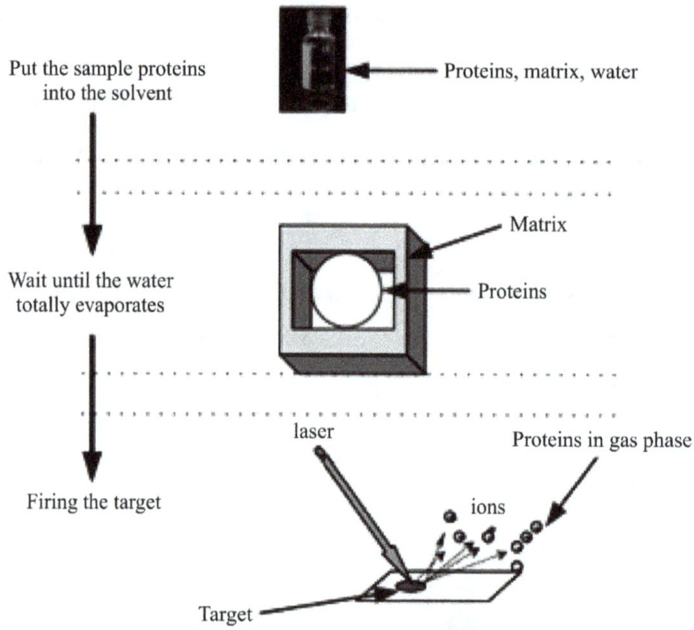

Fig. 10: Ion source of MALDI

TOF – Time Off Flight analyzer - An ion of known electrical charge and unknown mass enters a mass spectrometer and is accelerated by an electrical field of known strength. This acceleration results in any given ion having the same kinetic energy as any other ion given that they all have the same charge. The velocity of the ion will depend however on the m/z ratio (Fig.11).

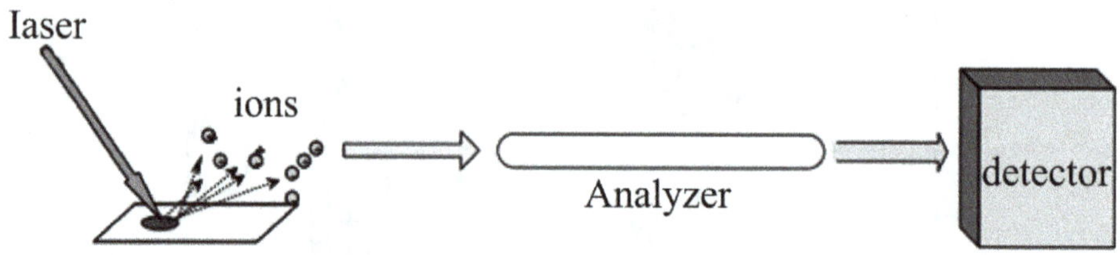

Fig.11: TOF Analyzer and Detector

The result given by the detector is in the form of a graph called as a mass spectrum based on the different molecules present in the sample (Fig. 12).

Fig. 12: Mass Spectrum

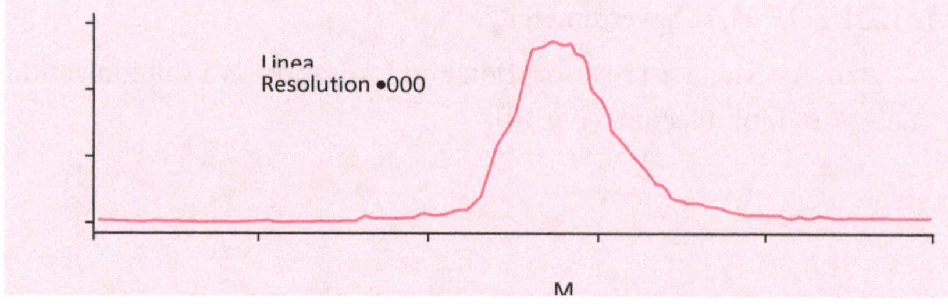

7.3.4 Uses

- Helps in identification of similar fragments of a molecule
- It can easily combine different ion source and mass analyzers.
- Has high resolution, friendly and robust; highly sensitive
- It is used for all kinds of chemical analyses, ranging from environmental analysis of petroleum products, trace metals and biological materials.

8

THEORY AND PRACTICE OF UHPLC AND UHPLC-MS

8.1 INTRODUCTION

Reversed-phase liquid chromatography (RPLC) is currently one of the most widely used separa- tion techniques. During the last few years, some substantial improvements, such as innovative supports and instrumentation, have been brought forth to perform high-throughput analyses and highly efficient separations in RPLC. Such advances were mainly driven by the need to handle either a growing number of analyses or more complex samples.

Regarding high-throughput separations, there is a growing demand in numerous fields, including toxicology, doping, forensics, clinical chemistry, and environmental analyses, where the delivery time response must be reduced as much as possible. The pharmaceutical field, with its need for enhanced productivity and reduced costs, is certainly the main driving force for faster separations. Because of the high number of analyses required for common pharmaceutical applications, such as purity assays, pharmacokinetic studies, and quality control, rapid analytical procedures (less than 5 min including equilibration time) are often mandatory.

Highly efficient separations are also necessary for many applications, including proteomics, plant extract analysis, and metabolomics, which all deal with very complex samples, such as biological samples, tryptic digests, or natural plant extracts. With such difficult samples, conventional high-pressure liquid chromatography (HPLC) systems present some obvious limitations.

Among the different strategies used to achieve fast and high-resolution separations, ultrahigh- pressure liquid chromatography (UHPLC) has been rapidly recognized as a powerful and robust analytical tool. As shown in Fig. 1, the number of articles published in the fields of UHPLC and UHPLCemass spectrometry (MS) has risen very quickly since 2004. Thus, in this chapter, the possibility to speed up and/or attain highly efficient separations will be demonstrated, using the UHPLC strategy in combination with UV as well as MS detectors. This chapter will also discuss the advantages and drawbacks of this approach, in comparison with other existing techniques.

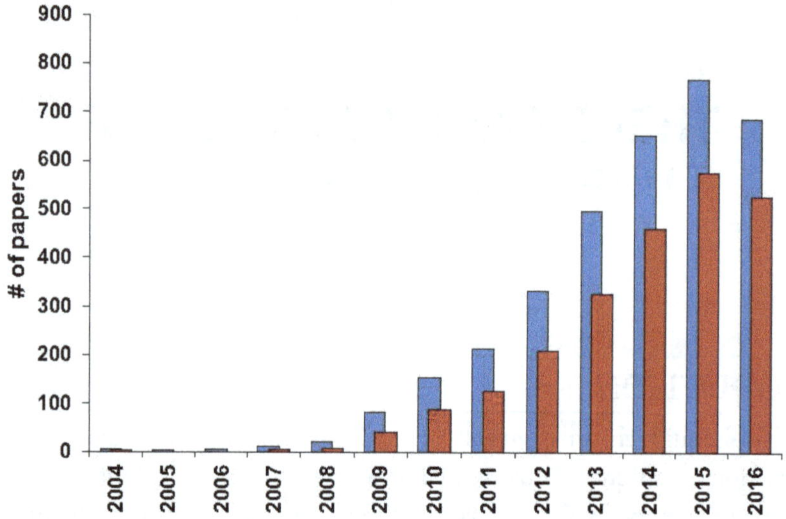

Fig. 1: Number of papers published each year in the field of ultrahigh-pressure liquid chromatography (UHPLC) and UHPLCemass spectrometry (MS), since 2004. Before this date, only few papers were published by Jorgenson's and Lee's groups. Blue bars (light gray in print versions) were obtained with keyword "UHPLC", whereas red bars (dark gray in print versions) were obtained with an additional filter (keyword "MS"). From Scifinder scholar 2016 search of the chemical abstracts database. Date of information gathering: August 2016.

8.2. BRIEF DESCRIPTION OF ULTRAHIGH-PRESSURE LIQUID CHROMATOGRAPHY AND SMALL PARTICLES IN LIQUID CHROMATOGRAPHY

In liquid chromatography (LC), it is well established that a reduction of particle size (dp) provides an important gain in chromatographic performance. Indeed, the reduction of dp allows faster separations as well as higher plate counts. Since the beginning of LC, there has been a continuing progression in the reduction of particle size that has culminated in the commercialization of columns packed with sub-2 mm particles (Table 1). According to Eqs. (1.1) and (1.2), these small particles lead to significant improvements in terms of (1) efficiency because N (number of theoretical plates) is inversely proportional to dp, and (2) time reduction because the optimal mobile phase linear velocity (u) is inversely proportional to particle diameter.

$$N = \frac{L}{h \cdot d_p} \tag{1.1}$$

Where, L is the column length, h the reduced plate height (generally between 2 and 3), and dp the particle size

$$u = \frac{v \cdot D_m}{d_p} \tag{1.2}$$

Table 1 Evolution of Particles Size in Liquid Chromatography

Years	Particle Size (mm)	Plates/15 cm
1950s	100	200
1967	50	1000
1972	10	6000
1985	5	12,000
1992	3e3.5	22,000
1996	1.5	30,000
2000	2.5	25,000
2004	1.7	30,000

[a]*Nonporous silica or resins.*

Adapted from Majors, R.E., 2003. HPLC column packing design. LC-GC Eur. 16, 8e13.

Where, *v* is the reduced linear velocity and Dm the diffusion coefficient of the solute into the mobile phase.

At the end of the 1960s, Horva´th et al. introduced columns packed with rigid pellicular particles (40e50 mm) compatible with high pressures. The thin porous coating allowed a rapid solute mass transfer into and out of the packing, producing a significant improvement in terms of column efficiency compared with the large porous particles commonly employed at that time. Nevertheless, this pellicular packing had a too limited surface area and therefore low sample capacity. The transition from large porous and pellicular particles to smaller particles (in the range of 10 mm) occurred in the 1970s. However, particles of silica smaller than 40 mm have demon- strated some difficulties with packing reproducibility, and irregular shapes of microporous particles were used, until spherical materials were developed and improved.

In 1977, Knox stated that ultrafast LC would require a new generation of particles and instru- mentation. Particles of 1 or 2 mm should be used to obtain to 10 s at reasonable pressures, in combination with column lengths of 20-40 mm. However, due to the dramatic reduction of the column volume, and the use of elevated mobile phase flow rate, the instrumentation was critical, and contributed strongly to additional peak broadening. Specifically, the injector (e.g., precise injection of small amounts) and detector devices (e.g., reduction of cell volume and improvement of electronics, such as acquisition rate and time constant) were not initially adapted. For this reason, 20e30 years have been spent to develop instrumentation compatible with short columns packed with sub-2 mm particles.

In the 1980s, columns packed with 5 mm particles became the standard, and in the early 1990s, 3e3.5 mm particle diameters were commercially available. The latter demonstrated 30%e50% faster analysis times and higher efficiencies compared with 5 mm particles. An additional advantage is that methods developed on columns packed with 5 mm can be easily transferred to a similar 3 or 3.5 mm stationary phase. To further improve chromatographic performance, small nonporous supports of 1.5 mm were introduced in 1996. These supports minimize the pore diffusion and mass transfer resistance effects. Therefore, nonporous silica columns can work in a much wider flow rate range without any loss of chromatographic performance and provide high efficiency (equivalent to 200,000 plates/m). However, due to the reduction in surface area, nonporous supports exhibit lower retention times and too limited loading capacity in comparison with porous columns. Nowadays, such materials are only proposed for separating very large molecules that slowly diffuse in the mobile phase.

In 2004, the first available porous silica with sub-2 mm particle size was commercialized (1.7 mm), allowing a better resolution compared with the current 5 or 3.5 mm. As shown in Fig. 2, the kinetic performance can be drastically improved when decreasing the particle size from the conventional columns packed with 5 mm particles to the sub-2 mm particles. With the latter, it is possible to attain an H value of less than 5 mm, meaning that a column of only 50 mm packed with sub-2 mm particles can provide an efficiency of 10,000 plates (i.e., equivalent to a 150 mm column length packed with 5 mm particles). In addition, the mobile phase linear velocity should be increased with smaller particles (see Eq. 1.2), and thus it is possible to work three times faster when using sub-2 mm versus 5 mm particles. Finally, the mass transfer resistance (right hand side of the Van Deemter curve) is drastically reduced, allowing columns to work at a linear velocity higher than the optimal one, with only a limited impact on efficiency. Because of these kinetic characteristics, the analysis time can be reduced by a factor of 9, for a similar efficiency between HPLC (150 mm, 5 mm) and UHPLC (50 mm, 1.7 mm) at the optimal linear velocity. On the other hand, it is possible to increase the column length to have a better efficiency without increasing the analysis time in comparison with an HPLC procedure.

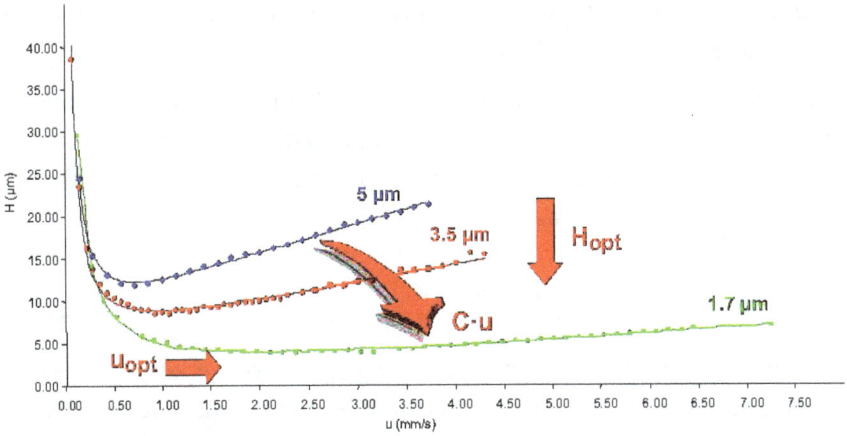

Fig. 2: Impact of particle size reduction on the Van Deemter curves. These experimental curves were obtained with a mobile phase: H_2O/MeCN (60:40 v/v), with butylparaben 25 ppm in H_2O, UV detection: 254 nm, columns: XTerra RP18 4.6 × 150 mm, 5 mm; XTerra RP 18 4.6 × 50 mm, 3.5 mm; Acquity Shield bridged ethylsiloxane/silica hybrid C18 2.1 × 50 mm, 1.7 mm.

In gradient mode, several authors have demonstrated that decreasing particle size improves peak capacity and thus chromatographic resolution. Indeed, particle size reduction has more impact than reducing column length, gradient time or flow rate to improve peak capacity in gradient mode.

However, these columns packed with small particles generate an elevated backpressure (>400 bar) often incompatible with conventional instrumentation. Darcy's law (Eq. 1.3) shows the dependence of column inlet pressure on the particle diameter *dp*.

$$\Delta P = \frac{u \cdot L \cdot \eta \cdot \phi}{d_p^2} \tag{1.3}$$

Where, L is the column length, h the mobile phase viscosity, and f the flow resistance.

At the optimal linear velocity (uopt), the pressure drop is inversely proportional to the cube of particle diameter (according to Eqs. 1.2 and 1.3). Thus, under optimal flow-rate conditions, 1.7 mm particles generate a pressure 25 times higher than that of 5 mm particles for an identical column length. Then, the backpressure limitation of conventional HPLC at 400 bar can turn out to be an issue, and there is a need to employ dedicated LC systems that withstand ultrahigh pressures (up to 1500 bar nowadays).

8.3 INTEREST IN VERY HIGH PRESSURES IN LIQUID CHROMATOGRAPHY

As described by Eq. (1.3), reducing the particle size rapidly increases the backpressure. Early in the sixties, Knox showed that the speed limitation of chromatographic systems would be related to the maximum operating pressure. In addition, when there is a need to improve efficiency with long columns, an increase in pressure drop is mandatory to work within acceptable flow-rate conditions.

Desmet et al. described a useful approach illustrating the influence of maximal backpressure (DPmax) on achievable N and t0, based on the Van Deemter data (u, H), column permeability (Kv,0) and mobile phase viscosity (h) with Eqs. (1.4) and (1.5).

$$N = \frac{\Delta P_{\text{max}}}{\eta}\left(\frac{K_{v,0}}{u_0 H}\right) \tag{1.4}$$

$$t_0 = \frac{\Delta P_{\text{max}}}{\eta}\left(\frac{K_{v,0}}{u_0^2}\right) \tag{1.5}$$

The kinetic plot representation (that corresponds to the kinetic performance of a chromatographic system, considering (1) mobile phase flow rate and column length, as variables and (2) mobile phase nature and composition as well as maximal pressure supported by the system, as constants) is advantageous because the performance of different LC supports can be directly compared whatever their lengths, particle sizes, and pressure drops. Fig. 3 shows kinetic plots representing the achievable efficiency on the X-axis as a function of the time required to attain a given efficiency on the Y-axis. The curves were constructed for a column packed with sub-2 mm particles and using different maximal operating pressures, namely 400, 600, 1000, and 1200 bar. This representation shows that, when dealing with high-throughput separations requiring short columns (N in the range 5000e15,000 plates), the corresponding analysis times were almost identical for systems with a maximal pressure drop of 600, 1000, or 1200 bar. However, the performance was reduced with a conventional HPLC system compatible with only 400 bar (logarithmic scale). These observations were related to the achievable mobile phase flow rate and that optimal linear velocity (uopt) cannot always be reached, depending on the upper pressure limit.

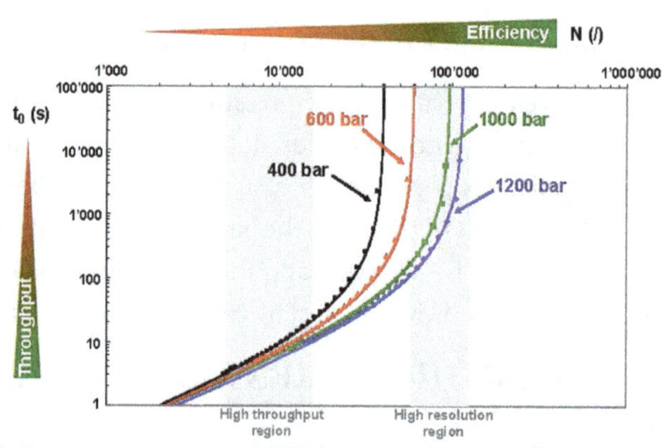

Fig. 3: Representation illustrating the importance of increasing the maximum pressure of the system in ultrahigh- pressure liquid chromatography. Kinetic plots were constructed for a Waters Acquity bridged ethylsiloxane/silica hybrid C18, 1.7 mm column, considering a mobile phase viscosity of 0.85 cp (mobile phase ACN:H2O 40: 60 at 30°C).

On the other hand, when a very high efficiency is required (right side of the kinetic curves, N ranging between 50,000e100,000 plates), the maximal operating pressure of the system becomes a critical parameter. Indeed, for a system that withstands 400 bar, the maximal achievable efficiency would only be equal to 40,000 plates, because a compromise should be found between column length and mobile phase flow rate, to maintain reasonable backpressure. Then, the N value increases proportionally with the maximal pressure drop, up to 120,000 plates for a system compatible with a maximal pressure of 1200 bar. To attain such efficiency, a column of 45e60 cm packed with sub-2 mm particles could be employed, as reported in the literature. Thus, there is a need to work with a dedicated system compatible with ultrahigh pressure, when using columns packed with sub-2 mm particles.

8.4 PRELIMINARY WORKS OF J.W. JORGENSON AND M.L. LEE IN ULTRAHIGH- PRESSURE LIQUID CHROMATOGRAPHY

The first report on UHPLC was published in 1972 by Bidlingmeyer and Rogers who employed long, thick columns packed with sub-2 mm particles. However, irreproducible results were obtained because of the inherent difficulty in packing such columns at that time.

The first remarkable separation performed at a pressure up to 4100 bar in a fused-silica capillary column (52 cm length, 30 mm I.D., 1.5 mm nonporous particles) was described in 1997 by Jorgenson et al. For an analysis time of less than 10 min, a number of plates in the range of 140,000-190,000 were obtained with small molecules. In 2003, the upper pressure limit of their system was extended and the particle size decreased to obtain outstanding chromatographic perfor- mance. As shown in Fig. 4A, a separation of four model compounds under isocratic conditions was obtained at a pressure of 7200 bar using a column of 43 cm length, 30 mm I.D., packed with 1.0 mm nonporous particles. For this separation, analysis time was reduced to less than 4 min for plate number values between 196,000 and 310,000.

Later, Lee et al. also investigated UHPLC with capillary columns but with the aim of reducing analysis times as much as possible. For this purpose, they developed a system able to work up to a pressure of 3600 bar, and a separation of benzodiazepines was performed in less than 60 s using a column of 125 mm length, 29 mm I.D., and packed with 1.5 mm nonporous particles (Fig. 1.4B) (Lippert et al., 1999). Their UHPLC setup was also successfully coupled to electrospray ionizationetime of flight (ESIeTOF) via a sheath liquid interface to carry out high-speed and high- resolution analyses of pharmaceutical compounds and herbicides. Columns of different geometries (29e100 mm I.D.) were used to obtain separations in less than 100 s with efficiencies ranging from 20,000 to 30,000 plates. Furthermore, elevated mobile phase temperature was employed to reduce the pressure drop and make the use of longer columns or smaller particles possible. Lee et al. (Xiang et al., 2003a,b) investigated this approach on a capillary column (145 mm length, 50 mm I.D., 1.0 mm nonporous particles) packed with zirconia material, due to its chemical stability at elevated temperature. A separation of various benzodiazepines was carried out in less than 1.2 min at 100°C using this column, at a pressure of 1480 bar. Five herbicides were also resolved with excellent efficiency in only 60 s at 90°C and 1800 bar.

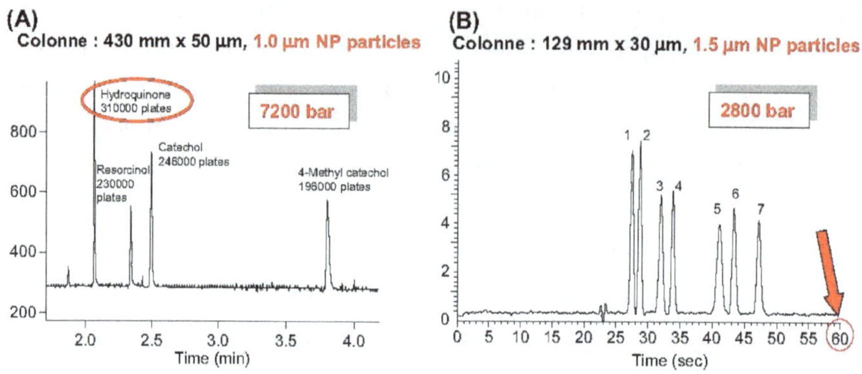

Fig. 4: Illustration of the first ultrahigh-pressure liquid chromatography experiments made by the groups of Jorgenson et al. and Milton Lee et al. (A) Ultraefficient separation of small organic test compounds obtained; conditions: 7200 bar inlet pressure, capillary column (430 mm × 50 mm I.D., packed with 1.0 mm nonporous particles) (B) Ultrafast separation of a triazine herbicides mixture. Conditions: 2800 bar inlet pressure; capillary column (129 mm × 30 mm I.D. packed with 1.5 mm Kovasil mass spectrometry-H nonporous particles

8.5 KINETIC COMPARISON OF ULTRAHIGH-PRESSURE LIQUID CHROMATOGRAPHY WITH OTHER EXISTING TECHNOLOGIES FOR FAST AND HIGH-RESOLUTION LIQUID CHROMATOGRAPHY

8.5.1 Brief Presentation of the Alternative Approaches to Ultrahigh- Pressure Liquid Chromatography

Table 2 summarizes the main characteristics (advantages and limitations) of alternative approaches to obtain fast and high-resolution separations.

Monolithic supports consist of a single rod of porous material that presents unique features in terms of permeability and efficiency. Silica-based monolithic supports are the most interesting ones for conventional RPLC and have been available from Merck and Phenomenex since 2000 and 2006, respectively. Their bimodal structure is characterized by 2- mm macropores and 13-nm mesopores, leading to efficiency equivalent to that of porous 3.5 mm silica particles. Tallarek et al. demonstrated that this material exhibits elevated permeability, equivalent to that of a column packed with 11 mm particles. The low generated backpressure and good mass transfer allow the application of elevated flow rates (3e10 times higher), thus enabling ultrafast separations, down to only a few seconds for the separation of several substrates and metabolites. Alternatively, it is also possible to use very long monoliths at a reasonable flow rate to reach an elevated resolution in a practical analysis time. For example, constructed an 11.4 m column by coupling numerous monolith columns, which provide 1,000,000 theoretical plates for an analysis time around 16 h. In 2011, a second generation of monolith has been released. The latter possesses smaller macropores of 1.2 mm and larger mesopores of 15 nm. As recently reported, because of a more dense packing, this second generation of silica monoliths provides kinetic performance close to that of columns packed with porous sub-2 mm particles. However, it possesses significantly lower permeability compared with the first-generation silica monoliths. Therefore, its use for high-resolution separations may be limited. Despite these exceptional properties, monoliths are not widely used and less than 1% of chromatographers routinely use silica-based monolithic columns (Majors, 2008). Several explanations for their limited use include patent exclusivity (limited number of suppliers), low range of column chemistry and geometry (columns are available in 2, 3, and 4.6 mm I.D. but only with a maximal length of 100 mm), and the limited resistance of the support in terms of pH and, more importantly, back pressure (DPmax ¼ 200 bar).

Table 2 Comparison of Selected Fast Chromatographic Approaches

Approaches	Advantages	Limitations
Monoliths	Very low backpressure due to the elevated permeability	Lack of chemistries (C18, C18 endcapped, C8) and providers
	Approach compatible with a conventional HPLC system	Direct method transfer impossible between conventional HPLC and monolithic supports
	Differnet geometries (e.g., 2.1 mm I.D.) are available	Limited resistance in terms of backpressure (<200 bar) and pH (2 < pH <8)
High-temperature liquid chromatography (HTLC)	Green chemistry: decrease of the organic modifier amount at elevated temperature	Stability of the solutes and silica-based stationary phases can be critical at T > 100°C
	Improvement of peak shape for basic drugs and large molecules (e.g., peptides and proteins)	Need to use dedicated instrumentation (preheating and cooling devices + backpressure regulator) Method transfer difficult because of changes in selectivity with temperature
	Possible to use this strategy with UHPLC system to further imporve perfomance	
Ultrahigh-pressure liquid chromatography (UHPLC)	Easy method transfer between HPLC and UHPLC	Need to use dedicated instrumentation (low $\sigma^2_{ext,}$ elevated acquistion rate, fast injection) and column compatible with ultrahigh pressures

Approaches	Advantages	Limitations
Superficially porous paticles (SPP)	Important decrease in analysis time Large variety of colums packed with sub-2 µm particles (more than 10 providers) Interesting approach to limit diffusion of large molecules in pores The quality of the packing is excelllent (h = 1.5) (h = 2-2.5) Approach potentially compatible with a conventional HPLC system	Cost of instrumentation and consumables higher than conventional HPLC Solvent compressibility and frictional heating are issues for ΔP close to 1000 bar Limited number of chemistries (C18, C8, HILIC) and providers Retention and loading capacity slightly lower than conventional HPLC (particularly for sub-2µm SPP) Lower resistance in terms of backpressure (<600 bar) and pH (2<pH<9) compared with fully porous paricles

HPLC, High-pressure liquid chromatography; HILIC, hydrophilic interaction liquid chromatography.

High-temperature liquid chromatography (HTLC) is a simple and elegant strategy to improve chromatographic performance. An elevated mobile phase temperature (beyond 60°C and up to 200°C) could be valuable for improving chromatographic performance. Indeed, a temperature increase results in a significant reduction of mobile phase viscosity, leading to higher diffusion coefficients and improved mass transfer, thereby increasing the optimal linear velocity toward higher values (uopt proportional to the ratio T/h). Temperature, which directly impacts solvent viscosity, also causes a significant reduction in column backpressure at a constant flow rate. Because of these features, it is possible to maintain resolution and accelerate separations by a factor 3e5 (at 90°C) and up to a factor of 20 (at 200°C), with hydro-organic mixtures containing MeOH and water. Alternatively, longer columns with acceptable backpressures can be employed at elevated temperatures, although it becomes difficult to work under optimal flow rate conditions. In addition to its kinetic performance, HTLC has some additional benefits, summarized in Table 2. First, both the dielectric constant and surface tension of water decrease at elevated temperature, and thus water can replace a significant proportion of the organic solvent in the mobile phase (5%e10% less organic solvent for each 30°C). Second, an improvement of the peak shape has been reported for basic compounds because of a decrease of their pKa values and thus a reduction of secondary interactions with residual silanol groups. For large molecules, such as peptides, the peak shapes can also be improved because diffusion coefficients strongly increase with temperature. Finally, as temperature is a thermodynamic parameter, it can be used to tune selectivity during method development. Although HTLC has been investigated in academic laboratories, it remains seldom used in an industrial environment. The major limitations of this approach are related to the limited number of stable stationary phases compatible with elevated temperature, the required modification of LC equipment to adequately control the mobile phase temperature, and most importantly, the putative thermal degradation of compounds.

The most recently introduced strategy to improve chromatographic performance is the use of superficially porous particles (SPP), also named fused-core or coreeshell, which became commer- cially available in 2007 under the trademark Halo from Advanced Materials Technology (Wilmington, DE, USA). This technology was originally developed by J.J. Kirkland in the 1990s to limit diffusion of the macromolecules into the pores, but smaller particle sizes (1.3, 1.7, 2.5-2.7 mm) have now been incorporated. Nowadays, this approach consists of using sub-3 mm SPP composed of a solid inner core and a thin porous outer core. In comparison with totally porous particles of similar diameters, the diffusion path is much shorter because the inner core is solid- fused silica, which is impenetrable by analytes (thus decreasing the C-term of the Van Deemter curve, mass transfer resistance). This obviously minimizes peak broadening, especially at elevated linear velocities. This characteristic is particularly important for the separation of large molecules (i.e., peptides or intact proteins) because slow mass transfer induces a loss of efficiency with rapid separations on porous particles. Additionally, this material presents an extremely narrow particle size distribution and high packing density compared to porous particles, leading to a smaller A term in the Van Deemter curve (i.e., Eddy diffusion). Additionally, the presence of a solid core within an SPP has a direct impact on longitudinal diffusion (B term of the Van Deemter equation). It decreases the B term contribution to the plate height by about 20%. Indeed, various authors have determined h values down to 1.5 or even 1 for such columns, in contrast to values of 2e3 for columns packed with porous particles. Thus, for an identical column length, the semiporous particles of 2.7 mm particles maintain around 80% of the efficiency of sub-2 mm particles but with a twofold lower backpressure. This promising approach has grown quickly in recent years, and today the number of suppliers for SPP particle columns is relatively large. However, even if the generated backpressure is two times lower than that of columns packed with sub-2 mm particles, the resistance of the support to pressure claimed by providers is also almost twice lower.

8.5.2 Best Liquid Chromatography Approach in Isocratic Mode theory and Applications

In isocratic mode, the throughput and resolving power of the different strategies were evaluated using a 2D map reported in Fig. 5. In this representation, two important parameters were selected. For the separation speed, the column dead time ($t_{10,000}$) required to attain an efficiency of 10,000 plates was estimated. Such a plate number is generally sufficient for high-throughput separations of conventional samples that contain a limited number of compounds. For the resolving power, the maximal achievable efficiency N30 min was calculated for a column dead time of 30 min (equivalent to an analysis time of 3 h for k = 5). This analysis time is quite long but not prohibitive when dealing with very complex samples (e.g., proteomics, metabolomics, etc.). Basically, the values of $t_{10,000}$ and N30 min were calculated using the Van Deemter data (H, u) and permeability values Kv,0 experimentally determined for each analytical strategy. Then, the data were computed considering the maximal pressure drop of the system (DPmax of 200 bar for monoliths; 400 bar for HPLC, sub-2 mm, and HTLC; 600 bar for SPP; and 1000 bar for UHPLC and HT-UHPLC). The corresponding column lengths and mobile phase flow rates required were calculated. Thus, the column dead times and plate numbers presented in Fig. 5 always correspond to a pressure drop equal to DPmax with variable column lengths and mobile phase flow rates. For a better understanding of the employed procedure to construct this 2D map, readers can refer to a few didactic papers on kinetic plot methodology.

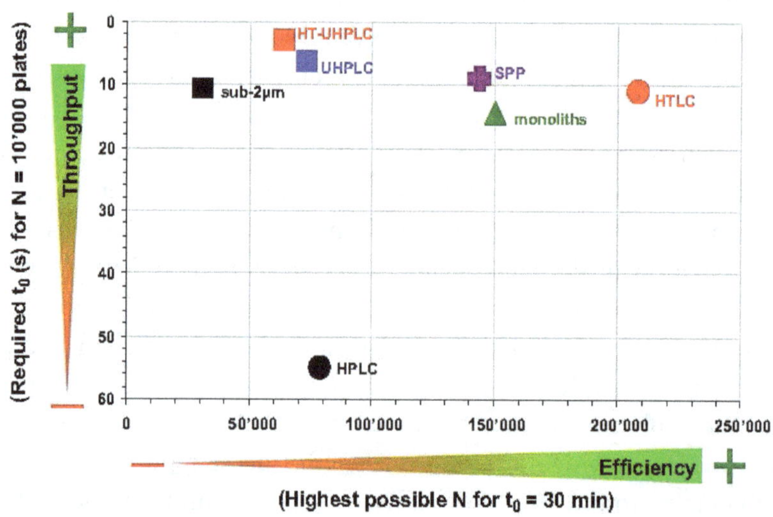

Fig. 5: Performance comparison of liquid chromatography strategies in isocratic mode in terms of throughput (t0 for N ¼ 10,000 plates) and maximal resolution (Nmax for t0 ¼ 30 min) for a model compound, butylparaben with MW of 200 g/mol. The data were gathered using the kinetic plot methodology, considering a maximal pressure of 200 bar for monoliths, 400 bar for high-pressure liquid chromatography (HPLC), high-temperature liquid chromatography (HTLC) and sub-2 mm, 600 bar for superficially porous particles (SPP) and 1000 bar for ultrahigh- pressure liquid chromatography (UHPLC) and HT-UHPLC.

According to Fig. 5, in terms of throughput (Y-axis), conventional HPLC ($t_{10,000}$ = 55 s for N = 10,000 plates) is clearly not competitive with the other strategies. As expected from theory, the analysis time can be significantly reduced with monoliths ($t_{10,000}$ = 15 s) because of the very low generated backpressure, but this material suffers from a too limited upper pressure limit compared with SPP columns or those packed with sub-2 mm particles. HTLC appeared beneficial for reducing the analysis time ($t_{10,000}$ = 11 s) because of the improvement of diffusion coefficients related to the mobile phase viscosity decrease with temperature. For example, it showed that ultrafast separations could be achieved at very high temperatures. A separation of five alkylphenones was carried out with a conventional 50 4.6 mm column packed with 2.5-mm particles in only 20 s (DP of 360 bar), instead of 20 min at ambient temperature. The use of columns packed with sub-2 mm particles remains the most efficient strategy to improve throughput, particularly when small particles are combined with high pressure and elevated temperature, where a 20-fold increase of throughput compared to that of conventional HPLC ($t_{10,000}$ = 3s for N = 10,000 plates) is possible. As shown in Fig. 6A, a mixture of four preservatives was separated in less than 10 s by HT-UHPLC using a 50 x 2.1 mm column packed with 1.7 mm particles at 1.8 mL/min and 90°C.

The maximal efficiency that can be reached for a t_0 = 30 min (X-axis, N30 min) is between 31,000 and 208,000 plates. Unfortunately, long columns operating in the B termedominated region of the Van Deemter curve are required (low mobile phase flow rate zone). To reach such elevated efficiencies, the required column was between 0.6 and 3.4 m in length, whereas the mobile phase flow rates ranged between 50 and 270 mL/min for a 2.1-mm I.D. column. Therefore, the strategies involving the use of small particles (i.e., sub-2 mm, UHPLC and HT- UHPLC) are less potent (31,000 < N30 min < 73,000 plates) because of the elevated generated backpressure. This observation has been experimentally confirmed by Sandra et al. Who demonstrated that an N_{max} of 74,000 plates was possible for a test mixture under isocratic conditions at 40°C, using a 450 mm column length packed with 1.7 mm particles at 1000 bar. In contrast, the maximal efficiency of columns packed with 5-mm particles reported in Fig. 5 was around 80,000 plates and can be increased by 2.6-fold at elevated temperatures. Because of their elevated permeability, monolithic supports also represent a valuable strategy for increasing the plate count compared to conventional HPLC (N30 min is twofold higher). However, the column length needs to be around 1.7 m, so numerous commercially available 100 mm columns length need to be coupled in series. Although some authors have coupled up to 10 columns, the cost becomes rapidly prohibitive. Despite the elevated maximal efficiency observed with monoliths, this approach is not competitive with HTLC because the lower maximal pressure drop capability limits the length of monolith that can be employed. In HTLC, more than 200,000 plates for a t0 30 min can be attained, but with a 3 m column, which can be very expensive (12 250-mm columns). Experimentally, Sandra et al. coupled in series eight 250-mm columns packed with 5-mm particles, corresponding to a total length of 2 m. With this configuration, efficiencies as high as 180,000 plates were attained at 80°C for a test mixture, with t0 values of 20 min and analysis times around 100 min. Finally, SPP technology provides N30 min values similar to monoliths but with more acceptable column lengths and mobile phase flow rates, such as 1 m and 90 mL/min for a 2.1-mm I.D. column. These results can be attributed to the elevated plates/m values of the SPP column in conjunction with a backpressure around twofold lower than that of columns packed with sub-2 mm particles.

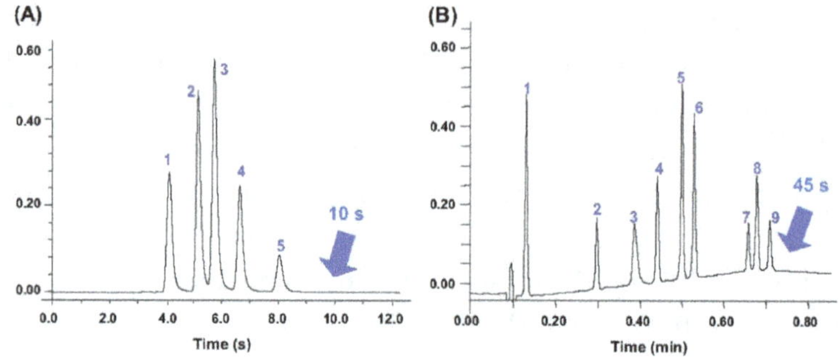

Fig. 6: Ultrafast separations carried out in HT-ultrahigh-pressure liquid chromatography. (A) Isocratic separation of various preservatives and uracil. Column: Acquity bridged ethylsiloxane/silica hybrid (BEH) C18 (50 × 2.1 mm I.D., 1.7 mm); mobile phase: watereacetonitrile (50:50%, v/v); flow rate: 1800 mL/min; temperature: 90°C; (B) Gradient separation of several doping agents. Column: Acquity BEH Shield RP18 (50 × 2.1 mm I.D., 1.7 mm); mobile phase: 0.1% formic acid in watere0.1% formic acid in acetonitrile; flow rate: 1800 mL/min; temperature: 90°C.

In conclusion, high-throughput separations require the use of columns packed with small particles (UHPLC) and should be carried out preferentially at increased temperatures (HT- UHPLC). It was also demonstrated that temperature and maximal system pressure drop should be increased as much as possible because both parameters are beneficial for increasing the plate count as well as the throughput. Finally, the use of sub-3 mm SPP allows an additional gain in performance.

8.6 BEST LIQUID CHROMATOGRAPHY APPROACH IN GRADIENT MODE THEORY AND APPLICATIONS

In the case of complex analyses, the separation is generally carried out in gradient mode, to handle compounds possessing very different physicochemical properties and/or improves the resolving power. In gradient mode, peak capacity must be calculated to estimate the performance. Peak capacity describes the number of peaks that can be separated with a resolution of 1 during the gradient duration and depends mainly on the isocratic efficiency, column dead time, and gradient time. Two parameters were selected to evaluate the throughput and resolving power in gradient mode, namely the gradient time required to attain a peak capacity of 100 (t_{100}) and the maximal peak capacity for a gradient time of 3 h (P3 h).

In recent papers, the strategy developed by Schoenmakers et al. was applied to construct kinetic plots in gradient mode. In the graphical representation proposed in Fig. 1.7, a similar approach was employed, using data previously obtained in isocratic mode (H, u, and Kv,0) for each analytical strategy. Both the lowest gradient times, t100 and highest peak capacities, P3 h, correspond to a pressure drop equal to DPmax and consequently to different column lengths and mobile phase flow rates.

According to Fig. 7, in terms of throughput, the ranking was similar for isocratic and gradient modes, and columns packed with small particles were clearly advantageous (i.e., sub-2 mm, UHPLC and HT-UHPLC). Indeed, it is theoretically possible to attain a P 100 in only 30 s with HT-UHPLC, whereas 7 min are required in conventional HPLC at 400 bar. Columns packed with superficially porous sub-3 mm particles (SPP technology) performed almost equivalently to columns packed with porous sub-2 mm particles (tgrad of 1.4 min for P = 100), but monoliths and HTLC were less powerful (tgrad around 2.5 min for P 100). These calculated values were in agreement with examples from the literature. For the separation of various pharmaceutical compounds, 50 mm columns packed with porous sub-2 mm or superficially porous sub-3 mm particles produce experimental P values higher than 70 in less than 2 min at ambient temperature. Fig. .6B shows the potential of HT-UHPLC, with a gradient separation of various steroids in less than 45 s, at a temperature of 90°C. Heinisch et al. also reported an impressive chromatogram of nine small aromatic compounds separated in less than 15 s by HT-UHPLC. This separation was performed with a 7.8 s gradient, time using a 50 2.1 mm, 1.7 mm column at 2 mL/ min and 90°C.

In terms of resolving power, monoliths, SPP, UHPLC, HT-UHPLC, and HTLC offer almost the same peak capacities, ranging between 415 and 480 for a gradient time of 3 h. Compared with the isocratic mode, UHPLC and HT-UHPLC generate higher peak capacities in gradient mode. This is because peak capacity is strictly related not only to the chromatographic efficiency but also to the column dead time in gradient mode. Because the latter is strongly reduced in UHPLC and HT-UHPLC compared to the traditional approaches, the maximal peak capacity for a 3 h gradient is enhanced. Conventional HPLC at 40° bar offers around 30% less peak capacity compared to the other approaches, demonstrating that 5 mm particles do not present any practical benefit in gradient mode.

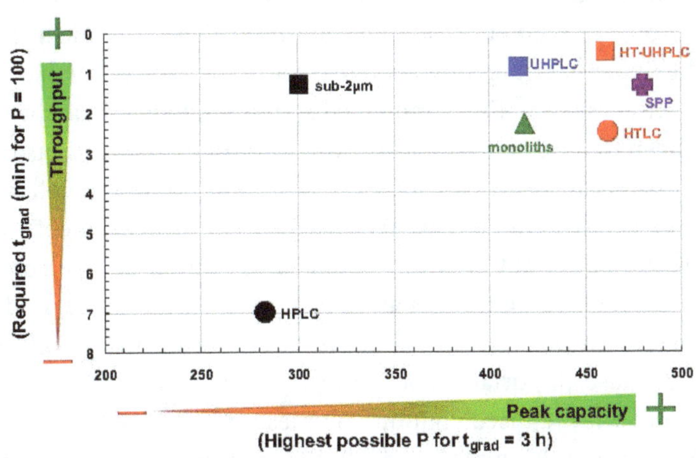

Fig. 7: Performance comparison of liquid chromatography strategies in gradient mode in terms of throughput (tgrad for P ¼ 100) and maximal resolution (Pmax for tgrad = 3 h) for a model compound, butylparaben with MW of 200 g/ mol. The data were gathered using the

kinetic plot methodology applied to gradient elution and considering the same maximal pressure as for isocratic mode. HTLC, high-temperature liquid chromatography; HT-UHPLC, high- temperature ultrahigh-pressure liquid chromatography; SPP, superficially porous particles; UHPLC, ultrahigh- pressure liquid chromatography.

Additionally, the sub-2 mm strategy, with a maximal pressure of only 400 bar, was also significantly less practical because of the low packing permeability and small backpressure limitation that reduces the column length and mobile phase flow rate that can be employed.

In conclusion, the SPP and UHPLC technologies are very attractive for maximizing both the throughput and resolution in gradient mode. Whatever the selected strategy, the use of elevated mobile phase temperature is an additional parameter to improve gradient performance.

8.7 PROBLEMS RELATED TO ULTRAHIGH-PRESSURE LIQUID CHROMATOGRAPHY

According to Table 2, the main drawbacks of UHPLC are (1) the need to acquire a dedicated system, optimized in terms of high backpressure pumps and injector, acquisition rate of the detector, injection cycle time, dwell volume, and system dead volume, (2) the need to work with a column that withstands ultrahigh pressures, (3) the elevated backpressure that can produce some important changes in solvent properties (frictional heating and solvent compressibility effects). All these limitations are discussed in detail below.

8.7.1 The Need to Work with a Dedicated Instrumentation

Currently, there is a wide choice of instruments that withstand pressures above 400 bar, from various suppliers. Regarding the selection of a UHPLC system, the cost is certainly a decisive consideration, but it is also important to look at specifications of all available instruments on the market, which are not equivalent. The most important feature is certainly the upper pressure limit and corresponding flow rate, which mostly defines the price of a UHPLC system. For commercial systems, DPmax varies between 600 and 1500 bar. It has been demonstrated that for fast or ultrafast separations of simple mixtures, the use of small particles was obvious, but there was no need to work with very elevated pressures. For high-throughput experi- ments, UHPLC instruments with pressure limits around 600-800 bar could provide a suitable so- lution at a reasonable price. On the other hand, for high-resolution separation, a system with an elevated maximal pressure is mandatory to work at acceptable flow rates. Beside the pressure capability of the apparatus, it is also important that the instrument was adapted to operate in fast and ultrafast modes with reduced column volumes. Indeed, small diameter columns (1 and 2.1 mm I.D.) commonly employed in UHPLC require limited extra-column volumes of detection, tubing, and injection. The following criteria have to be fulfilled to perform suitable separations.

The volumes of the connecting tubes should be drastically reduced. For this reason, the tubing length should be as short as possible and its diameter selected as a compromise between an acceptable pressure drop and a low volume. Hence, a system plumbed with 0.003 or 0.00500 I.D. stainless steel tubing and zero-dead volume fittings should be preferred for UHPLC experiments.

The injection volume should be selected in agreement with column geometry. A rule of thumb is to maintain the injected volume between 1% and 5% of the column dead volume in isocratic mode. As most of the experiments carried out in UHPLC are performed with a 50 × 2.1 mm column (V0 120 mL), the injected volume should be between 1 and 5 mL, to limit band broadening. In addition, a fast injection cycle time is mandatory for analysis times lower than 1 or 2 min.

Last but not least, the detector cell volume, time constant, and acquisition rate should be carefully selected. The detector used in UHPLC should ideally possess a low cell volume (<2 mL), while the sensitivity should not be lower compared to that of a conventional HPLC instrument (pathlength of 10 mm). The detector time constant has to be fast enough (s 100 ms) because peak widths are very small in UHPLC (only a few seconds). Finally, the UV detector sampling rate must be sufficiently high to acquire a suitable amount of data points for each peak (>20 Hz).

Finally, to reduce the detrimental effect of extra-column volumes and avoid unacceptable loss in efficiency, chromatographic conditions leading to sufficient retention factors (k or ke at least equal to 3) are recommended in UHPLC.

The last critical parameter to perform ultrafast separations in gradient mode is the system dwell volume, which corresponds to the time necessary for the mixed solvents to reach the column inlet. To work with ultrafast separations, a small gradient delay volume is obviously required. With a large dwell volume, fast separations are compromised because an isocratic hold is generated at the begin- ning of the gradient, inducing potential changes in selectivity and longer analysis times.

To check the compatibility of a UHPLC instrument with a given column geometry, it is recommended to characterize the chromatographic system by determining the extra-column and dwell volumes. For optimal compatibility with ultrafast separations, the former should be lower than 20 mL, whereas the latter should be reduced to a few 100 mL. Finally, a comparative study made by a pharmaceutical company on various UHPLC systems can be found elsewhere, for additional information.

8.7.2 The Need for Specific Columns Compatible with Ultrahigh Pressures

An important aspect when selecting a UHPLC setup is the selection of a stationary phase that should provide sufficient selectivity as well as acceptable performance and lifetime. Before the commercialization of UHPLC technology, only nonporous particles were used. Jorgenson et al. were the first to compare 1.5 mm porous hybrid particles packed into a

30 mm I.D. fused silica capillary column supplied by Waters (Milford, USA) to 1.0 mm nonporous silica material (Mellors and Jorgenson, 2004). Chromatographic performance was evaluated at pressures up to 4500 bar, and hybrid particles were similar to 1.0 mm nonporous silica particles, in terms of pressure resistance.

In 2004, Waters launched a new generation of hybrid columns packed with 1.7 mm particles that were stable up to 1000 bar. A bridged ethylsiloxane/silica hybrid (BEH) particle provides mechanical and chemical resistance in extreme conditions of pH (1e12), pressure, and temperature (up to 180°C). Numerous other providers have also made available stationary phases packed with sub-2 mm particles, demonstrating the opportunity to transfer almost all existing methods from HPLC to UHPLC. Indeed, the variety of phase chemistries can resolve almost all analytical issues: C8 and C18 for compounds of average polarity; C4 and cyano for the most apolar analytes; diol, amino, silica, and hydrophilic interaction liquid chromatography (HILIC) for polar molecules and biphenyl; perfluorophenyl or zirconia for alternative selectivity. All these stationary phases are not only equivalent in terms of pressure tolerance (from 600 to 1500 bar) and particle size (from 1.5 to 2 mm) but also pH and temperature ranges. Some performance comparisons between the different phases can be found in the literature, and data for column lifetime have also been published.

One of the main criticisms made by early UHPLC users has been the reduced lifetime of columns packed with sub-2 mm particles, compared to conventional columns. It is true that UHPLC columns are systematically exposed to very high pressures, but the packing pressure has been increased in the meantime. In our laboratory, we observed that lifetimes of UHPLC and regular HPLC columns were comparable. However, column lifetime could be expressed as the number of injections, number of column volumes or period of time used. With the latest generation of columns packed with sub-2 mm particles commercialized by different suppliers, it is possible to perform between 500 and 2000 injections or even more on a single column. Such values correspond to about 5000-20,000 column volumes, which are fully comparable to those obtained on standard HPLC columns. However, when considering the corresponding period of time, it is significantly reduced compared to conventional HPLC because of the higher throughput of UHPLC. For example, in a routine laboratory that performs a UHPLC analysis in 1e5 min, several hundreds of injections can be performed within a day.

8.7. 3 The Changes in Solvent Properties with Pressure

It is demonstrated that beyond 100 bar, experimental parameters that are considered as constant by chromatographers (e.g., column porosity, mobile phase density and viscosity, diffusion coefficient, retention factors, and efficiency) could depend on pressure to some extent. Consequently, irreproducible retention factors or loss in efficiency can be observed.

Frictional heating of the mobile phase is certainly the most critical effect that can be observed in UHPLC conditions and has been largely. Columns packed with small particles present low permeability, and thus generate a considerable amount of frictional heating under high pressure drops and elevated flow rates. The heat generation, or power dissipation, is the product of pressure drop (DP) and flow rate (F) as reported in Eq. (1.6).

$$Power = F \cdot \Delta P \tag{1.6}$$

Frictional heating and poor heat dissipation cause significant radial and axial temperature gradients across the column. An irregular radial temperature profile produces solute retention alteration, unexpected changes in selectivity and additional band broadening. Reducing the column diameter to capillary dimensions largely eliminates this effect. Therefore, only capillary columns (e.g., 30e150 mm I.D.) were used before commercialization of UHPLC systems, but Jorgenson et al. recognized that columns of larger diameter could be used with a reasonable value of generated power (i.e., <1 W). It also investigated in detail the power dissipation of columns with diameters in the mm range, and confirmed that the latter, packed with small particles, would generate acceptable frictional heating. A column of 150 1 mm I.D. packed with 1.5 mm particles size with a flow rate of 85 mL/min generated about 1380 bar. Under these conditions, the power dissipation was 195 mW, which corresponded to values encountered in conventional HPLC. Similar results were obtained with a 2.1 mm I.D. column packed with 1.7 mm particles at a linear velocity of 2.5 mm/s. Thus, the frictional heating generated with the 1e2.1 mm I.D. UHPLC columns employed nowadays was not detrimental to the separation performance.

Sandra et al. investigated the effects of the temperature gradient across the column length due to frictional heating. Experimental data obtained on a 2.1 mm I.D. column packed with 1.7 mm particles, at 1000 bar with a conventional still-air column heater showed no significant efficiency loss.

On the other hand, Guiochon et al. demonstrated the importance of the column external envi- ronment on its performance and particularly that frictional heating effects could be detrimental with a 2.1 mm I.D. column. This is illustrated in Fig. 8 where an experimental setup allowed the measure of the temperature at the surface of the column and the temperature of the liquid at the column outlet. With this strategy, complete temperature profiles along and across various UHPLC columns under still air conditions at room temperature were constructed. For this purpose, four different column lengths were considered, namely 150, 100, 50, and 30 mm lengths, in 1 and 2.1 mm I.D. Only the results for 2.4 mm I.D. columns of 50, 100, and 150 mm are presented in Fig. 8. The authors concluded that, with columns of 2.1 mm I.D., the difference between the inlet and outlet temperature (longitudinal gradient) was between 10 and 20°C, whereas it varied between 3 and 16°C with 1 mm I.D. columns, depending on the considered column length.

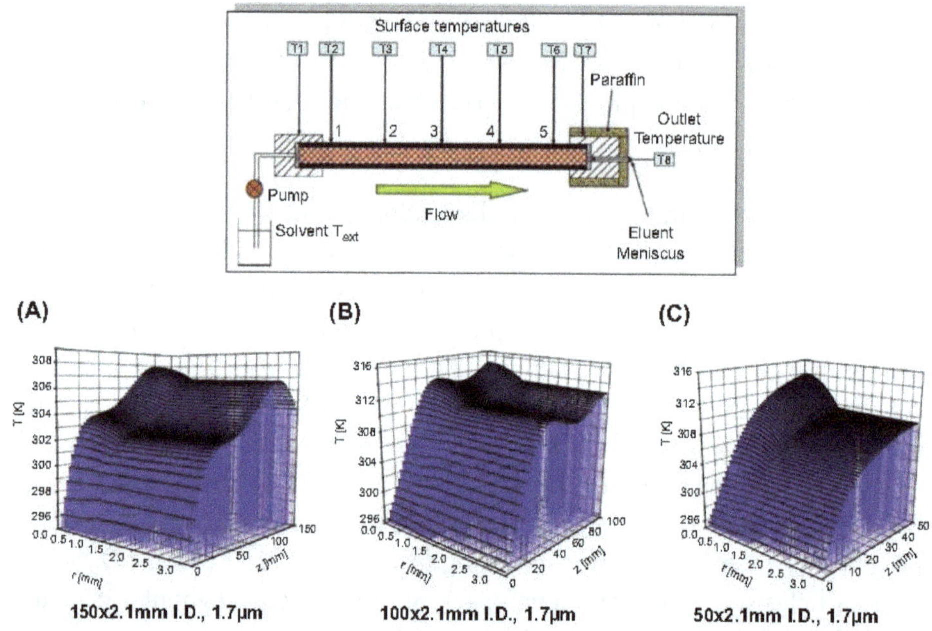

Fig. 8: 3D representation of the temperature profiles along and across various hromatographic columns (A) 150 mm × 2.1 mm (BEH) bridged ethylsiloxane/silica hybrid-C18 column, flow rate = 0.95 mL/min, Pinlet ¼ 979 bar. (B) 100 mm × 2.1 mm BEH-C18 column, flow rate = 1.45 mL/min, Pinlet = 1005 bar. (C) 50 mm × 2.1 mm BEH-C18 column, flow rate = 2 mL/min, Pinlet = 775 bar. The experimental setup used to measure the temperature of the surface of the column and the temperature of the liquid exiting the column outlet is also shown.

To reduce the effect of thermal gradient, and split up a column with a given length L into n segments with length L/n, while providing an active cooling to the capillaries connecting the segments. In this way, the viscous heat was removed at a location where the radial heat removal does not lead to an efficiency loss (i.e., in the thin connection capillaries), while the column segments can be operated under nearadiabatic conditions without suffering from an unacceptable rise of the mobile phase temperature.

Besides frictional heating effects, mobile phase viscosity is also dependent on pressure, and organic solvents such as methanol and acetonitrile are more compressible than water (Foley et al., 1989). Jorgenson et al. investigated the solvent compressibility effects in gradient mode and not only noticed a reduction of reproducibility in peptide retention times from run to run in UHPLC conditions (for pressures up to 5000 bar) but also indicated that this problem could be resolved with a 5.

8.8 METHOD TRANSFER FROM HIGH-PRESSURE LIQUID CHROMATOGRAPHY TO ULTRAHIGH-PRESSURE LIQUID CHROMATOGRAPHY

In various fields of application (i.e., pharmaceutical, environment, food, etc.), it is essential to be able to transfer existing methods (performed in conventional HPLC conditions) to faster separa- tions involving the use of columns packed with sub-2 mm particles. As most providers now offer equivalent stationary phases packed with 5, 3, and sub-2 mm particles, a geometrical transfer can be performed if the stationary phase chemistry remains identical between the original and final sets of conditions. For this purpose, some rules have to be strictly applied in both isocratic and gradient modes.

8.8.1 The Rules for Isocratic Mode Theory and Applications

For an isocratic transfer between conventional HPLC and UHPLC, two important parameters have to be adjusted, namely the injection volume and the mobile phase flow rate. To avoid a detrimental extra-column band broadening and maintain equivalent sensitivity, it is necessary to adapt the injection volume in agreement with the change of column dimensions. In LC, the injected volume should represent only 1%-5% of the column volume. The latter should be adjusted proportionally to the column internal diameter (dc) and length (L). Therefore, the injection volume is independent of the particle size and only proportional to the column volume. The new injected volume V_{inj2} can be determined simply by maintaining the ratio of column dead volume and injected volume constant between regular HPLC and UHPLC.

$$V_{inj_2} = V_{inj_1} \cdot \frac{d_{c2}^2}{d_{c1}^2} \cdot \frac{L_2}{L_1} \qquad (1.7)$$

In this equation, subscripts 1 and 2 are related to HPLC and UHPLC column dimensions, respectively. For example, from a conventional 150 4.6 mm, 5 mm column to a UHPLC 50 2.1 mm, 1.7 mm column, the injected volume should be decreased by 14-fold. To maximize sensitivity, it is possible to inject larger volume but in this case, the sample should be dissolved in a solvent of weaker eluent strength than the initial mobile phase composition (sample focusing).

Regarding mobile phase flow rate, this parameter should be adapted to remain close to maintaining a constant reduced mobile phase linear velocity (v). In HPLC, it is well known that the mobile phase linear velocity (u) is directly proportional to the square of column diameter and also depends on the particle size (d_p) of the support. It is however, completely independent of the column length. For a successful method transfer, it is mandatory to maintain the product $u \times d_p$ constant, to take into account simultaneous changes in column diameter and particle size of the support. Therefore, for a geometrical transfer, the UHPLC flow rate (F2) can be calculated with the following equation:

$$F_2 = F_1 \cdot \frac{d_{c_2}^2}{d_{c_1}^2} \cdot \frac{d_{p_1}}{d_{p_2}} \tag{1.8}$$

As an example, from a regular 150= 4.6 mm, 5 mm column to a UHPLC 50=2.1 mm, 1.7 mm column, the mobile phase flow rate should be decreased by 1.6-fold.

The expected analysis time of the transferred method tana2 is directly proportional to the change in column dead time and can be estimated according to:

$$t_{ana_2} = t_{ana_1} \cdot \frac{d_{p_2}}{d_{p_1}} \cdot \frac{L_2}{L_1} \tag{1.9}$$

The expected backpressure (DP2) can be calculated from Darcy's law, which shows that DP is inversely proportional to d3 (at the optimal linear velocity) and is strictly related to the column

$$\Delta P_2 = \Delta P_1 \cdot \frac{L_2}{L_1} \cdot \frac{d_{p_1}^3}{d_{p_2}^3} \tag{1.10}$$

Finally, the expected solvent consumption of the transferred method (V2) can be calculated by taking into account the change in internal diameter and column length.

$$V_2 = V_1 \cdot \frac{d_{c_2}^2}{d_{c_1}^2} \cdot \frac{L_2}{L_1} \tag{1.11}$$

Therefore, from a regular 150 4.6 mm, 5 mm column to a UHPLC 50 2.1 mm, 1.7 mm column, the analysis time is reduced by ninefold.

As an example, from a regular 150 4.6 mm, 5 mm column to a UHPLC 50 2.1 mm, 1.9 mm column, the gradient slope during the gradient process should be increased by ninefold.

When transferring a gradient method from regular HPLC to UHPLC, some changes in selectivity could occur during the gradient run because of differences in dwell volume between the original and the UHPLC configuration. The system dwell volume (Vd) refers to the volume between the mixing point of solvents and the head of the analytical column. After starting the gradient, it will take time until the selected percentage of solvent reaches the column. Because the gradient dwell volume may differ from one system to another, this extra isocratic step would be different and could result in retention time variations affecting resolution for early eluting peaks when transferring a method. To overcome this problem, the ratio of system dwell time (td) and column dead time (t0) must be held constant while changing column dimensions, particle size or mobile phase flow rate.

As the column dead time is reduced by around ninefold between a regular 150 4.6 mm, 5 mm column and a 50 2.1 mm, 1.7 mm column, the system dwell time should be reduced by the same factor. As mentioned previously, it is then mandatory to work with a UHPLC system possessing a low dwell volume (no more than a few 100 mL) to limit its influence.

Again, many applications of the above discussed approach can be found in the literature. One example has been selected, and it is presented in Fig. 10. The original separation of 12 pharmaceutical compounds was achieved using a 150 4.6 mm, 5 mm, C18 column and subsequently transferred to UHPLC with a 50 2.1 mm, 1.7 mm column possessing strictly similar chemistry. The original separation was performed in w27 min and was efficiently transferred to UHPLC, giving a separation that was complete in less than 3 min (reduction by a factor of 9, as expected from theory). In addition, both separations were equivalent in terms of sensitivity, peak capacities, and resolution, mainly because of an adequate reduction of system dwell volume (from 1 mL to 100 mL for HPLC and UHPLC, respectively). Finally, the mobile phase flow rate can be increased up to the maximal operating pressure of instrumentation to further reduce analysis time. As illustrated in Fig. 1.10C, at 1000 mL/min the analysis time was cut to only 1.6 min but with a slight loss (about 10%) of chromatographic performance.

8.9. FIELDS OF APPLICATION FOR ULTRAHIGH-PRESSURE LIQUID CHROMATOGRAPHY MASS SPECTROMETRY AND RELATED ISSUES

The coupling of UHPLC with MS appears as the ultimate approach in terms of sensitivity, selectivity, and peak assignment for the determination of analytes at low concentrations in complex matrices. Two main types of MS analyzers have been coupled with UHPLC, namely quadrupole-based and TOF-MS instruments. The first, operating in the selected ion monitoring (SIM) or selected reaction monitoring (SRM) modes, was preferentially selected for targeted analysis (e.g., bioanalysis, multiresidue screening), whereas the second was particularly useful for nontargeted analysis (e.g., drug metabolism studies, metabolomics).

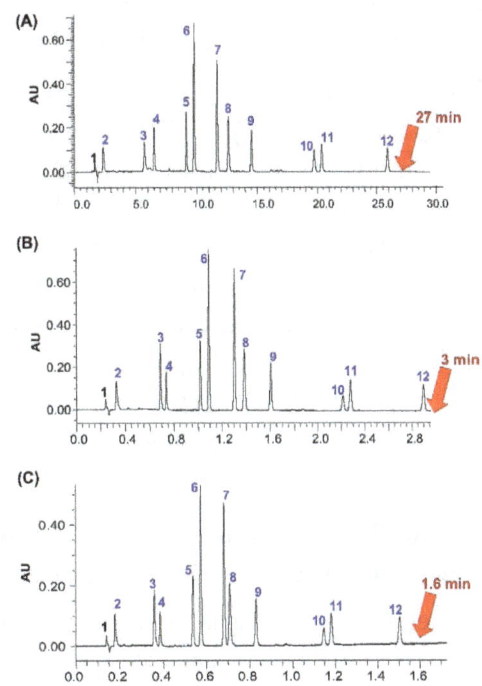

Fig. 10: Gradient method transfer from regular high-pressure liquid chromatography (HPLC) to ultrahigh-pressure liquid chromatography (UHPLC). Separation of a pharmaceutical mixture containing the main product (6) and 11 impurities in gradient mode with HPLC and UHPLC systems: (A) original HPLC method: column: XBridge C18 150 × 4.6 mm,

5 mm; flow rate: 1000 mL/min; injected volume: 20 mL; total gradient time: 45 min. (B) Trans- ferred UHPLC method: column: Acquity bridged ethylsiloxane/silica hybrid (BEH) C18 50 × 2.1 mm, 1.7 mm; flow rate: 610 mL/min; injected volume: 1.4 mL; total gradient time: 5.1 min. (C) Transferred and optimized UHPLC method: column: Acquity BEH C18 50 × 2.1 mm, 1.7 mm; flow rate: 1000 mL/min; injected volume: 1.4 mL; total gradient time: 3.1 min..

Regarding quadrupole-based analyzers, the sampling rate can be an issue and modern instruments possessing improved acquisition rates should be selected for hyphenation with UHPLC. With the new generation of analyzers, dwell times have been reduced, down to less than 1 ms for some providers, in SIM as well as in SRM modes. TOF instruments are also well adapted to record and store data over a broad mass range without compromising sensitivity. With the latest generation of TOF-MS, high mass resolution (e.g., up to 100,000 FWHM) can be attained at speeds of up to 100 full spectra per second.

Aside from the acquisition rate, it has been demonstrated that MS instruments could represent a no negligible source of extra-column band broadening in UHPLC compared to a UV. In addition, even if fast polarity switching (i.e., 15e20 ms) and/or fast ESI/ atmospheric pressure chemial ionization mode switching (i.e., 20 ms) are available from several providers to increase productivity, it always compromised sensitivity, peak width, and sampling rate in UHPLC, and should then be used with caution.

Fig. 1 shows that UHPLCeMS(/MS) is the detector of choice for more than 70% of separations carried out by UHPLC. The current state of the art of UHPLCeMS(/MS) is reported in this chapter, with emphasis on the throughput increase and/or resolution improvement afforded by UHPLC tech- nology, and its impact on MS detection capabilities. For this purpose, three promising application fields of the technique were considered: (1) bioanalysis and drug metabolism with tandem MS operating in the SRM mode on triple quadrupole mass spectrometer (QqQ) analyzers; (2) rapid multiresidue screening, using either quadrupole-based instruments in SRM mode or quadrupole time-of-flight mass spectrometer (QqTOF) analyzers; and (3) metabolomics, taking advantage of the very high chromatographic resolution of UHPLC in combination with the exact mass measurement of QqTOF/MS.

8.9.1 Ultrahigh-Pressure Liquid Chromatographyemass Spectrometry/Mass Spectrometry for High Throughput in Bioanalysis

Challenges in bioanalytical laboratories include the development of fast LCeMS methods able to separate closely related compounds, such as analytes and metabolites, from endogenous components. Because UHPLC greatly enhances the separation throughput and resolution, peaks as narrow as (or less than) 1 s can be obtained. This induces practical issues for bioanalytical applications using MS because sufficient data points (e.g., >15 points per peak) are essential to ensure reliable quantitation. Several critical applications can be found in the literature. For example, Petsalo et al. (2008) published a

UHPLCetandem mass spectrometry (MS/MS) procedure for analyzing nine drugs and their respective metabolites in urine, using a 4 min gradient. An ESI source operating sequentially in positive and negative polarity modes was employed, and dwell times (DT) of 20e30 ms were applied for each SRM transition. Because peak widths of 4 s were experimentally obtained, only six points were acquired to define peaks, which could limit performance, particularly at the lower limit of quantitation. To accommodate the small UHPLC peak widths, DT can be reduced when many SRM transitions have to be monitored, but this can result in sensitivity loss. An alternative approach consists of using various time windows during the acquisition, as proposed by Berg et al. (2009). Finally, to circumvent DT reduction, Li et al. (2008) suggested a useful "peak parking" strategy, which consisted of reducing the flow rate during peak elution, and thus extending the MS acquisition window for quantitative bioanalytical assays. However, the latter strategy is only suitable when a limited number of targeted analytes are analyzed.

In addition, it has been demonstrated in various studies that a significant reduction of matrix effects was brought about by UHPLC technology, compared to regular HPLC. As an example, Chambers et al. (2007) proved that polymeric mixed-mode solid-phase extraction (SPE), combined with UHPLC technology and appropriate mobile phase pH, provided significant benefits for reducing matrix effects from plasma matrix components, and improving the ruggedness and sensitivity of bioanalytical methods.

Considering the analysis time reduction offered by UHPLC technology, the sample preparation procedure becomes the limiting step in terms of total analysis time. Numerous UHPLC bioanalytical applications still involve traditional sample preparation procedures, which drastically increase the total analysis time. A few authors have suggested solutions to this issue, while maintaining sufficient sample preparation selectivity. To date, solideliquid extraction (SLE) and SPE based on a 96-well plate format were used prior to UHPLCeMS/MS bioanalysis, allowing for selective, sensitive, and, above all, high-throughput analyses.

8.9.2 High Resolution Drug Metabolism by Ultrahigh-Pressure Liquid Chromatography Mass Spectrometry Using Quadrupole Time-of- Flight Mass Spectrometer Analyzers

In drug metabolism experiments, there are also many challenges because of the complex nature of biological matrices and the large diversity of produced metabolites. To identify unknown metabolites, high chromatographic resolution and mass accuracy for fragmentation patterns are key requirements. UHPLC with high resolution analyzers such as QqTOF are particularly useful to fulfill both tasks in a high-throughput environment.

In the year (2005) were the first to report the use of UHPLC technology with a QqTOF analyzer in the early drug discovery process. This study emphasized improved resolution, in terms of chromatographic and mass spectral quality, and the associated gain in sensitivity afforded by the UHPLCeQqTOF/MS system. These features were explained by the combination of reduced peak width and low ion suppression due to the enhanced resolution of metabolites and endogenous compounds.

Walles et al. have investigated the benefits and drawbacks of three UHPLCeQqTOF/MS methods for fast metabolite identification using alternative MS/MS experiments (MSE). The high efficiency attributable to UHPLC was the key to the successful identification of isobaric metabolites. In fact, they could not be distinguished with the accurate mass of QqTOF, as they had an identical elemental composition and often similar MS/MS fragmentation patterns. It can be noted that the time spent for structure elucidation created additional bottlenecks and becomes the limiting step when UHPLCeQqTOF experiments are performed.

8.9.3 Ultrahigh-Pressure Liquid Chromatography Mass Spectrometry for Multiresidue Screening

Multiresidue screening techniques are generally developed to quickly assess the presence of contaminants in a complex sample. Thus, the developed method should be able to detect as many components as possible in a single analytical run. In this context, UHPLC coupled with tandem MS or TOF/MS remains the gold standard. Multiresidue screening methods with UHPLC technology have been applied to a wide variety of analytes and matrices, including: (1) doping agents and veterinary drugs in biological matrices; (2) drugs, pesticides, and herbicides in environmental matrices; and (3) veterinary drugs drugs, and pesticides in food samples.

Because of the high number of investigated compounds, conventional HPLC runs can be relatively long, particularly to avoid peak coelution leading to matrix effects. It is indeed important to attain sufficient chromatographic resolution, to minimize coelution of compounds with close m/z ratios and similar fragmentation pathways. As shown in Fig. 11, for the separation of 103 doping agents in a urine sample, gradient conditions were selected to elute analytes presenting a wide polarity range, and formic acid is generally the preferred additive in the mobile phase. To screen compounds by UHPLC, 50e100 mm column lengths were used with gradient times ranging between 5 and 15 min, followed by a reequilibration time of 2e4 min. The column length should be selected in agreement with the gradient time because the longest column did not always provide the highest peak capacity in UHPLC. Indeed, a 50 mm column was found to be optimal for gradient times shorter than 7 min, whereas a 100 mm column was only beneficial for longer gradients. Finally, the 2.1 mm I.D. column was often preferred, to limit extra-column band broadening contributions. Only reported the successful screening of about 50 pharmaceuticals in wastewater using a 1 mm

I.D. column. Even if the consumption of mobile phase and analyte was drastically reduced, peaks were notably broader and distorted with a 1 mm I.D. column, as expected from the influence of external volume contributions.

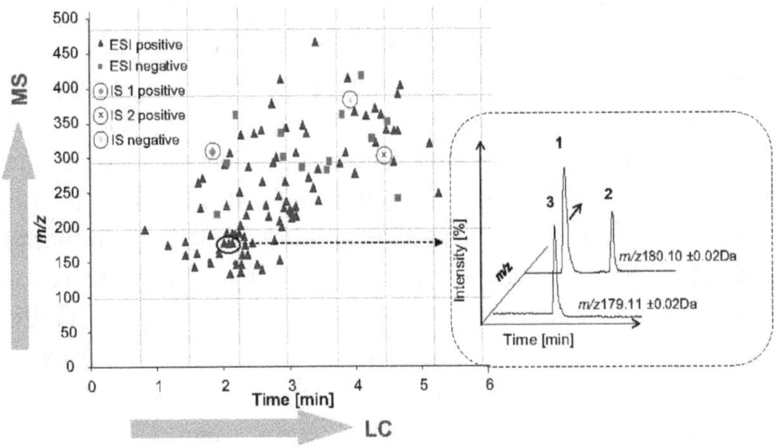

Fig. 11 shows that UHPLCeMS(/MS) Separation of 103 doping agents in urine sample according to m/z and tR values. Data from electrospray ionization (ESI) positive and negative mode are plotted together. The three I.S are circled with a continuous line. A zone (dashed line) is magnified to show the selectivity of coupling ultrahigh-pressure liquid chromatography to the quadrupole time-of-flight mass spectrometer. In the magnified zone, the compounds (1) methylephedrine, (2) 3,4- methylenedioxyamphetamine, and (3) nikethamide are separated as a function of time, intensity, and m/z.

From the above papers, an approximate increase in throughput by three- to fivefold was observed in UHPLC compared to conventional HPLC methods. In addition to the analysis time decrease, an equivalent or higher chromatographic resolution was reported. Such improvements were attributed not only to an increase in peak capacity but also to column selectivity changes. Because of different column chemistries, strict comparisons were not always possible.

8.10 ULTRAHIGH-PRESSURE LIQUID CHROMATOGRAPHY MASS SPECTROMETRY IN METABOLOMICS

Due to the inherent complexity of biological samples and because metabolites can be found at low concentrations, there is a need for analytical systems providing high resolution and increased sensitivity. For this reason, the value of UHPLCeTOF/MS and QqTOF/MS platforms has been demonstrated in a number of studies. The strong reduction of analysis time provided by UHPLC versus HPLC opens up the possibility of high-throughput screening for metabolomic fingerprinting. On the other hand, a longer UHPLC run can be employed to increase the amount of information, essentially for metabolomic profiling. The UHPLC platform has been applied for the global metabolic profiling of (1) human and animal biological fluids, including rat urine, human urine, and human serum, as well as (2) plant extracts, such as Arabidopsis thaliana and Panax herbs.

Wilson et al. have used UHPLC for the profiling of rat and mouse urine since 2005. Initially, biological fluids were analyzed on a 50 mm column packed with 1.7 mm particles in combination with TOF/MS. In terms of chromatography, the average peak widths were around 1 s, thus generating a peak capacity of 60 for UHPLC runs of only 1 min. With the additional TOF/MS information, a total of 1000 features (i.e., signals observed with specific m/z and retention times that can be considered as a variable for data treatment) were determined in rat urine. This number was equivalent, or even better than, that achieved on conventional HPLC instrumentation, but with a 10-fold reduction in analysis time. The study performed by Nordstrom et al. (2006) on the quantitative analysis of endogenous and exogenous metabolites in human serum confirmed these results. Indeed, UHPLC provided 20% more detected components in comparison with HPLC. Finally, it was demonstrated that UHPLC displayed some additional advantages over HPLC, such as better retention time repeatability and signal-to-noise ratios. Finally, two interesting approaches, namely, the application of elevated temperature in UHPLC (up to 180C) and the use of HILIC columns packed with sub-2 mm particles, were proposed by Wilson et al. to further extend the applicability of UHPLC in metabolomics.

Because of the complexity and chemical diversity of metabolites present in natural plant extracts, metabolomics is also gaining interest in the field of phytochemistry. The use of UHPLCeTOF/MS as well as QqTOF/MS for untargeted metabolic profiling has been reported by two research groups. Jia et al. reported the profiling of several medicinal Panax herbs, while Wolfender et al. evaluated a UHPLCeTOF/MS platform for the analysis of a model plant, A. thaliana. As shown in Fig. 12, they proposed a useful and sensitive multistep strategy for the detection, isolation, and identification of stress-induced metabolites in A. thaliana after leaf wounding, which mimicked herbivore attack. In the first step, a rapid screening gradient was carried out by UHPLCeTOF/MS using a short column of 50 mm. This metabolite fingerprinting was performed on numerous plant specimens to evaluate the intrasample variability and achieve adequate pool formation. The second step consisted of high resolution metabolite profiling of selected pool samples using a UHPLC column of 150 mm. Gradient conditions used in the metabolomic fingerprinting were adequately transferred to the new column geometry, and analysis times were increased up to 100 min. This profiling allowed confirmation of the presence of different stress-related compounds. The high peak capacity afforded by long columns packed with sub-2 mm was indeed essential to obtain a complete deconvolution of the biomarkers, and resolution of numerous closely related isomers. The last step of the process was the complete structural determination of minor biomarkers in plants using LCeMS triggered preparative isolation. For this purpose, the UHPLC separation obtained during the metabolic profiling was transferred to semipreparative conditions, using a 19 mm I.D. column packed with 5 mm particles of the same chemistry. Based on the use of a capillary-nuclear magnetic resonance probe, 1D and 2D spectra of good quality were obtained at the mg level, allow- ing unambiguous structural elucidation of the isolated wound-biomarkers (including known signaling molecules, as well as original oxylipins and jasmonates). This generic analytical platform can be used to screen various other plant extracts.

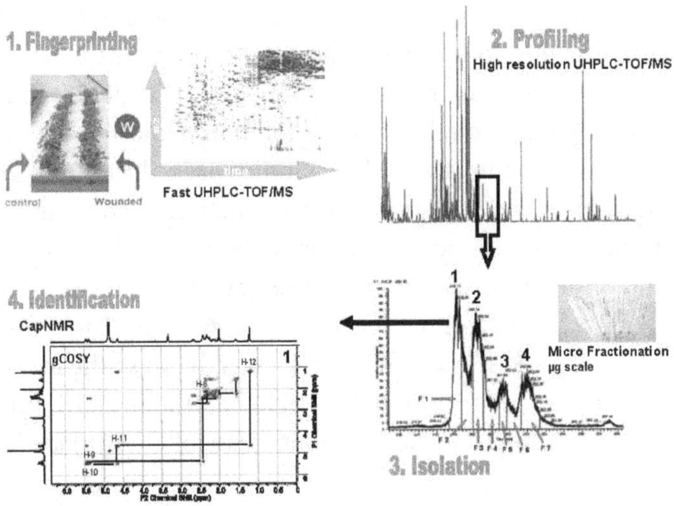

Fig. 12 Shows that UHPLCeMS(/MS)Plant metabolomics based on a four-step strategy: fingerprinting, profiling, isolation, and identification of stress biomarkers. UHPLCeTOF/MS, ultrahigh-pressure liquid chromatographyetime of flight mass spectrometry; cap NMR, cap nuclear magnetic resonance spectroscopy.

8.11 PERSPECTIVES

As shown in this chapter, UHPLC is a powerful technology, which is quite easy to implement. With this strategy, it is possible to increase drastically the throughput, while maintaining equivalent per- formance and/or to increase the resolution within an acceptable analysis time. However, to take full advantage of this platform, it is important to keep in mind that the performance of UHPLC is based not only on the column packing but also on the quality of the chromatographic system. For this reason, UHPLC experiments should be performed on a system compatible with ultrahigh pressure, which possesses reduced system and dwell volumes.

The coupling of UHPLC with MS appears to be the ultimate approach, in terms of sensitivity, selectivity, and peak assignment for the determination of analytes at low concentrations in complex matrices. This strategy has become very popular and has now been applied in numerous fields of application, such as bioanalysis, drug metabolism, multiresidue screening, and metabolomics. How- ever, it is recommended to work with an MS device of the latest generation (quadrupole-based and TOF instruments are the most appropriate) that possesses a sufficient data acquisition rate.

In the future, if 2.1 mm I.D. column becomes the standard dimension for UHPLC, there is certainly not much interest in reducing even more the particle size of the support. Indeed, the backpressure generated by such packing would be detrimental for the chromatographic separation because of the axial and longitudinal temperature gradients within the column (frictional heating effects). To limit these negative effects, the solution would be to work with reduced I.D. (500 mme1 mm) columns. But in this case, the current chromatographic system needs to be strongly improved in terms of extra- column and dwell volumes. From our point of view, the use of small particles, in conjunction with ultrahigh pressure and elevated temperature, up to 90°C (HT-UHPLC), is certainly more promising both not only for throughput and resolution but also for selectivity and peak shape. Finally, monolithic columns as well as columns packed with sub-3 mm SPP should be neglected as they could represent some good alternatives to columns packed with fully porous sub-2 mm particles. Second generation system. In another study performed in isocratic mode, the surge in linear velocity caused by mobile phase compression at the start of isocratic UHPLC runs (with pressures up to 6300 bar) was found to induce a 50% increase in the measured Knox C-term. However, because the maximal pressure of commercial UHPLC instrument is limited to 1500 bar, the negative impacts of solvent compressibility (i.e., lack of reproducibility and increase of C-term) are less critical.

9

RECENT ADVANCES IN HYDROPHILIC INTERACTION: LIQUID CHROMATOGRAPHY

9.1. INTRODUCTION: AN OVERVIEW

In reversed-phase (RP) systems, most frequently used in contemporary high-performance liquid chromatography (HPLC) practice, the stationary phase is nonpolar, usually an alkyl-silica type bonded phase, whereas the mobile phase is a mixture of one or more organic solvents and water or an aqueous buffer. As a rule, sample retention increases for more lipophilic samples and stationary phases and in more polar mobile phases; polar solutes are less strongly retained than the nonpolar ones. Very hydrophilic samples such as carbohydrates or small strongly polar compounds are weakly retained in RP-LC systems and often elute close to the column hold-up volume, so that their separation from one another and from polar matrix interferences may be difficult to accomplish, even in highly aqueous mobile phases. On the other hand, the stationary phase in normal-phase (NP) chromatography is more polar than the mobile phase, anddopposite to RP-HPLCdretention increases with increasing polarity of samples and of the stationary phase and also in less polar mobile phases. In nonaqueous mobile phases traditionally used in conventional NP (adsorption) chromatography, the retention mechanism is based on the competition between the sample and the mobile phase for localized polar adsorption centers on the adsorbent surface. However, strongly polar compounds are often excessively retained in nonaqueous NP systems or are poorly soluble in nonpolar or in weakly polar organic solvents. Often, their separation on polar stationary phases can be improved by adding water to the mobile phase. Some water accumulates close to the polar surface, where it forms a diffuse layer more polar than the bulk aqueousorganic mobile phase, which is less rich in water (Fig. 1). This approach had been occasionally used a long time before Alpert introduced the name "hydrophilic interaction liquid chromatography" (HILIC) for this separation mode. The term "hydrophilic" refers to affinity for water. Essentially, HILIC systems can be understood as a "normal-phase stationary phase" in combination with a "reversed-phase mobile phase," usually containing 50% or more of water. The HILIC technique provides appropriate retention and resolution for many polar compounds, often with better separation efficiency in comparison to the RP chromatography.

The diffusion coefficients of ionized basic compounds in less viscous organic-rich mobile phases under HILIC conditions are approximately twice those under RP conditions, leading to improved separation efficiency (lower height equivalent of a theoretical plate, HETP).

Another reason for the increasing popularity of HILIC is its excellent suitability for coupling to mass spectrometry (LC/MS). Originally, HILIC was applied to carbohydrate and peptide analysis in proteomics and glycomics. Later, HILIC has gradually been used for separations of drugs, toxins, plant extracts, and other small polar compounds in clinical, food, and pharmaceutical analysis. Growing interest in HILIC separations resulted in several recently published review articles. An excellent overview on recent progress in the development of polar stationary phases until 2006 was written by more recent advances are included in reviews. Other reviews have focused on the effects of the operating conditions on HILIC separations, separation efficiency, HILIC method development coupling HILIC systems with MS and MS/MS implementation of HILIC systems in two-dimensional (2D) separation modes, and on HILIC applications in biological, pharmaceutical and metabolite analysis. An excellent book on HILIC has appeared recently, covering various aspects of the technique: separation mechanism, stationary phases, method development, and applications.

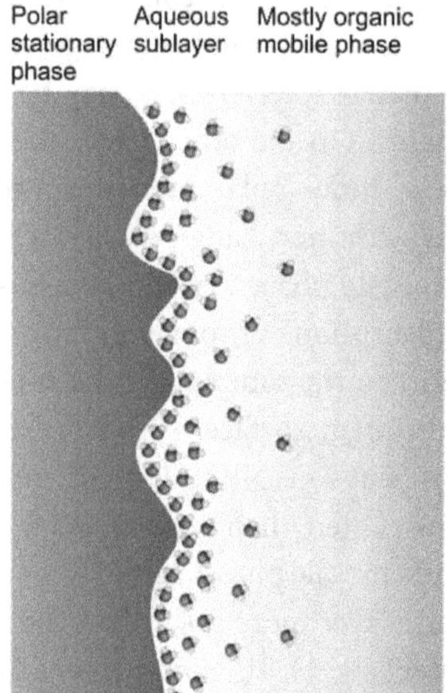

Fig. 1: shows that UHPLCeMS(/MS) Schematic representation of a diffuse water layer at the surface of a polar stationary phase in a highly organic environment.

9.2 COLUMNS FOR HYDROPHILIC INTERACTION LIQUID CHROMATOGRAPHY SEPARATIONS

Properties of HILIC separation systems strongly depend both on the stationary phase and composition of the mobile phase. Polar, hydrophobic, or ion-exchange interactions potentially contribute to sample retention, often giving rise to a mixed retention mechanism. In acetonitrile-rich mobile phases, hydrophilic interactions control the retention (HILIC systems), whereas in highly aqueous mobile phases the column may show essentially RP behavior with the major role being hydrophobic interactions. NP separations on bare silica and amino-silica columns in aqueous-organic mobile phases were reported as early as in 1970s; since then, a large variety of silica-based columns suitable for HILIC separations have been developed, providing excellent separations of peptides, proteins, oligosaccharides, drugs, metabolites, and various natural polar compounds. Bare silica gel, hydrosilated or hybrid silica, inorganic oxides, or organic polymers carrying various polar groups, but also ion exchangers and mixed-mode or zwitterionic-bonded ligands, have proved useful as HILIC materials.

9.2.1 Silica Gel and Hybrid Inorganic Sorbents

Highly purified "sil-gel" spherical silica particles (type B silica) formed by the aggregation of silica sols in the air, stable to at least pH 9, are still the most frequently used column materials in HILIC chromatography. Bare silica gel shows suitable separation of many types of polar compounds. The efficiency of silica particles improves with their decreasing size. Recently, fully porous sub-2 mm particles were introduced for UHPLC (ultra-high performance liquid chromatography) in the HILIC mode. Better separation efficiency for polar and basic compounds on sub-2 mm coreeshell silica particles was reported, in comparison to fully porous particles of the same size, because of a reduced diffusion path inside the particles and consequently reduced sample transfer resistance (van Deemter C-coefficients) under HILIC conditions. On bare silica columns, the retention mechanism in aqueouseorganic mobile phases may be attributed to partitioning, adsorption, and ion exchange, depending on the nature of the analytes, the composition of the mobile phase, and the characteristics of the analytical support. Separations of polar drugs can often be accomplished on bare silica gel columns in aqueouseorganic mobile phases with acetonitrile concentrations ranging from 75% to 95%, usually containing 10e100 mmol/L ammonium acetate or formate buffers. HILIC applications of silica gel columns in pharmaceutical analysis have been reviewed in detail. At higher pH, ionization of the silanol groups on the silica gel surface increases, and cation exchange may play a significant role in retention, especially of positively charged basic compounds, which are generally strongly retained on silica gel by hydrogen bonding and ion-exchange interactions. The peak symmetry of basic compounds on "sil-gel" columns in HILIC mobile phases (acetonitrileeammonium formate buffer) often improves in comparison to RP-HPLC. Addition of trifluoroacetic acid to HILIC mobile phases generally suppresses ion-exchange interactions on the silica surface but may promote an ion-pairing mechanism.

Continuing efforts have been reported to develop new types of sorbents with improved thermal and pH stability, such as 1.7 mm bridged ethylene hybrid (BEH) organicesilica gel particles. BEH columns with bonded amide, diol, or cyanophenyl polar ligands differ from underivatized BEH hybrid particles in separation selectivities for various polar test probes under HILIC conditions.

Silica gel supports coated with metal dioxides or graphene oxide (GO) also show improved stability in comparison to conventional silica-based HPLC materials. For improved stability, SiO_2 particles were wrapped by GO sheets. The GO surface is covered with a monolayer of oxidized carbon atoms containing polar hydroxy-, epoxy-, and carboxy groups.

Materials with ZrO_2 nanoparticle multilayers immobilized on a SiO_2 surface show the advantages of the high chemical and temperature stability of the ZrO_2 surface (up to 200_C and in the mobile phase range of pH 1e14), while preserving the inner pore structure of SiO_2. Bare ZrO_2 and ZrO_2/SiO_2 monoliths provide separation of some compounds with different spatial configurations in the HILIC mode.

A ZrO_2eSiO_2 stationary phase was prepared by coating silica particles with ZrO_2 "layer by layer," modified by surface bonding adenosine-5-monophosphate. The stationary phase with the bonded nucleotide is suitable for HILIC separations. By adjusting the pH of the mobile phase, the stationary phase can also show some ionic interactions resulting in changes in selectivity. Basic compounds showed a mixed HILIC/RP mechanism, whereas weak acids were retained only in the HILIC mode, most strongly in the pH range from 4 to 5. Above pH 5, electrostatic repulsion weakened the retention and improved the peak symmetry and resolution of nonsteroidal anti-inflammatory drugs (florfenicol, flufenamic acid, mefenamic acid, ibuprofen, ketoprofen, sulindac, and indoprofen).

9.2.2 Chemically Bonded Silica-Based Stationary Phases

Silica gel materials modified with various polar functionalities have been prepared for improved retention and separation selectivity of various sample types in HILIC. These mainly include diol, amino, amide, cyclodextrin (CD), ion exchange, zwitterionic, poly(2-hydroxyethyl aspartamide) or poly(succinimide), and other bonded phases prepared by chemical modification of the silica gel surface. The preparation and applications of some of these were covered in earlier reviews. Structures of some bonded ligands frequently used as HILIC stationary phases are shown in Figs. 2.2e2.5: Fig. 2A presents simple bonded ligands, Fig. 2B amide and polysuccinimide phases, Fig.3A cyclodextrine and sugar bonded ligands, Fig. 3B mixed-mode stationary phases providing HILIC and ion-exchange retention mechanisms, Fig. 4 some silica hydride bonded phases, and Fig. 5 zwitterionic ligands.

9.2.2.1 Amino, Amide Amino Acid, and Peptide Bonded Stationary Phases

Aminopropyl silica has been traditionally used for NP separations of sugars, amino acids, peptides, carboxylic acids, nucleosides, and some pharmaceuticals in aqueouseorganic mobile phases. Bonded amino stationary phases show increased affinities for acidic compounds, due to weak ion-exchange interactions. The contribution of the ion-exchange mechanism to the retention may cause weaker retention at increasing ionic strength of the mobile phase. The primary amino group is relatively reactive, which may cause bleeding of the bonded phase by selfdecomposition, or formation of Schiff bases with aldehydes, which may cause problems in HILIC separations of carbohydrates on amino-bonded silica. Aminopropyl-silica columns provide an increased rate of anomer mutarotation, eliminating the formation of doublet peaks of monosaccharide diastereomers and hence better peak shapes in the separation of sugars. Stationary phases containing secondary or tertiary amine groups, such as the YMC-Pack Polyamine II, cannot form Schiff bases with carbonyl compounds, which may improve the column stability and lifetime with respect to aminopropyl silica.

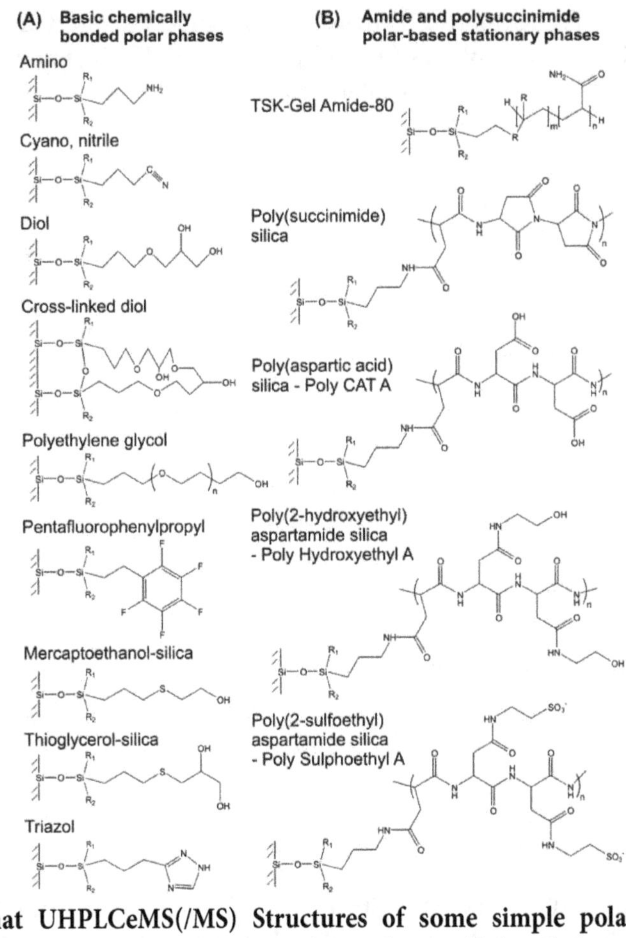

Fig. 2 Shows that UHPLCeMS(/MS) Structures of some simple polar (A), amide and polysuccinimide (B) HILIC stationary phases chemically bonded on a silica gel surface.

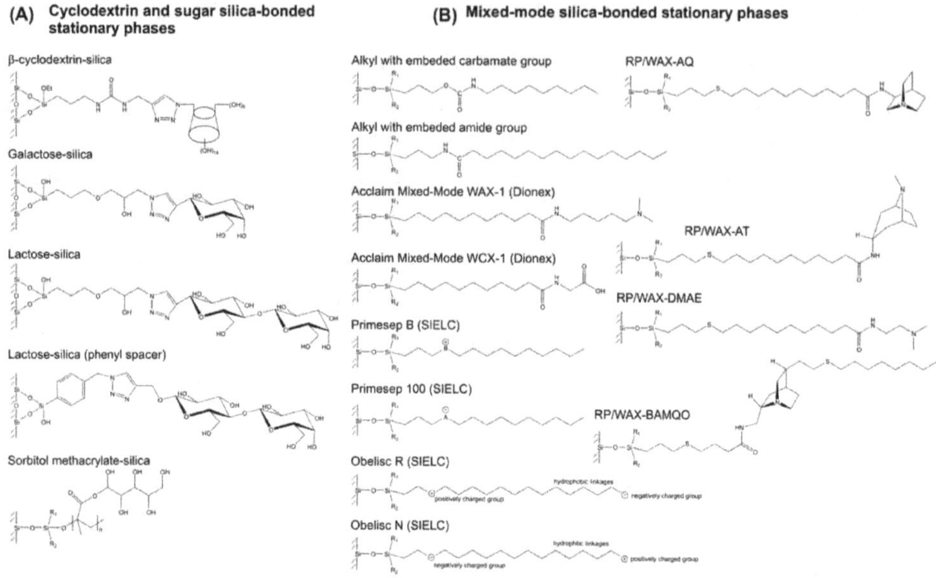

Fig. 3: Shows that UHPLCeMS(/MS) Structures of some cyclodextrin and sugar silica-bonded HILIC stationary phases (A) and mixed-mode HILIC/ion-exchange stationary silicabonded phases (B).

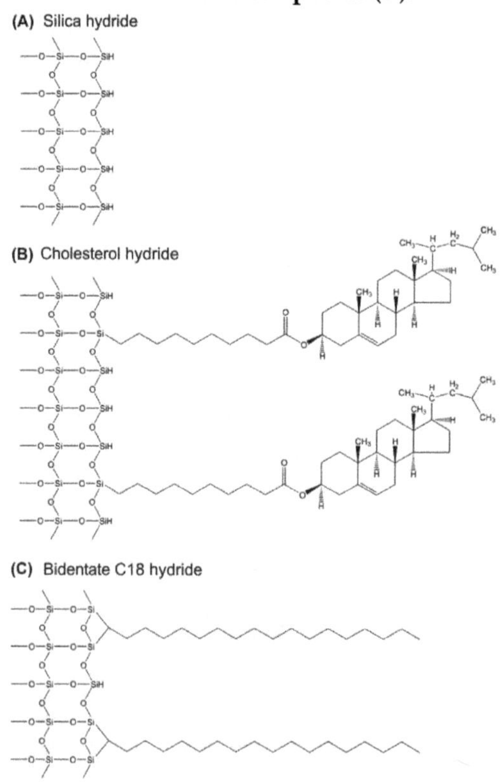

Fig. 4: Shows that UHPLCeMS(/MS) Structures of unmodified silica hydride (A) and hydrosilated silica modified with nonpolar cholesterol B) and bidentate octadecyl (C) ligands.

Amide stationary phases contain a carbamoyl or an amide group bonded to the silica gel surface via a short alkyl spacer and are generally recommended for efficient and fast separations of highly polar samples (Fig.2B). Amide-silica columns are often applied to the separation of peptides but are also suitable for HILIC separations of other hydrophilic or amphiphilic high-molecular weight samples, such as oligosaccharides, glycoproteins, or glycosides In contrast to the aminoesilica or poly(2-sulphoethyl aspartamide) columns, the retention of ionizable analytes is unaffected by ion-exchange interactions. Consequently, buffer additives to the mobile phases usually are not necessary for the separation, resulting in improved stability over a long period

Some amide stationary phases have been specifically designed for HILIC applications; e.g., carbamoyl-silica HILIC TSKgel Amide-80, which shows stronger retention of many polar compounds in comparison to silica gel and has been used for HILIC separations of mono- and oligosaccharides, sugar derivatives, peptides, and amino acids. 1.7 mm BEH amide columns strongly retain polar basic pteridines, for which they provided improved separation under HILICeUHPLC conditions versus underivatized BEH silica, especially at a high pH. Important contributions of ionic interactions and adsorption to retention are generally reported for the amino acid bonded phases. "Thiol-ene" click reactions (see more on reactions below) were employed for preparation of amino acid bonded stationary phases. By reactions between cysteine or cystine and the silica gel surface modified by horizontal polymerization of vinyl trichlorosilane, selective and efficient stationary phases were prepared for separations of nucleosides, sulfadiazine, sufisoxazole, and derivatives of benzoic acid in mobile phases containing salts.

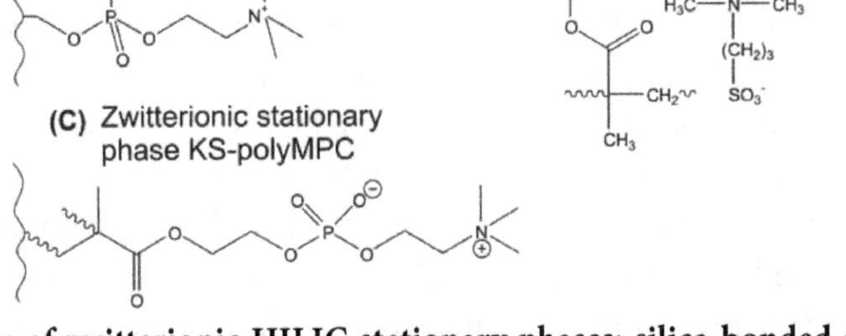

Fig. 5: Structures of zwitterionic HILIC stationary phases: silica-bonded sulfobetaine ZIC-HILIC (A), phosphorylcholine ZIC-cHILIC (B), poly amino-phosphate (C), and organic polymer N,N-dimethyl-N-methacryloxyethyl-N-(3- sulfopropyl)ammoniumeethylene-dimethacrylate (D).

Linear peptides immobilized on silica surfaces possess multiple amidic bonds and have traditionally been applied to LC of structurally similar compounds, such as for chiral separations, and for purification, fractionation, and isolation of peptides, proteins, etc., in HILIC systems. At a lower pH and higher ionic strength of the mobile phase, only the retention of acidic compounds decreased because of an ion-exchange mechanism; the retention of basic compounds was barely affected.

9.2.2.2 Diol, Polyethylene Glycol, Thioglycerol, Cyclodextrin, and Sugar Bonded Phases

Chemically bonded diol phases are prepared by bonding glycidoxypropyltrimethoxy silane to the silica gel surface and contain neutral hydrophilic 2,3-dihydroxypropyl ligands, which show high polarity and hydrogen bonding properties (Fig. 2A) and have been used for HILIC separations of proteins and low-molecular phenolic compounds in the HILIC mode. Diol columns may resolve anomers and different cyclic forms of monosaccharides, which allow monitoring the rate of the transition between the individual forms (mutarotation).

The Luna HILIC 200 cross-linked diol stationary phase contains both oxyethylene and hydroxyl bonded groups and combines to a certain extent the properties of the polyethylene glycol (PEG) and the diol columns (Fig. 2A). It shows improved stability against hydrolysis and better peak shape and resolution in comparison to the noncross-linked diol-silica phases. For example, nicotine and its principal metabolites cotinine, trans-3- hydroxycotinine, nicotine-N-oxide, and cotinine-N-oxide in urine were separated on a Luna HILIC 200 column using gradient elution from 98% to 35% acetonitrile at pH 3.0 with 10 mmol/L ammonium formate buffer. Six basic hydrophilic sympathomimetic drugs were separated by nano-LC under HILIC conditions on aminopropyl-silica and Luna HILIC 200 columns. Hydrophobic interactions are more pronounced with less polar polyethylene glycol (PEG) bonded stationary phases than with diol columns.

The separations on diol and PEG stationary phases show highly different selectivities in comparison to bonded alkyl-silica columns, both in RP mode and under HILIC conditions. CD stationary phases bonded on silica were prepared via click chemistry by immobilization of alkyne-modified b-cyclodextrin on the surface of azide-modified silica particles (Fig. 3A). Bonded CD stationary phases show chiral recognition properties under HILIC conditions, which have been used for chiral separations of sugar alcohols, monosaccharides, and oligosaccharides with 1e8 monosaccharide units. The retention increases with increasing number of monosaccharide units, which is attributed to interactions with the hydrophilic aqueous layer on the exterior of the CD molecules rather than to the penetration of sample molecules inside the cavity. CD-bonded columns retain amino acids and some other polar samples more strongly in comparison to the TSKgel Amide-80 phase and are more stable and reproducible than the aminopropyl silica stationary phases.

2-mercaptoethanol and oxidized 1-thioglycerol groups were attached onto vinylized silica, followed by oxidation with excess hydrogen peroxide in aqueous medium to yield stationary phases that can be used both in HILIC and in the RP mode (Fig. 2A). The separation selectivity and even the elution order of various sample types such as vitamins, nucleosides, and bases on these columns differ significantly from the behavior of polar compounds on diol bonded phases.

Avariety of sugar stationary phases were prepared using the "click" chemistry approach, in which sugar alkynes are covalently coupled to the azido-activated silica gel surface in the presence of a copper catalyst. The bonded carbohydrates retain their configurations, so that these highly polar stationary phases can be applied to stereoselective separations of monosaccharides and for HILIC separations of highly polar amino acids, glycopeptides, oligonucleotides, and natural products such as flavonoids. Separation selectivity differing from that of other polar-bonded columns and bare silica was also found for a hydrophilic stationary phase prepared by graft polymerization of sorbitol methacrylate on the surface of silica particles. Columns packed with carbohydrate-modified silica particles may show fair efficiencies for separation of oligosaccharides and chiral separations in the HILIC mode.

Several other polar stationary phases bonded on silica gel were used for separations under HILIC conditions, such as the Cosmosil HILIC column with 1,2,4-triazol groups bonded on silica, which has basic character and was reported to provide highly efficient separations of water-soluble vitamins, carboxylic acids, amino acids, peptides, and polar pharmaceuticals.

9.2.2.3 Polysuccinimide Bonded Stationary Phases

Alpert prepared a series of HILIC stationary phases by reaction of aminopropyl-silica with poly (succinimide) followed by a second reaction step such as alkaline hydrolysis yielding poly(aspartic acid)-silica, reaction with 2-aminoethanol providing poly(2-hydroxyethyl) aspartamide-silica, which may be followed by further modification with 2-aminoethylsulfonic acid to poly (2-sulfoethyl) aspartamide-silica (Fig. 2.2B). A wide range of poly(succinimide)-silica bonded materials available commercially have been applied for separations of various highly polar compounds such as peptides, proteins, nucleic acid constituents, oligosaccharides, carbohydrates, etc., in buffered aqueouseorganic mobile phases.

PolySulfoethyl A shows a mixed-mode HILIC/cation-exchange mechanism and was used for separations of hydrophilic peptides in aqueouseorganic mobile phases containing 50% or more acetonitrile, with increasing salt gradients. The separations show complementary selectivity to RP systems. The separation efficiency may be poor in mobile phases containing less than 90% acetonitrile; hence the PolySulfoethyl A column is recommended for not too strongly retained samples.

PolyHydroxyethyl A can be applied for separations of phosphorylated and nonphosphorylated amino acids, peptides and glycopeptides, proteins, oligonucleotides, carbohydrates and glycosides, and small polar solutes. Plant metabolites elute in the order: polar lipids, flavonoids, glucosinolates, saccharides, and amino acids. Successful separations with symmetrical peaks require mobile phases containing at least 7e10 mM electrolyte (such as triethylammonium buffer), especially for ionized solutes.

Poly Glycoplex columns are suitable for separations of monosaccharides, oligosaccharides, and sialyl sugars or their p-nitrobenzyloxy derivatives.

9.2.2.4 Other Polymer Coated and Bonded Silica Stationary Phases

The popularity of polysaccharide bonded silica stationary phases (Fig. 2.3A) for HILIC separations has steadily increased. Commercially available native cyclofructan-6 chemically bonded to porous silica particles (FRULIC-N) or to superficially porous silica (coreeshell) particles has been employed for separations of nucleic bases, nucleosides, nucleotides, xanthines, b-blockers, phenolic acids, carbohydrates, and other compounds in HILIC and in purely organic NP modes.

The dextran-bonded stationary phase prepared by reacting dextran with carbonyldiimidazole (CDI)-activated silica was reported to preserve the hydrophilic interactions of dextran and the high mechanical strength of the silica better than earlier prepared dextran-bonded phases. The retention can be adjusted by varying the salt concentration in the mobile phase. Efficient separations of sugars, sugar alcohols, oligosaccharides, and glycosilated peptides were achieved, with promising application possibilities in glycobiology and proteomics.

A hydrophilic polymer layer improves the pH stability of the silica-based stationary phases in comparison to unmodified silica. A polyvinyl alcohol (PVA)ecoated silica stationary phase (PVASil) was prepared by dipping silica particles into a hot PVA solution and then settling the suspension, providing a thin layer of PVA, which shields silanol groups from interactions with basic organic compounds and suppresses unwanted electrostatic interactions. The method can be extended to prepare different stationary phases with additional functionalities by doping desired ingredients in a PVA solution. Stronger retention of some nucleosides and similar performance with an efficiency of w57,000 theoretical plates/m of PVA-Sil were observed relative to the commercial bare silica. Simultaneous separation of eight cephalosporins was achieved with a gradient of decreasing content of acetonitrile in buffered mobile phases. Recently, a highly stable stationary phase (up to pH 12) was prepared by coating porous graphitic carbon with PVA. The material shows characteristic HILIC properties, but the separation selectivity differs from other HILIC columns.

9.2.2.5 Zwitterionic and Mixed-Mode Silica Stationary Phases

Silica-bonded zwitterionic stationary phases containing both positively and negatively charged moieties (Fig. 5) are widely used in the practice of HILIC and often show a dual HILIC/RP retention mechanism, depending on the mobile phase composition.

Mixed-mode anionecation exchange/hydrophilic interaction liquid chromatography (ACE/HILIC materials, Fig. 5) shows improved separation selectivity for small molecule drugs by combining ionic interactions with polar hydrophilic interactions.

Irgum's group introduced sulfoalkylbetaine stationary phases for HILIC separations. The active zwitterionic layer grafted on wide-pore silica gel or a polymer support contains both strongly acidic sulfonic acid groups and strongly basic quaternary ammonium groups, separated from each other by a short alkyl spacer. The two oppositely charged groups are present in a 1:1 Mratio, so that there is only a very low net negative surface charge of the bonded layer, attributed to the larger distance of the sulphonic groups from the silica gel surface.

Sulfobetaine-bonded ZICeHILIC silica (Fig. 5A and D) is suitable for separation of a wide range of small polar compounds, metabolomes, glucosinolates, aminoglycosides , peptides, or glycopeptides. Polar (hydrogen-bonding and dipoleedipole) interactions in the stationary phase are of primary importance, even though weak electrostatic interactions may affect the separation of partially ionized analytes. The HILIC separation of peptides on ZIC-HILIC columns strongly depends on pH. At pH ¼ 3, it resembles separations on strong cation exchangers (SCX). At a higher pH (7e8), chromatographic resolution improves, especially for prevalent þ2 and þ3 charged peptides in comparison to SCX separations. The resolution was the best at pH 6.8, but the orthogonality against a C18 phase was better at pH = 3.

HILIC on a capillary zwitterionic ZIC-cHILIC column in combination with electrospray ionization mass spectrometry has been applied to the detection and identification of more than 100 N-glycopeptides and O-glycopeptides in a single run. Glycopeptides could be distinguished by their glycan composition on a monolithic silica gel capillary surface modified with sulfobetaine stationary phase, in contrast to RP-HPLC. Purine and pyrimidine bases and nucleosides can be separated using gradient elution with a decreasing concentration of acetonitrile in buffered aqueouseorganic mobile phase on sulfobetaine ZIC-HILIC columns. The compounds elute in order of decreasing hydrophobicities, in agreement with generally observed HILIC behavior.

Phosphorylcholine stationary phases for HILIC separations (Fig. 5B and C) are prepared by graft polymerization of 2-methacryloyloxyethyl phosphorylcholine onto the surface of silica gel support. They differ from the sulfobetaine ZIC-HILIC material not only by the nature of the negatively charged group but also by the charge arrangement, causing significant differences in the elution order and in the separation selectivity for peptides and other samples, e.g., free amino acid and carboxylic acid ligands and their metal complexes in plant samples.

Amino-phosphate zwitterionic stationary phases contain a negatively charged phosphate group bonded via an ester spacer and a positively charged quaternary amine group. Successful separations of vitamins, nucleosides, deoxynucleosides, nucleobases, and aromatic acids were accomplished in aqueous acetonitrile containing ammonium formate. The retention of basic compounds increased at higher buffer pH, whereas the retention factors of uncharged analytes, such as nicotinamide, are almost independent of pH. reported a novel amino-phosphate zwitterionic stationary phase with phospho amine and diamine groups for HILIC applications.

Mixed-mode HILIC/ion-exchange columns possessing a ligand with a long alkyl chain and a hydrophilic polar terminal group bonded on the silica gel support (such as the Acclaim mixed-mode columndFig. 3B) have been introduced to enable separations of a wide range of strongly and weakly polar, or even nonpolar, compounds in organic-rich mobile phases. A weak anion exchanger, PolyWAX LP, prepared by modifying silica with a cross-linked coating of linear poly (ethyleneimine) can be also used for mixed-mechanism HILIC/ion-exchange separations in highly organic mobile phases.

A combined anion-exchange/cation-exchange/HILIC mechanism on silica-based weak ion exchangers was found useful for the analysis and purification of dipeptides and other compounds in biological samples, and of synthetic combinatorial chemistry products, using combined gradients of pH and increasing concentration of water.

A trimodal stationary phase prepared by coating porous spherical silica particles with charged organic polymer nanobeads (0.1 mm) has the inner pore area covalently modified with an organic ionexchange layer. The outer surface is modified with strong cation-exchange groups; hence the spatial separation of the anion-exchange and cation-exchange regions in the stationary phase provides possibilities for simultaneous separations of acids, bases, and neutral compounds using a mixed HILIC/ion-exchange retention mechanism at high acetonitrile concentrations. The material was used for simultaneous separation of ionized hydrophilic drugs and their counterions such as penicillin G and its potassium salt.

A mixed-mode glutamine silica-bonded stationary phase containing an amino alcohol group and two different amide groups, one being a polar head and the other embedded in an aromatic phenyl ring, retains polar, moderately polar, and nonpolar analytes. The column shows similar properties to zwitterionic stationary phases and provides a HILICeion exchange dual retention mechanism. Six nucleotides were resolved in 10 min in the HILIC mode with phosphate buffer (pH 3.25); phytohormones and phenolic compounds were separated under RP conditions. A mixed-mode glutathione HILIC/cation-exchange stationary phase was used to analyze peptides varying both in hydrophobicity/hydrophilicity and charge. Neutral fructosan with a high degree of polymerization was separated from basic chitooligosaccharides. Strongly acidic carrageenan oligosaccharides could also be resolved. The retention mechanism was attributed to the ERLIC mode (HILIC combined with electrostatic repulsion interactions) in mobile phases containing low amounts of buffers.

Neomycin is a hydrophilic aminoglycoside that contains six amino groups, seven hydroxyl groups, and six glycosidic oxygen functions, which has been grafted onto the surface of silica gel to act as a suitable stationary phase for HILIC separations of nucleosides, cytokinins, and sulfonamides in buffered aqueous acetonitrile. It offers additional selectivity for charged compounds, such as organic acids by mixed-mode hydrophilic/ion-exchange interactions.

Thiol-ene click chemistry, based on the reaction between a thiol and an alkene (alkene hydrothiolation), was used for coupling thioglycolic acid onto vinyl-bonded silica to provide a Thiol-Click-COOH column with high reaction yields and stereoselectivity. The material, which is more hydrophilic than pure silica columns, was reported to provide selective separations of 13 nucleosides and bases, and four water soluble vitamins, because of a combined HILICeion exchange mechanism.

The steviol glycoside Rebaudioside A (RA), an ent-kaurene-type diterpene glycoside, contains a hydrophobic aglycone and four hydrophilic sugar units. RAwas immobilized on silica through the "thiol-ene" click chemistry reaction between the terminal alkene group in RA and thiol groups on the silica surface to provide a HILIC/RPLC mixed-mode stationary phase, which enables separation of several sugars whose retention decreases with increasing content of water in the mobile phase, in agreement with the HILIC mechanism. However, the RP mechanism could be applied for separation of glycosides in highly aqueous mobile phases, too.

9.3 AQUEOUS NORMAL-PHASE CHROMATOGRAPHY ON HYDROSILATED SILICA PHASES

Through the hydrosilation process, ordinary silica gel (B-type) can be transformed to C-type, "hydride silica," by transforming up to 95% of the original surface silanols (SieOeH) to nonpolar silicon hydride SieH groups. The silica hydride material is less polar and shows less attraction for water and improved pH stability in comparison to sil-gel columns; hence the authors suggested to denote the retention mechanism on hydrosilated silica as "aqueous normal-phase chromatography" rather than HILIC. However, HILIC separations are also essentially conducted in an NP system. Water adsorbed on hydrosilated materials at the point of column saturation fills only 2.6%e5.5% of the pore volume, which corresponds to approximately half a monomolecular adsorbed water layer. According to recent studies, water uptake on the hydrosilated surface could probably be explained by the effects of the electrical interfacial double layer at the adsorbent surface, characterized by the zeta potential, rather than by adsorption on residual silanol groups.

Hydrosilated silica can be chemically modified by low-polarity bonded groups, such as alkyls or cholesterol, to provide a mixed-mode HILIC-RP retention mechanism and improved selectivities for separations of moderately polar compounds, using either the RP mode in highly aqueous mobile phases or the NP mode in buffered mobile phases containing more than 50%e70% acetonitrile. The UDC cholesterol and C18 bidentate columns (Fig. 2.4) show significant mixed hydrophobic and NP retention mechanisms, whereas weak retention is observed on silica hydride and Diamond hydride columns (containing w2.5% carbon) in the RP mode. Hence, a single column can be used with different selectivities and elution order in successive experiments in RP and NP modes, switching between different RP compatible water-rich and NP compatible highly organic mobile phases. Strong retention of both hydrophobic and hydrophilic peptides in high-organic mobile phases was used for their simultaneous separations on a Diamond hydride column using a gradient of decreasing acetonitrile concentration in water (from 80%-90% to 30%-50%) with addition of 0.1% acetic or formic acid. This column also provided better separation and suppressed peak tailing of acidic lipids such as (lyso-) phosphatidic acids with an acetonitrile concentration decreasing from 99.7% to 75% at pH 4, in comparison to bare or chemically modified silica columns modified the silica hydride surface with undecanoic acid (UDA silica). This stationary phase showed increased separation selectivity for mono-, di-, and triphosphate nucleotides because of enhanced ion-interaction (ion repulsion) properties with respect to the Diamond hydride column. The unmodified silica hydride provides similar elution order but stronger retention of more polar phenolic acids in comparison with hydrosilated silica modified by attaching cholesterol or bidentate C18 ligands under comparable HILIC conditions. The elution order of phenolic acids is identical on various hydrosilated columns and the retention increases with increasing polarity of the stationary phase surface: Bidentate C18 < cholesterol < Diamond hydride < silica hydride. The Diamond hydride column provides the best resolution and baseline separation of eight phenolic acids in a HILIC gradient with decreasing acetonitrile concentration.

9.4 MONOLITHIC COLUMNS FOR HYDROPHILIC INTERACTION LIQUID CHROMATOGRAPHY SEPARATIONS

9.4.1 Silica Gel and Hybrid Monoliths

Bare silica monoliths generally show low HILIC retention. The domain size (i.e., the sum of the average sizes of the through-pores and skeleton) controls the efficiency of a silica gel monolithic column. Coating the monolithic silica surface with cationic latex nanoparticles by electrostatic attachment provides separation media that retain the high efficiency and permeability of the native silica monolith, but show a significantly larger amount of adsorbed water, leading to stronger HILIC/ion-exchange retention. Fast separations of benzoates, nucleotides, and amino acids at high acetonitrile concentrations in the mobile phase were reported on latex-coated monoliths, based on mixed-mode electrostatic repulsionehydrophilic liquid interactions (ERLIC).

Hybrid monolithic silica capillary columns for HILIC were prepared by on-column polymerization of acrylic acid on monolithic silica in a fused silica capillary modified with anchor groups. The products maintained their high permeability and provided a theoretical plate height (H) of 10e20 mm for polar solutes, including nucleosides and carbohydrates.

Zwitterionic silica-based monolithic capillary columns were prepared by grafting a layer of zwitterionic monomer ([2-(methacryloyloxy)ethyl]-dimethyl-(3-sulfopropyl)-ammonium hydroxide or 2-methacryloyloxyethyl phosphorylcholine) on the silica monolithic surface. Two types of efficient capillary zwitterionic organic-silica hybrid monolithic columns were synthesized in a single-step thermally induced polymerization, by using [2-(methacryloyloxy)-ethyl] dimethyl-(3-sulfopropyl)ammonium hydroxide or [2-(methacryloyloxy)-ethyl] phosphorylcholine as the organic monomers and [3-(methacryloxy)-propyl]trimethoxysilane. The columns were suitable for HILIC separations of various low-molecular-weight neutral, basic, and acidic analytes, as well as small peptides in tryptic digests.

9.4.2 Organic Polymer Hydrophilic Interaction Liquid Chromatography Columns

Monolithic (poly)methacrylate diol or (poly)hydroxymethacrylate columns were successfully employed for HILIC separations of oligonucleotides. Unfortunately, organic polymer monolithic columns traditionally have shown rather low separation efficiency for low-molecular-weight compounds. Organic polymer-based monoliths have a heterogeneous structure, which rather resembles a net of interconnected nonporous cauliflower-like microglobules with low surface area. This morphology is suitable for separation of polymers requiring large pores (15e100 nm) with relatively low specific surface area (10e150 m^2/g). Large molecules of biopolymers cannot diffuse into the small pores in the microglobules of organic monoliths and can access only the large pores by convection; hence they provide narrow peaks. On the contrary, small molecules can penetrate into the narrow pores by slow diffusion, generally resulting in a poor separation efficiency. Essential improvement in the performance of monoliths was achieved by controlling the inner pore morphology of the monolithic bed structure. The proportion of mesopores should be increased while preserving a sufficient amount of through-pores. Several strategies were followed to this end:

- Chemistry and proportions of the functional monomers, cross-linkers, and porogen solvents have major impacts on the morphology of the final monolith.
- The initiation conditions and the temperature of polymerization should be carefully adjusted.
- Termination of the polymerization reaction at an early stage, in less than 1 h, provided incomplete polymerization, and consequently, less cross-links and larger porosity of the final monoliths. However, nanostructural heterogeneity in the polymer network and a void volume can be formed during polymerization because of different local compositions of monomer and cross-linkers. This may cause poor reproducibility of monoliths prepared using this approach.
- The chromatographic performance of the organic polymer monoliths could also be improved via adjustment of the pore morphology using post polymerization modifications or two-step monolith fabrication, hypercrosslinking, or

incorporation of additional structural elements into the monolithic skeleton, such as carbon nanotubes.

Organic polymer monoliths with zwitterionic groups incorporated into (poly) methacrylate monolithic structures are suitable for HILIC separations of neutral, basic, and acidic polar compounds in aqueouseorganic mobile phases with 60% or more acetonitrile. Polymer monolithic columns containing sulfoalkylbetaine moieties were prepared by photo-induced or thermal copolymerization of ethylene dimethacrylate (EDMA) and N,N-dimethyl- N-methacryloxyethyl-N-(3-sulfopropyl)ammonium betaine (MEDSA) inside fused silica capillaries (Fig. 2.5). Capillary monolithic columns prepared by copolymerization of methacryloxyethyl-N-(3- sulfopropyl) ammonium betaine (MEDSA) with EDMA or with 1,2-bis(p-vinylphenyl) ethane (BVPE) were employed for separations of nucleic bases and other neutral, basic, and acidic polar analytes in aqueouseorganic mobile phases with 60% or more acetonitrile. Their performance compared favorably to particle-packed silica-based zwitterionic columns.

Porosity, permeability, selectivity, and retention characteristics of monolithic sulfobetaine columns improve when using water-containing porogen solvents in the polymerization process. Long-chain alkyl monomers and crosslinkers increased the proportion of the mesopores in polymethacrylate monolithic columns, resulting in improved separation efficiency up to 70,000-80,000 theoretical plates/m for small molecules.

Monolithic stationary phases prepared by copolymerization of 3-chloro-2-hydroxypropyl methacrylate and ethylene dimethacrylate in a fused silica capillary could be further functionalized by reaction with triethanolamine ligands and used for separations of nucleobases and substituted benzoic acids.

Dimethacrylate cross-linkers with relatively long polar (poly)oxyethylene chains generally improve the performance of polar monolithic columns for HILIC applications. By copolymerizing MEDSA with N,N0-methylenebisacrylamide (MBA), 1,4-bis(acryloyl) piperazine (PDA), or bisphenol A glycerolate dimethacrylate (BIGDMA) cross-linkers, columns with significantly improved efficiency for small molecules were prepared. For the last type of monolithic column, excellent long-term stability, reproducibility of preparation within 1%-3% (relative), and efficiency up to 60,000 theoretical plates/m were reported at the optimum flow rate of 1-3 mL/min. After correction for extra-column contributions to band broadening, the efficiency increased to 120,000 theoretical plates/m.

Recently, investigated the effect of three charged hydrophilic groups introduced into hydrophilic sulfobetaine stationary phases, namely N,N-dimethyl-N-acryloyloxyethyl-N-(3- sulfopropyl) ammonium betaine (SPDA), [2-(acryloyloxy) ethyl] trimethylammonium chloride (AETA), and 3-sulfopropyl acrylate potassium salt (SPA), on the chromatographic properties of zwitterionic monolithic capillary columns. Depending on the combination of stationary phaseemobile phaseesolute, hydrophilic interactions and electrostatic and hydrogen-bonding interactions, together with molecular shape, contribute to the retention and separation selectivity of analytes. Because of the weak anion exchange properties, the zwitterionic poly(SPDA-co-MBA) hydrophilic monoliths exhibit the best separations for phenols, b-blockers, and small peptides, in terms of selectivity, peak shape, and analysis time. On the other hand, the cationic poly(AETA-co-MBA) monolith provided the best separation for nucleobases and nucleosides.

A two-step surface modification of a poly (styrene-co-vinylbenzyl chloride-co-divinylbenzene) nonpolar monolithic layer, including hypercross-linking and thermally initiated surface grafting of [2- (methacryloyloxy) ethyl] dimethyl (3-sulfopropyl) ammonium hydroxide, has been used to form a second zwitterionic layer for efficient separation of small polar compounds.

A primary monolithic poly(lauryl methacrylate-co-tetraethyleneglycol dimethacrylate) layer could be used as a stable scaffold for a second, polar monolithic layer with zwitterionic functionality by in situ copolymerization of poly(N,N-dimethyl-N-methacryloxyethyl-N-(3-sulfopropyl)ammonium betaineco- bisphenol A glycerolate dimethacrylate) on the primary monolith; FriedeleCrafts grafting onto the scaffold layer was not necessary. The column was suitable for HILIC separations of phenolic acids, flavones, nucleosides, and bases of nucleic acids, with similar efficiencies, but with selectivities differing from zwitterionic methacrylate monolithic columns prepared by a single step polymerization.

9.5 HYDROPHILIC INTERACTION LIQUID CHROMATOGRAPHY COLUMN SURVEY

Usually, 100e150 mm long, 4.6 mm I.D. columns packed with 3e5 mm particles are used for standard HILIC separations; smaller diameter columns are preferred for HILIC coupled online with MS detection. The column length and flow rate can be changed for increasing resolution or decreasing run time, with pressure as the limiting factor. Working at very high pressures considerably increases the speed of separation; however, secondary impacts on HILIC separations were observed. The retention of sugars in hydrophilic interaction chromatography decreases at high pressures, probably because of changes in the hydration of strongly polar analytes as they move from the bulk highly organic mobile phase into the water-rich layer at the surface of the polar adsorbent.

The retention of selected probes was used as an indicator of specific interactions: the relative retention of cytosine versus uracil as the probe of the "hydrophilicity" of the HILIC phases; the relative retention of adenosine versus adenine as a probe of the hydrogen bonding activity of the stationary phase; and the relative retention of benzyltrimethylammonium versus cytosine as the probe of the cation-exchange/anion-exchange character of the column.

HILIC stationary phases with various chemistries provide different retention times and selectivities for individual analytes. Very similar selectivities were found for neutral, acidic, and basic analytes on bare silica and on diol stationary phases. Cyanopropyl columns provide different selectivity for basic compounds, whereas aminopropyl stationary phases show a strong retention preference for acidic compounds, due to the ion-exchange effect, whereas bare silica stationary phases show a strong preference for basic compounds attributed to interactions between the surface silanols and ionized analytes compared the separation properties of some types of particle-packed and monolithic polar columns (bare silica, aminoesilica, amideesilica, diolesilica, cyanoesilica, polysuccinimideesilica, sulfobetaineesilica, triazolesilica, and CDesilica) frequently used for HILIC separations of carboxylic- and amino-acids, peptides, amines, amides and substituted urea derivatives, nucleic bases and nucleosides, poly-alcohols, and oligosaccharides and carbohydrates.

Some HILIC columns may require lower flow rates for optimum performance. Similar to RP-HPLC, the efficiency and column back-pressure increase with decreasing particle size; organic polymer columns usually show higher heights equivalent to theoretical plate (HETP) in comparison to silicabased columns. Compared the retention of representative carboxylic acids, nucleosides, and nucleotides on bare silica and polar silica-bonded amide, amino, aspartamide, and sulfobetaine stationary phases commonly used in the HILIC mode. Silica gel showed the least retention but the highest selectivity differences with respect to other polar bonded stationary phases. In another study, the chromatographic behavior of 10 model peptides on bare silica, amide, poly-hydroxyethyl aspartamide, diol, and zwitterionic HILIC columns was investigated. The AlltimaeAlltech silica column showed the best performance for the selected peptide set, followed by the zwitterionic ZIC-HILIC column. A generic screening protocol was suggested for the selection of the best column packed with BEH organic silica gel 1.7 mm particles, suitable for fast HILIC separations. The selection from among three stationary phases, namely bonded amide, diol, and cyanophenyl polar ligands, which differ from one another and from underivatized BEH HILIC, is based on the selectivity correlations obtained in experiments with 28 polar test probes in weakly acidic (pH= 3) and weakly basic (pH= 9) mobile phases.

Applied the linear solvation energy relationship (LSER) model using 68 test compounds comprising small neutral, basic, acidic, zwitterionic, and amphoteric molecules with a broad range of solvatochromic parameters to characterize 22 silica-based polar stationary phases into basic, neutral, or acidic classes at pH ¼ 3. The "acidic" modified columns exhibit only weak hydrogen bond acceptor properties and are less capable of achieving an even distribution of diverse analytes along the available retention window, in contrast to "basic" (bonded amino) and "neutral" columns. Of the "neutral" columns, amide and zwitterionic columns show a broader retention window for a wide range of analytes in comparison to diol columns and therefore are preferred as the first choice for HILIC separations. The results of these studies confirmed major effects of hydrogen bonding and ionic interactions on the retention.

9.6. SEPARATION MECHANISM AND EFFECTS OF THE ADSORBED WATER AND MOBILE PHASE

9.6.1 Adsorption of Water on Polar Columns

The mechanism of separation on polar columns in aqueous organic mobile phases is complicated. HILIC is traditionally understood as the partition process between the aqueous layer accumulated close to the solid surface and a bulk mobile phase containing high concentrations (usually more than 60%) of a polar organic solvent in water. However, the actual separation mechanism is obviously more complex, and differences in the chromatographic selectivities of various polar compounds indicate that adsorption on polar functionalities on the solid phase surface may also play an important role in the retention. Both polar and hydrophobic groups on the structure of adsorbents may cause excess adsorption of either acetonitrile or water, depending on the mobile phase composition. In organic solvent normal-phase adsorption chromatography, the retention is primarily based on competition between the solute and the polar solvent for the localized adsorption sites on the surface of a polar adsorbent, usually bare silica gel. Polar solvents, especially water, are strongly adsorbed on polar adsorbents from mixed aqueouseorganic mobile phases. In the original HILIC model, considered a partition-driven retention mechanism in which the analytes distribute between the stagnant water-rich layer adsorbed onto the polar sorbent on one hand and the water-poor bulk mobile phase on the other, without contribution of the sorbent backbone.

However, the solid adsorbent cannot be considered as an inert support for the adsorbed "pseudo stationary" aqueous phase in mobile phases containing 2%-40% water in organic solvent (usually acetonitrile). Although the water uptake is often characterized in terms of "adsorbed water layers," there is an essential difference between the water retained in HILIC and that in classical adsorption organic solvent normal-phase chromatography. A compact water layer can be accumulated on the adsorbent surface from traces of water present in no polar or weakly polar organic solvents. On the other hand, water is miscible at any proportion with acetonitrile, acetone, methanol, or other polar organic solvents. Hence, an adsorbed "water layer" is rather diffuse, without sharp boundaries, and the concentration of water progressively decreases from the solid surface toward the bulk organic-rich mobile phase outside (and, possibly even partly inside) the pores of the stationary phase (Fig. 1).

Specific solid phase effects such as hydrogen-bonding and ionic interactions with polar functional groups such as ionized silanol groups on bare silica, or residual silanols on silica-bonded phases, may Contribute to the sample retention and also affect the amount of water adsorbed on the solid support; even nonpolar siloxane groups may adsorb some water at very low organic solvent concentrations.

The plethora of polar stationary phases used in the contemporary practice of HILIC separations show large variability in the amount of adsorbed water in aqueouseorganic mobile phases, depending on the column type. Computer molecular dynamics simulation studies indicate that the relative proportion of the amount of water contained in the pores of silica-based phases to the water concentration in the bulk mobile phase increases at low total water concentrations in the column. The water molecules close to the silica surface are almost immobilized by the hydrogen bonds to the silanol groups.

NMR studies of bare silica and silica with bonded zwitterionic sulfobetaine groups suggested that three types of water can be distinguished inside the 6-10 nm pores: (1) free water molecules, (2) "freezable" bound water, and (3) water bound within the polymeric stationary phase network that does not freeze at the regular water freezing temperature. This may be the reason why separations are often irreproducible or fail in mobile phases containing less than 2% water in acetonitrile. Water adsorption can often be described by the Langmuir model, Eq. (1), which enables measuring the adsorbed amount of water over a varying composition of the bulk mobile phase:

$$q_i = \frac{ac_m}{1 + \frac{c_m a}{q_s}} \qquad 1$$

Where qi is the adsorbed water concentration, cm is the water concentration in the bulk mobile phase, and a is the distribution constant of water in the pores of the stationary phase, forming a layer between the column and acetonitrile at very low cm. The plateaus of the isotherms characterized by qs in Eq. (1) give the saturation capacities for water adsorption, which can be calculated from the experimental parameters of the Langmuir equation determined by the frontal analysis technique, where a sample solution is pumped through the column and the breakthrough volume of the sample from the column is usually monitored online in the column effluent using a nonspecific (refractive index or low-wavelength UV) detector. As the boundaries of the adsorbed water diffusion layer are difficult to determine and depend on the composition of the bulk mobile phase, the adsorbed water amount may be characterized in terms of the percentage of the water-occupied pore volume.

The water breakthrough volumes, VB, at varying concentrations of water in the column feed (from 0% to column saturation) could be measured using off-line coulometric Karl Fischer titration of small collected column effluent fractions. This method does not rely on the online nonspecific detection, which may be affected by interfering compounds. The amount of the water retained in the stationary phase can be conveniently expressed in terms of the volume fraction of the excess water contained in the inner pore volume, qex, (over the bulk mobile phase water concentration, cm) and was calculated from Eq. 2):

$$q_{ex} = \frac{(V_B - V_M) \cdot c_m}{V_i} \qquad 2$$

V_M is the total column hold-up volume, including the inner pore volume, Vi ¼ εiVM, and the outer (interparticle) volume, V0 = ε0VM; εi and ε0 are the inner pore and the inter-particle porosities, respectively. Uracil or thiourea are often used as column hold-up volume, VM, markers in RPLC, which are quite strongly retained on polar columns under HILIC (NP) conditions.

Benzene and toluene are more suitable nonretained markers of column hold-up volume in acetonitrilerich mobile phases. Their elution times on silica columns slightly increase as the water content in the mobile phase grows from 0% to 30%, which means that the water amount adsorbed close to the polar adsorbent surface depends on the water concentration in the bulk mobile phase. Water adsorption isotherms on nonmodified and monomeric functionalized silica phases show monolayer formation followed by multilayer adsorption, whereas water uptake on polymeric functionalized silica stationary phases may lead to the formation of hydrogels. The water uptake correlates with the retention factors of neutral analytes, supporting the idea of the coexistence of adsorption and partitioning of neutral solutes in the water concentration regime normally encountered in HILIC.

Frontal analysis with Karl Fischer titration was used for the determination of water uptake on 19 stationary phases, including fully porous and coreeshell silica gel and silica-bonded phases with different polarities: octadecyl and cholesterol, phenyl, nitrile, pentafluorophenylpropyl, diol, zwitterionic sulfobetaine and phosphorylcholine ligands bonded on silica, hybrid organic-silica, and hydrosilated silica stationary phases, together with a few nonpolar alkyl-bonded phases. All columns adsorb water from acetonitrile, and even though there are large differences between the individual stationary phases, the adsorption can be described by Langmuir isotherms, Eq. 1). Six isotherm examples are shown in Fig. 2.6. Nonpolar long alkyl chain (YMC Carotenoid) and hydrosilated (Cogent Silica C, Cogent C18 Bidentate) columns show steep isotherms with low plateau water concentrations (low water uptake), whereas the water adsorption isotherms on polar columns (YMC Triart Diol HILIC, ZIC HILIC, and TSKgel Amide-80) have a much less steep slope and at least one order of magnitude higher water saturation concentrations, which is often not achieved even in mobile phases containing 20% water in acetonitrile. Fig.7 compares the column saturation capacities, qs, calculated from the Langmuir isotherms. The TSKgel amide, ZIC-HILIC and ZIC-cHILIC, and TSKgel NH_2 columns have high saturation capacities, corresponding tow45% (vol/vol) water in the bulk mobile phase, which is consistent with their high affinity to water reported elsewhere. A relatively high saturation capacity, 35.4% (v/v) water, was observed for the Ascentis Express OH column, whereas the Xbridge HILIC, Atlantis HILIC, Ascentis Express ES-CN, and Ascentis Express F5 columns have lower saturation capacities, less than 9% v/v. At full column saturation, the excess adsorbed water, Vex, fills up to 45.3% of the pore volume of normal silica-based columns but only 2.6%e5.5% of the pore volume of hydrosilated silica columns. Because of the low affinity of the hydrosilated silica material to water, saturation capacities are as low as 0.2%-0.4% water in the inner pore volume, which are achieved in mobile phases containing 3%-6% v/v water (for the hydrosilated bare silica or bonded C18 Bidentate columns), in agreement with the Pesek ANP model. This low water saturation capacity is similar to the long nonpolar alkyl chain (C30) columns. On the other hand, polar columns used frequently in HILIC show much less steep water isotherms and relatively high saturation capacities, which are not approached even in mobile phases containing 20% v/v water in acetonitrile.

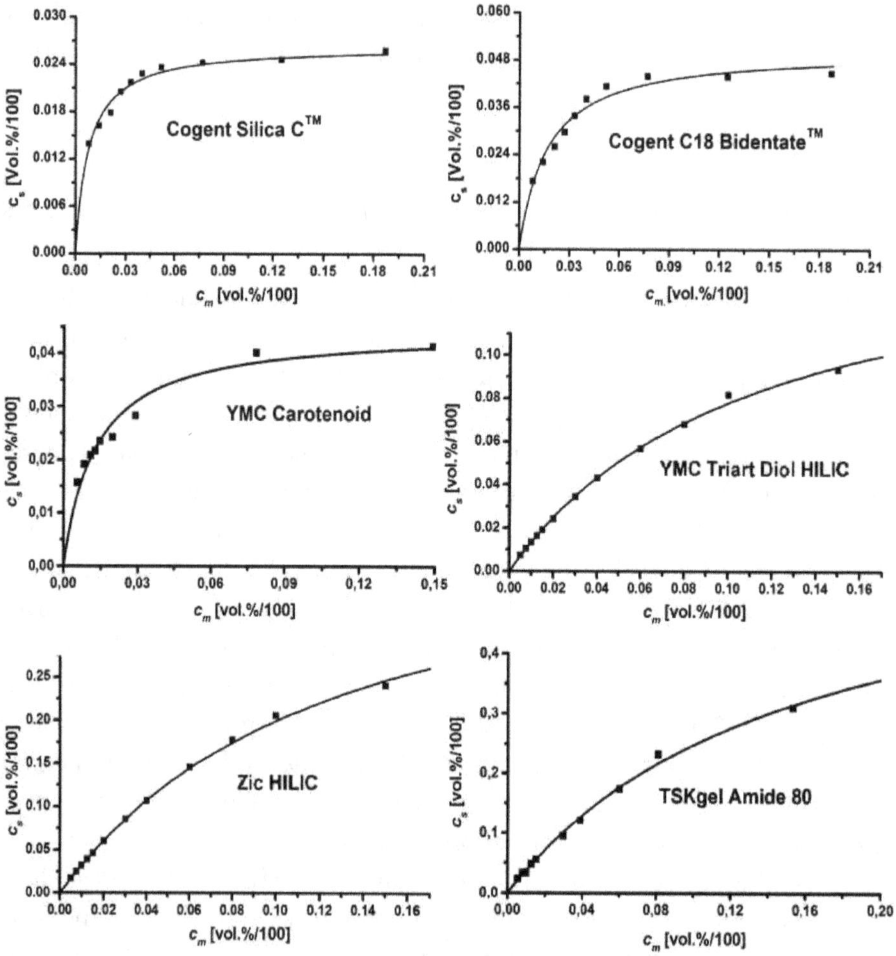

Fig. 6: Langmuir isotherms of water adsorbed on two hydrosilated (Cogent Silica, Cogent Bidentate) and four silica gel (C-30 YMC carotenoid, YMC Triart Diol HILIC, zwitterionic sulfobetaine ZIC-HILIC, and TSkgel Amide-80) bonded stationary phases. cm, volume fraction of water in the mobile phase in equilibrium with the stationary phase; cs, volume fraction of excess water contained in the pores of the stationary phase.

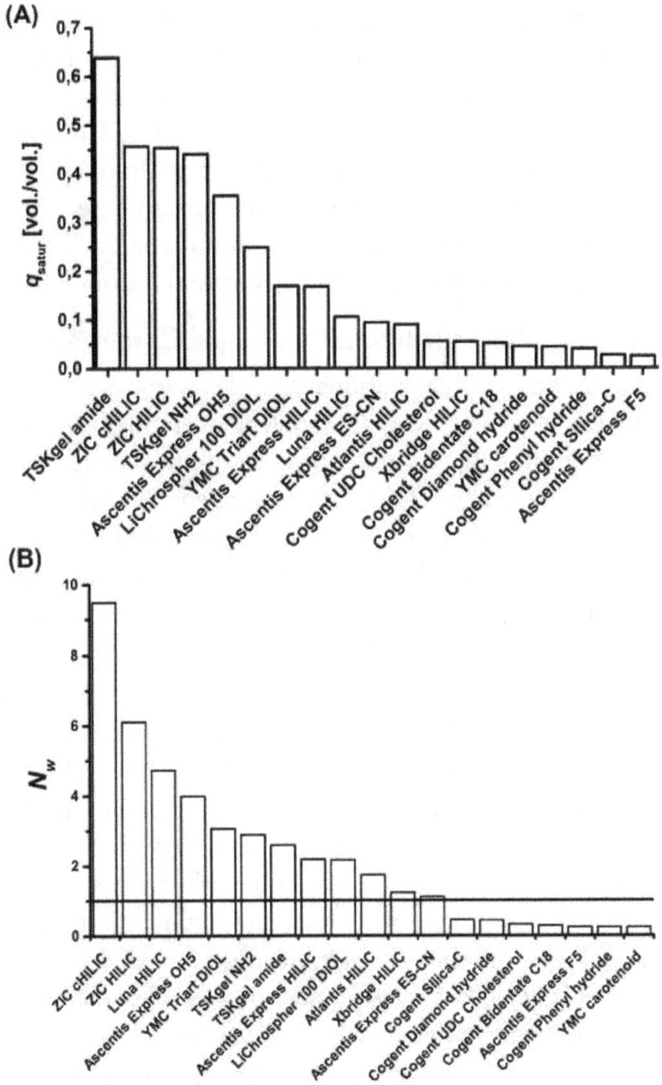

Fig. 7: Water adsorption on high-performance liquid chromatography columns: (A) excess water saturation capacities, qsatur, calculated from Eq. (2.1) and (B) the equivalent number of adsorbed monomolecular water layers, Nw, inside the pores at full saturation capacity of the columns.

Fig. 7(B) shows the water uptake in terms of the number of "hypothetical" monomolecular water layer equivalents, Nw, at full saturation capacity of the 19 columns tested. The number of adsorbed water layer equivalents generally agrees with the order of column sorption capacities in Fig. 7(A) but with some exceptions from the rule. Less than one monomolecular water layer equivalent (full horizontal line) was adsorbed on the silica hydridebased stationary phases and on moderately polar coreeshell columns (Ascentis Express F5 and Ascentis Express CN) at the column saturation capacity. On strongly polar stationary phases, several water layer equivalents are captured from the mobile phase. The sample partition between the bulk mobile phase and a submonomolecular layer of adsorbed water lacks a physical meaning. Rather, competition between the adsorbed water and polar solutes based on a NP adsorption mechanism is more realistic. Hence, a low number of the adsorbed monomolecular equivalents, Nw < 1, may be used to distinguish between aqueous normal phase (ANP) and traditional HILIC. The strongest affinity to water was observed on the ZICcHILIC stationary phase, where more than nine water layer equivalents were adsorbed at the saturation capacity. Columns with bonded hydroxyl and diol ligands show stronger water adsorption in comparison to bare silica.

9.6.2 Mobile Phase in Hydrophilic Interaction Liquid Chromatography Separations

HILIC of polar compounds on polar stationary phases employs aqueouseorganic mobile phases, usually containing 60%e95% organic solvent (acetonitrile) in water or an aqueous buffer (often volatile ammonium formate or ammonium acetate, compatible with MS detection and online identification). It is recommended to perform initial experiments at a relatively high concentration (e.g., 40%) of aqueous buffer and then increase the concentration of acetonitrile until acceptable sample retention and resolution are achieved. A scouting gradient of decreasing acetonitrile concentration starting from 95% acetonitrile can be helpful for this purpose.

Selection of the organic solvent has a strong effect on the retention and overall separation performance in HILIC mode. Increasing polarity and the ability to participate in protonedonor/ protoneacceptor interactions enhance the solvent strength, which generally decreases in the order: water > methanol > ethanol > 2-propanol > tetrahydrofuran > acetonitrile. Acetonitrile, an aprotic solvent, which does not show protonedonor interactions, shows the best retention and separation selectivity among all water-miscible organic solvents in HILIC (or in ANP) chromatography. Methanol and other lower alcohols are protic solvents that provide significant hydrogen bonding interactions and compete with water for solvation of the polar adsorbent surface, decreasing the amount of adsorbed water layer on polar stationary phases. Theoretically, it is possible to substitute water by an organic solvent and to employ, e.g., a methanoleacetonitrile mobile phase in "nonaqueous HILIC chromatography," but usually the retention is weaker than that in conventional HILIC separations.

Mobile phases containing methanol (and to lesser extent ethanol, 2-propanol, or other protic solvents) often provide insufficient sample retention and broad or nonsymmetric peak shapes. The poor performance of alcohols as organic components of HILIC mobile phases is attributed to their similarity to water in providing strong hydrogen-bonding interactions. The retention is obviously promoted by increasing differences in selective polar interactions between the highly aqueous liquid layer occluded on the surface of polar adsorbents and the bulk organic-rich mobile phase. These differences are higher with acetonitrile-rich mobile phases in comparison to the mobile phases containing protic organic solvents.

The attempts to replace acetonitrile with a less toxic solvent have not been very successful so far. Acetone has similar polarity as acetonitrile but shows lower sample retention under HILIC conditions. Moreover, it cannot be used with UV detection and provides lower intensity MS signals. At high concentrations of carbon dioxide added to ethanol-water mobile phases, HILIC separations of nucleic bases and other simple polar compounds on bare silica were reported to resemble those in acetonitrile-water mobile phases. However, the routine use of such mobile phases can be subject to serious practical problems.

HILIC mobile phases usually contain a buffer, whose pH and ionic strength are usually selected to enhance the sample ionization, retention, and separation selectivity, i.e., pH > 7 for acids and pH < 7 for bases. The retention in HILIC systems with uncharged stationary phases usually increases with increased salt (buffer) concentrations, probably because of enhanced hydrogen-bonding interactions between the analyte and the stationary phase. On the other hand, the retention of ionic samples on a stationary phase with ionic or ionizable functionalities may decrease with increasing ionic strength because the salt counterions displace the retained ionized sample molecules by ion-exchange interactions. That is why the retention of acidic compounds on bonded amino phases decreases at higher salt concentrations, in contrast to other simple polar bonded silicabased stationary phases).

High ionic strengths often reduce peak tailing in HILIC systems. Some samples may require a buffer concentration of 100 mmol/L or higher for acceptable peak shapes. The addition of trifluoroacetic acid (TFA) to the mobile phase may also improve peak shapes of basic compounds by formation of ion associates. However, high ionic strengths or TFA additives are not recommended in HILIC/MS applications because they may suppress sample ionization in the mass spectrometer.

In some applications, water was partly replaced by a weaker polar protic solvent. For example, the retention of methacrylic acid, cytosine, nortriptyline, and nicotinic acid on a BEH-HILIC column increased considerably and the separation improved when substituting 5% of water with a second organic solvent (methanol, ethanol, or 2-propanol) in the original buffered mobile phase containing 90% acetontrile. Water can even be fully replaced by a polar organic solvent in so-called "nonaqueous HILIC chromatography" (NA-HILIC), employing mixed polar organic mobile phases. There is no water in the mobile phase; instead, the liquid diffuse layer adsorbed on the stationary phase contains an increased concentration of an organic "protic modifier."

NA-HILIC may fill the gap between traditional nonaqueous NP (employing a nonpolar and one or more polar organic solvents) and aqueouseorganic HILIC systems. It can potentially be useful in the analysis of some poorly soluble oligomers or weakly polar compounds, which may precipitate in water-containing HILIC mobile phases. Selecting the type of protic solvent added to acetonitriledethylene diol, methanol, or ethanold(with elution strength decreasing in that order), the retention and separation selectivity of nucleobases, nucleosides, and deoxynucleosides could be adjusted in HILIC on bare silica, diol (Luna HILIC), bonded thioglycerol, and oxidized thioglycerol polar bonded phases.

In HILIC mode, the retention on polar columns increases with decreasing concentration of water in a water/organic mobile phase. The effect of water in the mobile phase on the sample retention factor, k = tR/tM, can sometimes be described by Eq. (3), or by Eq. (2.4), in the limited range of high water concentrations:

$$\log k = a_1 - m_{\mathrm{HILIC}} \cdot \log \varphi_{\mathrm{H_2O}} \qquad 3$$

$$\log k = a' - m_{\mathrm{HILIC}} \cdot \varphi_{\mathrm{H_2O}} \qquad 4$$

Where $4H_2O$ is the volume fraction of water, "a1" is the (extrapolated) logarithm of the retention factor in pure water, and "a0" is the logarithm of the retention factor in pure organic solvent. (tR is the sample retention time and tM is the elution time of a nonretained standard, i.e., the column hold-up time.) The parameter "mHILIC" characterizes the rate of decreasing retention with increasing content of water in the highly aqueous mobile phases. Traditionally, it has been believed that Eq. (3) characterizes the retention in adsorption systems and Eq. (4) in partition systems. In fact, various experimental HILIC systems show better data fit for either of the two equations. However, a good fit of any model equation to the experimental data does not guarantee the correctness of the model. Furthermore, the original idea of the validity of Eq. (4) was applied to immiscible stationary and mobile phases, such as RP-LC with long-chain chemically bonded alkylsilica materials, or organic NP-LC. The correct application of theoretical models relies on knowledge of the exact volumes of both the stationary and mobile phases in the column, which can hardly be guaranteed in HILIC systems, where the amount of adsorbed water changes with the composition of the bulk aqueouseorganic mobile phase.

In many cases, neither Eq. (3) nor Eq. (4) fits the HILIC experimental data satisfactorily. For example, this was the case with 35 acidic, basic, and neutral polar compounds on bare silica and bonded aminopropyl, amide, diol and cyanopropyl columns, including fully porous, hybrid, and coreeshell types of particles. Combining Eq. (3) and Eq. (4), we obtain Eq. (5), which takes into account a dual HILIC-RP retention mechanism and provides improved fit to the retention data over a broader range of aqueouseorganic mobile phase compositions, in comparison to polynomial empirical equations:

$$\ln k = a + b \ln \varphi_{\mathrm{H_2O}} - c\varphi_{\mathrm{H_2O}} \qquad 5$$

The ratio of terms b and c of Eq. (5) was tentatively employed to estimate the importance of the relative contributions of partitioning and the surface adsorption mechanism. Based on the results of application of Eq. (5) to the retention of water-soluble vitamins, it was concluded that the transition from a partitioning to a surface adsorption mechanism for neutral compounds occurs at more than 75%-80% acetonitrile on diol, silica, and amino columns, depending on the different degree of hydration of the stationary phases.

For electrostatically attracted compounds, surface adsorption remains the dominant retention mechanism even at lower acetonitrile concentrations.

In spite of differences in the conclusions of some studies of HILIC mechanisms, the adsorption and the partition retention mechanisms most probably actually coexist in many HILIC systems, depending on the solute, the stationary phase polar functional groups, and the eluting conditions. For example, less hydrophilic nortriptyline was reported to be retained by a partition-like mechanism and cytosine by a more hydrophilic mechanism, rather than by an adsorption-like mechanism, which was attributed to slower diffusion of the more retained polar species in the viscous diffuse water layer, even though diffusion is not a thermodynamic phenomenon.

Many ionized compounds can be separated in HILIC systems. Very often, either attractive (ion exchange) or repulsive (ion exclusion or ion repulsion) electrostatic interactions participate in the retention mechanism, especially on strong (SAX) or weak (WAX) anion-exchange columns. The addition of salts, weak acids, bases, or ion-pairing reagents as mobile phase additives usually significantly improves the separation in the mixed HILIC/ion-exchange mode. Adjusting the pH and salt (buffer) concentrations may significantly improve the retention selectivity, peak profile, and separation, however, with very different selectivity effects for acids and bases. On bare silica columns, acids show much stronger retention in mobile phases containing trifluoroacetic acid than that in ammonium formate buffers, wheredon the contrarydbases are better retained.

Some mixed-mode silica-based HILIC/IEX stationary phases can be used for separations of polar and ionic solutes under HILIC conditions in organic solvent-rich mobile phases and for separations of less-polar compounds under RP conditions in more aqueous mobile phases. The RP/WAX phases differ from the typical HILIC stationary phases, TSKGel Amide-80, ZIC-HILIC, or polysulfoethyl A, towhich they provide a certain degree of complementary application possibilities.

At increased concentration of acetonitrile, adequate retention and satisfactory resolution of both basic and acidic peptides can be achieved in a single run, on either SAX or WAX columns at low pH, where HILIC and electrostatic repulsion retention mechanisms superimpose to produce the ERLIC mode, which offers possibilities for independently adjusting the HILIC and the ion-exchange selectivity's in highly organic mobile phases.

9.6.3 Dual Hydrophilic Interaction Liquid Chromatography/Reversedphase Retention Mechanism

Increasing the water concentration in a mobile phase is connected with a decrease in sample retention over a more or less limited range of high organic solvent concentration in the mobile phase, where the retention is controlled essentially by the sample's polar interactions in the HILIC mode. When a certain concentration of water in the bulk mobile phase is exceeded, the sample is not retained and elutes close to the column hold-up (dead) volume. However, by further increasing the concentration of water in the mobile phase, some compounds may become retained again because of the solvophobic interactions with nonpolar moieties contained in the stationary phase, which is typical for RP behavior. Hence, many polar columns show a dual HILIC-RP retention mechanism: nonpolar (RP) at high water concentrations and polar (HILIC) in mobile phases with high concentrations of organic solvents. In this case, the graphs of the sample retention factors, k, versus the volume fraction of water, 4(H_2O), show characteristic U-profiles over a broad range of binary aqueouseorganic mobile phases (see examples in Fig. 8). The dual HILIC-RP behavior depends both on the column and sample, and it probably increases for polar stationary phases containing significant nonpolar moieties in the structure.

In the presence of a dual HILIC-RP mechanism, the effects of the volume fraction of the more polar solvent, water, 4ðH_2OÞ, on the retention factor, k, can often be described by Eq. (6) over a broad range of compositions of aqueouseorganic mobile phases _at 4ðH2OÞ > 0:02

$$\log k = a + m_{\mathrm{RP}} \cdot \varphi_{\mathrm{H_2O}} - m_{\mathrm{HILIC}} \cdot \log\left(1 + b \cdot \varphi_{\mathrm{H_2O}}\right) \qquad 6$$

The parameter mRP characterizes the effect of the increasing concentration of water in the mobile phase on the retention because of the RP mechanism in water-rich mobile phases, whereas the parameter mHILIC is a measure of the water contribution to the decrease of retention in highly organic mobile phases (i.e., the HILIC range). The system constant, a, depends on the solute and type of organic solvent. The empirical term, b, improves the description of the retention at low water concentrations.

Eq. (6) describes the U-shaped experimental graphs, where the "U-turn" transition from the HILIC to the RP mode can be localized at the minimum retention corresponding to the 4min concentration of water in the mobile phase:

$$\varphi_{\min} = 0.434 m_{\mathrm{HILIC}}/m_{\mathrm{RP}} \qquad (2.7) \qquad 7$$

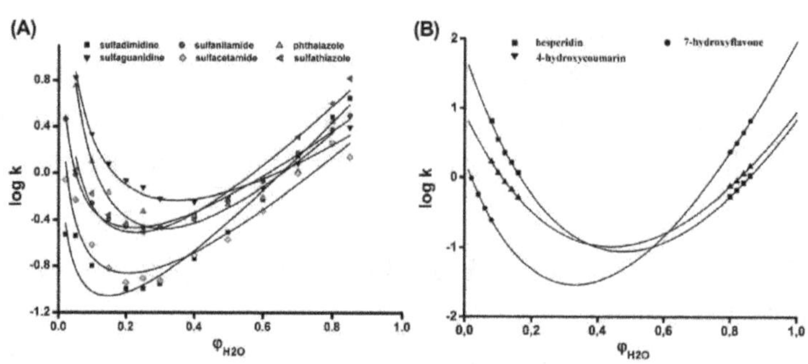

Fig. 8: Effect of the volume fraction of aqueous buffer, 4H2O, in aqueous-acetonitrile mobile phase on the retention factor, k, of sulfonamides on monolithic column BIGDMA-MEDSA (A) and flavonoids on carbamoyl bonded silica column TSKgel Amide-80 (B). Points are the experimental data and lines the best fit plots of Eq. (5).

The transition between the HILIC and the RP behavior is observed at a lower volume fraction of the aqueous buffer on less PEG columns (such as bonded polyethylene glycol in 70%-90% acetonitrile), in comparison to more polar columns, e.g., diol, zwitterionic, etc., in 40%-70% acetonitrile.

Zwitterionic polymethacrylate monolithic columns show dual HILIC and RP mechanisms. Orthogonal selectivity in the HILIC and the RP ranges may provide excellent separations of flavonoids and phenolic acids (as well as others) on a single column in both retention modes. Silica hydride columns with bidentate or cholesterol nonpolar ligands bonded on the surface also show dual retention mechanisms, despite very low affinity to water.

Some columns may provide practically useful separations only either at high organic solvent concentrations (HILIC) or in highly aqueous mobile phases (RP), ruling out from practical use the major part of the medium mobile phase composition range, where the retention is too low. The useful mobile phase range depends both on the stationary phase and on the analyte. For example, organic polymer zwitterionic columns provide relatively broad mobile phase ranges both for HILIC and RP separation modes. Fig. 8A shows examples of U-turn retention plots of sulfonamides on a homemade capillary zwitterionic column and Fig. 8B the U-turn retention plots of flavonoids on a commercial bonded amide column.

In a recent study, the retention of 35 small polar compounds was compared among aminopropyl, amide, diol, and cyanopropyl columns with silica-based, hybrid, and fused-core particles . Most acidic and neutral compounds experienced greater retention under RP conditions than that under HILIC conditions. Only with the beta-blockers atenolol and propranolol were typical U-profile retentionemobile phase graphs observed. The aminopropyl stationary phase strongly acids because of electrostatic interactions, even under RP conditions. Bare silica exhibits a preference for basic compounds by interactions with the acidic surface of silanols.

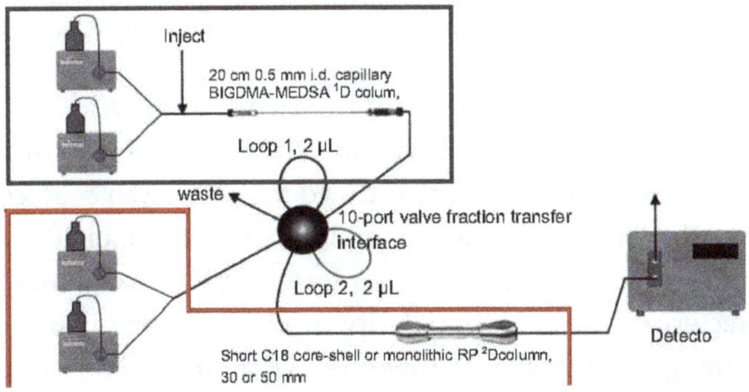

Fig. 9: Instrumental two-dimensional comprehensive HILIC _ RP HPLC setup with a zwitterionic polymethacrylate capillary column in the first dimension and a short efficient C18 column in the second dimension.

Because the separation selectivities on a dual-mechanism column are highly complementary (orthogonal), a single column can be used to obtain useful information on the sample injected subsequently in acetonitrile-rich (HILIC) and highly aqueous (RP) mobile phases. The dual mechanism is best used with alternating increasing (HILIC) and decreasing (RP) concentration gradients of water in buffered mobile phase at pH ¼ 3.1, both in unidimensional and 2D HPLC separations.

9.6.4 Sample Structure and Selectivity in Hydrophilic Interaction Liquid Chromatography

Correlations between the retention factor, k, and sample hydrophobicity are often used in RP chromatography to predict the effect of sample polarity on retention. The original approach was based on simple correlation of log k with the logarithm of the sample partition coefficient between octanol and water, log P. This correlation may give a useful first indication of the effect of polarity on the retention of neutral compounds; for ionizable compounds, the partition coefficient P should be substituted by the distribution ratio, D, taking into account the retention of both the nonionized and ionized solute forms, which of course depends on pH.

Combined effects of the solute structure and the mobile phase composition on the HILIC retention were employed in a model based on five predictors, including the %ACN in the mobile phase, the log D, the number of hydrogen bond donors, the desolvation energy in octanol, and the total absolute atomic charge structural descriptors. The model was applied to predict the retention of adrenoreceptor agonists and antagonists on a PVA-bonded stationary phase under HILIC conditions. Also, a two-parameter model employing the %ACN and the solute local dipole index was used to predict the HILIC retention of seven ginsenosides on a polyamine-bonded stationary phase.

The effects of the sample structural contributions in RP-LC are described in detail by the LSER model, which employs multiple correlations between the retention (log k) and so-called solvatochromic parameters (molecular structural descriptors), characterizing the solubility and salvation of the solute and the stationary phase.

$$\log k = c + v \cdot V + s \cdot S + a \cdot A + b \cdot B + d^{-} D^{-} + d^{+} D^{+} \qquad 8$$

The structural descriptors in Eq. (8) characterize the properties of the sample: the volume of the solvated solute molecule, V, is called the McGowan characteristic volume, which is proportional to the size of sample molecule (in cm3/mol/100) and the energy necessary to disrupt solventesolvent bonds to form a cavity accommodating a sample molecule; S characterizes the sample polarity and polarizability; B and A represent the hydrogen-bonding basicity and the hydrogen-bonding acidity, respectively. Ionic molecular descriptors D_ and Dþ were later added by to the LSER model, to take into account the degree of dissociation of weak acids and weak bases, depending on pH. For neutral species, the D parameters are zero; fully dissociated acids or bases have D values close to 1.

The parameters v, s, a, b, d_, and dþ of Eq. (8) characterize the response of the separation system (the specific combination of the stationary and the mobile phases) to the solute structural parameters. The system response coefficients are determined by multiple linear regression applying Eq. (2.8) to a representative group of sample compounds. The original LSER model was also later applied in HILIC chromatography. Relatively good fit of the LSER model was found for the silica hydride, bonded polyethylene glycol, diol, and zwitterionic sulfobetaine columns in the HILIC range up to 95% acetonitrile. The LSER model does not allow direct comparison of column selectivity, but it is a useful tool in column design, which helps to verify the expected effects on retention of various ligands immobilized on the silica surface.

9.6.5 Temperature Effects

Elevated temperature often improves the HILIC bandwidths, and consequently the efficiency of separation, by decreasing the mobile phase viscosity and accelerating the diffusion. Furthermore, temperature affects the Gibbs molar free energy of retention, DG0, and consequently the distribution constant, K, and the retention factor, k, in chromatography. The temperature effects in chromatography can often be described by the Van't Hoff equation :

$$k = \ln K + \ln\frac{V_S}{V_M} = -\frac{\Delta G^0}{RT} = \frac{\Delta S^0}{R} + \ln\frac{V_S}{V_M} - \frac{\Delta H^0}{RT} = A_i + \left(\frac{B_i}{T}\right) \qquad 9$$

DH0 is the standard partial molar free enthalpy of retention and DS0 is the corresponding standard partial molar free entropy; T is thermodynamic temperature (in Kelvins); VS and VM are the volumes of the stationary phase and mobile phases in the column, respectively. Hence, temperature usually affects the retention and selectivity of a separation through the decrease in enthalpy. Good fit of Eq. (9) to the experimental data is considered as proof of an enthalpy-controlled exothermic process within the experimental temperature range. In addition to changing enthalpy, the retention can be connected with entropic changes, namely conversion between different sample configurations. For example, an enthalpy-controlled retention process (positive Bi terms in Eq. (9) and improved separations at elevated temperature were found in HILIC separations of polar compounds on mercapto-ethanol and 1-thioglycerol columns (DH0 from 4.5 to 19.5 kJ/mol). Phenolic acids and flavonoids show linear Van't Hoff log k versus 1/T plots both on silica hydride and UDC cholesterol silica hydride columns, both in the highly organic (HILIC) and highly aqueous (RP) mobile phase ranges. Hence, enthalpic effects control the retention in both retention modes, but the energy of interactions with the stationary phase is 3e5 times stronger in the HILIC range, which suggests a stronger effect of adsorption with respect to partition distribution in the HILIC (NP) mode. Optimum HILIC mode separations on hydrosilated silica and UDC cholesterol silica hydride columns have been achieved at ambient temperature.

Most silica-based and organic polymer monolithic zwitterionic columns yielded Van't Hoff Bi parameters close to zero, indicating low or negligible enthalpic effects on retention. The retention of acids on dual-mechanism zwitterionic columns in acetonitrile-rich mobile phases (HILIC mode) may show significant entropic effects, whereas the same column is subject to a "regular" enthalpycontrolled exothermic retention mechanism in more aqueous mobile phases (in the RP mode).

9.7 HYDROPHILIC INTERACTION LIQUID CHROMATOGRAPHY MODE IN TWO-DIMENSIONAL LIQUID CHROMATOGRAPHY SEPARATION SYSTEMS

The number of peaks that can be separated in a single HPLC run can hardly exceed 100. However, in clinical, pharmaceutical, food, natural and environmental analysis, molecular biology, and elsewhere, we encounter samples that may contain even millions of compounds, in concentrations possibly spanning over 10 orders of magnitude. The peak capacity, i.e., the maximum number of peaks that can be separated in a sample, can be significantly increased by combining two or more different separation mechanisms in multidimensional separation systems. To quantify the differences in selectivity of the individual HILIC columns, the correlation coefficient, r2, between the retention factors, k, of a suitably selected group of test compounds, or of their logarithms, on various polar columns and a reference column (such as bare silica) may be measured to calculate the coefficients s2, characterizing the relative separation selectivities of the individual polar columns:

$$s^2 = 1 - r^2 \qquad (10)$$

s^2 = 1 for fully noncorrelated (orthogonal) systems (r2 ¼ 0), whereas s2 ¼ 0 for equivalent separation systems. The 2D peak capacity can theoretically reach the product of the peak capacities of the individual systems in orthogonal 2D separation systems, where the retention in the first dimension is not correlated with the retention in the second dimension. In practice, most 2D systems show a lower number of resolved peaks because ofmore or less significant retention correlation in the two dimensions. Coupled HILIC and RP separation systems offer two completely different retention mechanisms and a very high degree of orthogonality in comparison to other 2D LC systems. Hence, 2D separation systems combining HILIC and RP modes allow the number of compounds resolved in complex samples to be significantly increased.

The expected increase in the number of really separated compounds (the practical peak capacity) that can be achieved in various off-line and online multidimensional setups and the price to be paid for it in terms of the analysis time and sample dilution (i.e., decreased sensitivity) have been reviewed in detail. Later, the performance of off-line, online stop-flow, and comprehensive HILIC chromatography combined with RPLC was compared in terms of peak capacities, analysis times, and peak production rates. An online comprehensive LC _ LC system is best for samples requiring peak capacities up to 600. The off-line or stop-flow systems provide higher peak capacities, at the cost of long separation times. The contribution of the stop-flow to band broadening was found negligible.

Generally, the peak capacity is higher in gradient mode than that under isocratic conditions. Hence, the number of resolved peaks increases when simultaneous gradients are used in the first dimension and in the second dimension of a comprehensive online 2D setup. Optimization of the gradient range and gradient profile, especially in the second dimension, can significantly increase the practical 2D peak capacity.

The off-line approach using two (or more) separate columns is very simple, does not necessitate any special instrumentation, and has been practiced for many years. It allows independent optimization of the two separation systems; the fractions collected from the first column can be pretreated before injection onto the second column, such as using evaporation to dryness of manually collected HILIC fractions before introduction to a C18 column in the second dimension. Off-line coupling of HILIC and RP-LC was shown to provide a powerful separation system for procyanidins and flavonoids. Another recent example reports an off-line RP-HILIC method coupled with MS for impurity profiling of infusion solutions. Unfortunately, off-line procedures are labor demanding and time demanding.

Online combinations of HILIC and RP-HPLC (Fig.9) are subject to compatibility problems originating from the differences in the mobile phase elution strengths in the HILIC and RP modes. High concentrations of the organic solvents used for HILIC separations usually provide weak retention in the RP systems, whereas the mobile phases rich in water used in RP-HPLC are usually too strong as HILIC eluents. If the mobile phase from the first (HILIC) dimension is used for the fraction transfer to the second dimension, a significant decrease in retention and unsymmetric or even split peaks may appear, with detrimental effects on the separation.

The compatibility of the mobile phases used in 2D HILIC-RP systems can be improved in several ways:

1. Transferring only small volume fractions online from the first dimension to the second dimension column (for example, 2 mL onto a 50 _ 3 mm I.D. C18 column) often minimizes the sample solvent effects in the second dimension, however, sometimes with possible negative impacts on the sensitivity of 2D separations.
2. The mobile phase strength in the acetonitrile-rich fractions transferred from the HILIC column can be modulated by diluting with water before introducing the fractions onto an RP column in the second dimension using a make-up flow mediated by an additional auxiliary pump.
3. Fractions from the first column can be trapped on a small column and another solvent can be used for the transfer to the second dimension.

Approaches 1 and 2 can be combined. For example, 2D HILIC _ RP separation of phenolic compounds in green tea was reported using a diol HILIC column in the first dimension and a gradient of water and methanol in acetonitrile with acetic acid additive.

50 mL fractions were collected with 1 min frequency and evaporated under nitrogen to 2 mL before introduction onto a C18 column in the second dimension.

Online 2D separations can employ either serial or parallel column setups. Serial column coupling does not require complex instrumentation, and allows only a moderate increase in peak capacity, because of the additivity of the contributions of the individual columns, in contrast to comprehensive online 2D LC _ LC providing multiplicative effects on peak capacity and considerably higher peak production rate (i.e., the number of resolved compounds in a preset separation time). A set of serially coupled columns containing different stationary phases behaves like a new column with modified selectivity, which may enable separations of samples with widely differing properties. An octadecylsilica column in the first dimension serially coupled with a HILIC column in the second dimension, to which a gradient of acetonitrile in water up to the end concentration of 80% or more was applied, allowed separation of a broad range of pharmaceuticals in a single run. A serial combination of a ZIC-HILIC sulfobetaine column and an amide column with a HILIC gradient run from 95% to 35% aqueous buffer in acetonitrile allowed simultaneous separations of polar and nonpolar metabolites in a mouse serum sample. Tandem coupling of a C18 column with a zwitterionic column was employed for separation of polar and nonpolar phenolic compounds in wine, with a single gradient of simultaneously increasing concentration of acetonitrile and decreasing salt concentration.

A 2D capillary liquid chromatographye Fourier transform mass spectrometry method, employing a 100 mm _ 150 mm I.D. C18 RP column serially coupled with a 250 mm _ 200 mm I.D. (polyhydroxyethyl aspartamide) HILIC column, was used for the analysis of 11 quaternary ammonium compounds in brain extracts, including acylcarnitines of low polarity. To overcome the mobile phase compatibility problem on the two columns, they were connected via a T-piece allowing addition of a make-up flow of 95% acetonitrile to the effluent from the RP column before transfer to the HILIC column.

An interesting solution to the problem of different elution strengths required to elute polar and nonpolar compounds employed a column-switching setup, where the sample was first injected onto two serially coupled 100 mm _ 2.1 mm BEH columns. A plug of early eluting weakly polar compounds was directed onto a trapping RP column, where the analytes were stored until the separation of polar compounds on the HILIC columns had been finished. Then the configuration of switching valves was changed to redirect the weakly polar compounds onto a Phenyl Hexyl RP column, where they were separated using a gradient of increasing acetonitrile concentration in 0.02% aqueous formic acid.

A crucial point affecting the separation time in comprehensive 2D liquid chromatography is the performance of the column used in the second dimension, which should allow highly efficient fast chromatographic separations. For this purpose, UHPLC with a short column packed with sub-2 mm particles can be used at a very high operating pressure. A coreeshell column represents another possibility that can be used with conventional liquid chromatographic instrumentation. Online connection of a capillary HILIC column in the first dimension and an ultra-high performance RP-LC in the second dimension, coupled with MS, was used for highresolution separation and detailed characterization of anthocyanins and related pigments in berries and aged red wine. Because of lower flow resistance, monolithic columns can be used in the second dimension at higher flow rates in comparison to particle-packed columns operated at the same operating pressure. Very fast second-dimension gradient separations on coree shell columns were achieved at ambient temperature without excessive backpressure and without compromising optimal first dimension sampling rates. Very good band symmetry and retention time repeatability in gradient separations of phenolic compounds and flavonoids could be achieved in optimized comprehensive HILIC _ RP on a 0.5 mm I.D. monolithic sulfobetaine HILIC capillary column coupled with various 2.5e5 cm long, 3 mm I.D. monolithic and coreeshell C18 columns. A flow rate of a few microliters per min was used on the capillary column in the first dimension; the short C18 columns were operated at flow rates of 3e5 mL/min in the second dimension.

Polar columns showing a dual RP-HILIC mechanism allow a combined 2D RP _ RP and HILIC _ RP setup to be used. In the first dimension, the RP mode in a highly aqueous mobile phase alternates with the HILIC mode in a mobile phase with a high acetonitrile concentration. In the second dimension, the RP mode can be used. A recently introduced zwitterionic polymethacrylate sulfobetaine BIGDMA-MEDSA capillary column shows a dual HILIC/RP mechanism at high concentrations of acetonitrile and RP behavior in water-rich mobile phases.

The BIGDMA-MEDSA column in the first dimension coupled online with a short monolithic or coreeshell C18 column was used for combined alternating HILIC _ RP and RP _ RP comprehensive 2D separations of polyphenolic compounds. During the HILIC _ RP period, a gradient of decreasing acetonitrile gradient was used for the separation in the first dimension. At the end of the gradient, the polymeric monolithic microcolumn was equilibrated with a highly aqueous mobile phase and was ready for the second sample injection in the RP _ RP period. This time a gradient of increasing concentration of acetonitrile was used in the first dimension. Fig. 2.10 presents 2D chromatograms of flavones and related polyphenolic compounds, acquired with a single first-dimension BIGDMAMEDSA capillary column in two experiments with consecutive injections of the sample, the first one into a decreasing and the second into an increasing acetonitrile gradient. The automated dual LC _ LC approach allows obtaining three-dimensional data in a relatively short time.

9.8 PERSPECTIVES FOR FURTHER DEVELOPMENT

A number of new types of polar stationary phases have been reported during the past years, based either on silica gel, other inorganic supports, or on organic polymers. Many column types have been successfully used for separations of various classes of strongly and moderately polar, or even ionic, inject compounds. Large differences in retention and separation selectivity, depending on the chemistry of stationary phases, indicate that the retention mechanism is more complex than the originally suggested partition between the water layer adsorbed on a solid support and the bulk mobile phase. Probably, a sample-dependent mix of nonpolar and selective polar [hydrogen bonding, dipoleedipole, ionexchange, and ion-repulsion (ERLIC)] interactions participates in separation.

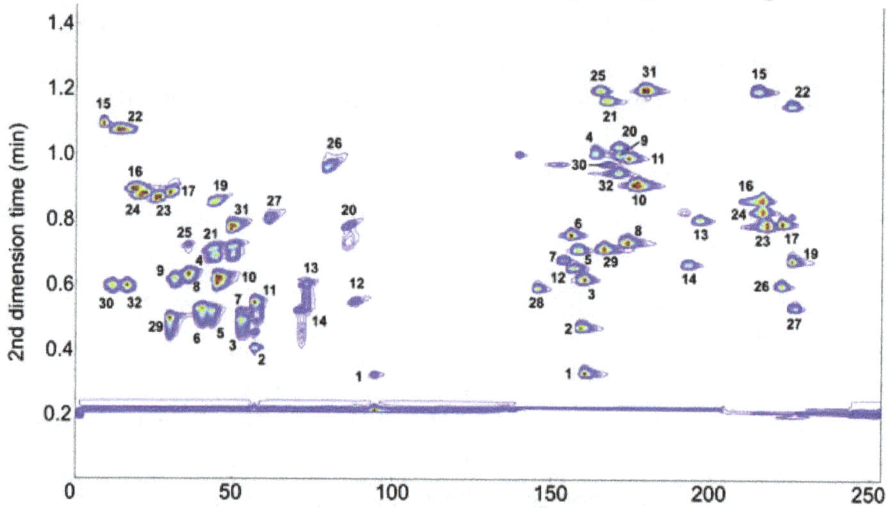

Fig. 10: Dual comprehensive two-dimensional chromatogram of polyphenolic and flavonoid compounds recorded in subsequent HILIC mode (left, gradient of decreasing acetonitrile concentration) and reversed-phase mode (right, gradient of increasing acetonitrile concentration) on a single monolithic capillary zwitterionic polymethacrylate BIGMA-MEDSA column in the first dimension coupled online with a Kinetex XB-C18, 50 _ 3.1 mm I.D., coreeshell column in the second dimension.

The column types used in HILIC strongly differ in the adsorption capacity for water. The columns based on hydrosilated silica showless than amonomolecular water layer capacity, which suggests that the adsorption process (i.e.,NP-LC) characterizes the retention more appropriately than the partition process. We can expect future development of tailor-made HILIC stationary phases intended for specific types of samples. Another almost unexplored promising field is better utilization of mobile phases for adjusting the retention mechanism (HILIC-RP) on a single column to extend separation selectivity by alternating use of different separation conditions. These possibilities can probably be used in future 2D combinations implementing the HILIC mode in off-line, serial, stop-and-go, and comprehensive setups, and their combinations, to improve peak capacities for complex samples containing hundreds of compounds.

10

ADVANCED CHROMATOGRAPHIC TECHNIQUES IN THE ANALYSIS OF LUTEOLIN AND ITS DERIVATIVES

Luteolin is a flavonoid often found in various medicinal plants that exhibits multiple biological effects such as antioxidant, anti-inflammatory and immunomodulatory activity. Commercially available medicinal plants and their preparations containing luteolin are often used in the treatment of hypertension, inflammatory diseases, and even cancer. However, to establish the quality of such preparations, appropriate analytical methods should be used. Therefore, the present paper provides the first comprehensive review of the current analytical methods that were developed and validated for the quantitative determination of luteolin and its C- and O-derivatives including orientin, isoorientin, luteolin 7-O-glucoside and others. It provides a systematic overview of chromatographic analytical techniques including thin layer chromatography (TLC), high performance thin layer chromatography (HPTLC), liquid chromatography (LC), high performance liquid chromatography (HPLC), gas chromatography (GC) and counter-current chromatography (CCC), as well as the conditions used in the determination of luteolin and its derivatives in plant material.

10.1 INTRODUCTION

Luteolin is a yellow dye commonly found in fresh plants. It is a flavonoid of the flavone type that is distributed widely throughout the plant kingdom. Similar to other derivatives of 2-phenylbenzo-γ-pyrone, its basic skeleton has a characteristic C6-C_3-C_6 system, containing two benzene rings and a bridge with a C_2-C_3 double carbon bond and an attached oxygen atom. Structure-activity studies have demonstrated that the pharmacological effects of luteolin and other flavonoids are strongly related to the presence of hydroxyl groups at the C_5, C_7, C_3' and C_4' carbons as well as to the presence of the double bond in the C_2-C_3 position. The presence of the -OH group at position C_3' distinguishes luteolin from apigenin, and the lack of this group at the C_3 carbon is an element that places luteolin in the flavone group.

Luteolin exhibits multiple biological effects such as antioxidant, anti-inflammatory and immunomodulatory activity. Plants rich in luteolin are often used in traditional medicine for treatment of various diseases such as hypertension, inflammatory disorders, and even cancer.

Because luteolin bears four hydroxyl groups (at the C5, C7, C3' and C4' positions), many derivatives of luteolin can be created. Various types of functional groups and/or sugar molecules can be attached to those positions, creating a huge number of different but structurally similar molecules. The most common are methyl derivatives, as well as *C*- and -*O*-glycosides.

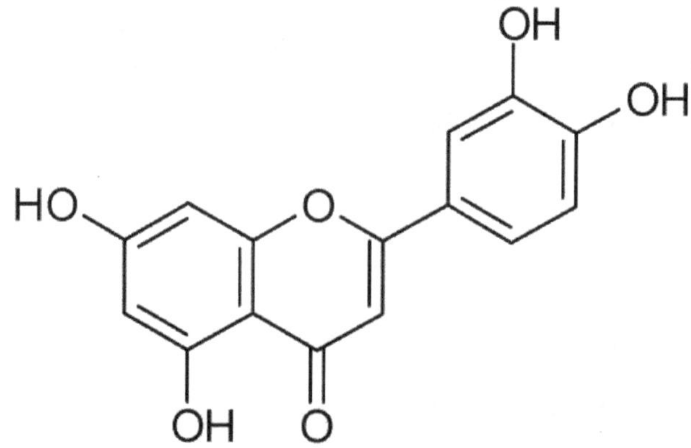

Fig. 1: Chemical structure of luteolin. Orientin, an 8-C-glucoside derivative of luteolin, displays an array of health-related biological properties, such as antioxidant, anti-ageing, antiviral, antibacterial, anti-inflammatory, vasodilatation, cardioprotective, radiation protective, neuroprotective, antidepressant-like, anti-adipogenesis, and antinociceptive effects. It may be found in different medicinal plants such as Ocimum sanctum (holy basil), Phyllostachys nigra (bamboo leaves), Passiflora sp. (passion flower), Linum usitatissimum (flax), Euterpe oleracea (Acai palm) and many others. Another luteolin derivative, isoorientin (luteolin-6-C-glucoside) acts as an antioxidant, photoprotective, skin lightening, hepatoprotective and anti-inflammatory agent. O-glucosides of luteolin also display biological activities. For example, luteolin 7-O-glucoside alleviates skin lesions in murine models of atopic dermatitis and protects cells against apoptosis induced by hypoxia/reoxygenation.

An increasing number of herbal preparations on the market contain luteolin and its derivatives, either as single-ingredient products or in mixtures with other phytochemicals, e.g., in form of medicinal plants extracts. To establish the quality of such products, it is important to use appropriate analytical methods. However, there is a lack of quality reviews of the available methods for quantification of luteolin derivatives, and the information on their comparison is lacking. Therefore, the aim of this article is to systematize knowledge and information in the field of chromatographic analytical techniques used for quantification of luteolin and its derivatives. The presented review is the first description of this type and provides a systematic overview of chromatographic analytical techniques including thin layer chromatography (TLC), high performance thin layer chromatography (HPTLC), liquid chromatography (LC), high performance liquid chromatography (HPLC), gas chromatography (GC) and counter-current chromatography (CCC), as well as the conditions used to assess luteolin and its derivatives.

10.2. CHROMATOGRAPHIC TECHNIQUES FOR THE ANALYSIS OF LUTEOLIN DERIVATIVES

Chromatography occupies a leading position among other instrumental methods in the analysis of chemical compounds. As a physicochemical method of separation and analysis of mixtures of chemical compounds, it allows detection and identification as well as quantitative determination of the test substance with high accuracy. The coupling of chromatography with other methods of analysis contributes to a more accurate detection and expansion of analytical capabilities, especially for complex mixtures of organic compounds.

Chromatographic techniques are based on the interaction of the mixture components with the mobile and stationary phases of the chromatographic system. This results in the division of the mixture components between the two phases. In addition, the interaction of the mobile and stationary phases is also important in the separation process. According to the aggregation state of the mobile phase, chromatographic techniques are divided into gas, liquid and supercritical chromatography. Another criterion for classification of chromatographic techniques is the type of stationary phase. If the stationary phase is a liquid, the chromatography technique is referred to as partition chromatography. In the case of a solid, the technique is referred to as adsorption chromatography. Another example of the classification of chromatographic methods is their division depending on the chromatographic processing method. This classification allows distinguishing column chromatography and planar techniques, which include TLC and paper chromatography. Many different chromatographic techniques are used in the analysis of luteolin derivatives. These include TLC, HPTLC, LC, HPLC, GC and CCC.

10.2.1. Thin Layer Chromatography in the Analysis of Luteolin Derivatives

Thin layer chromatography (TLC) is a rather simple but relatively popular method used in the analysis of flavonoids since 1960. It is a variation of LC that is carried out on a plane and is therefore referred to as planar chromatography. Despite the dynamic development of other chromatography techniques, TLC has not lost its importance in phytochemical analysis. Its values are still recognized as is the basic tool in the qualitative analysis of natural products, and as such, it is still recommended by most modern Pharmacopoeias. In addition, the TLC technique is currently being improved, and the scope of its application is widening, while the results are becoming comparable to those obtained by GC or HPLC. The stages of chromatographic analysis consist of placing the sample on the stationary phase and developing the chromatogram, followed by its visualization. In the final step, qualitative and/or quantitative determinations of the tested components are made. General guidelines for flavonoid separation on TLC plates are presented in Table 1.

Table 1 Recommended combinations of solvents / adsorbents for identification of different flavonoid types by thin layer chromatography (TLC).

Flavonoid Type	Adsorbent Type/Mobile Phase		
	Cellulose	Polyamide	Silica gel
Polar flavonoid aglycones, e.g., flavones	BuOH:AcOH:H_2O (3:1:1 *v/v/v*) [a] $CHCl_3$:AcOH:H_2O (30:15:2 *v/v/v*) [b]	MeOH:AcOH:H_2O (18:1:1 *v/v/v*)	To:Py:FA (36:9:5 *v/v/v*)
Non-polar flavonoid aglycones, e.g., methylated flavones	10–30% AcOH	---	$CHCl_3$:MEOH (15:1 to 3:1 *v/v*)
Flavonoid glycosides	BuOH:AcOH:H_2O (3:1:1 *v/v/v*) [a] BuOH:AcOH:H_2O (4:1:5 *v/v/v*) [a]	H_2O:MeOH:MEK:methyl acetylacetone (13:3:3:1 *v/v/v/v*)	EtOAc:Py:H_2O:MeOH (80:20:10:5 *v/v/v/v*) (especially flavone C-glycosides)

Abbreviations: AcOH, acetic acid; BuOH, butanol; $CHCl_3$, chloroform; EtOAc, ethyl acetate; FA, formic acid; MeOH, methanol; MEK, methyl ethyl ketone; Py, pyridine; To, toluene. a The mobile phase is thoroughly mixed in the separating funnel and the upper phase is used. b The mobile phase is thoroughly mixed in the separating funnel and the excess water is discarded.

The main advantage of TLC is that it is a relatively simple and inexpensive technique that allows for rapid qualitative and quantitative analysis of the tested compounds. Samples analyzed with this method usually do not require pre-treatment, such as purification or concentration. In addition, several dozens of samples can be analyzed simultaneously on one plate. A large amount of the diluted sample can be applied to the stationary phase because the solvent evaporates during the application to the plate. Furthermore, due to the evaporation of the solvent phase after the development of the chromatogram, the detection method does not depend on the type of mobile phase used for separation. In TLC, it is possible not only to compare the analyzed components with the standards, but also to differentiate between substances bearing specific functional groups using the appropriate reagents for detection.

Thin layer chromatography and column chromatography (CC) are interchangeable techniques that may be combined, which significantly reduces costs and analysis time. To achieve this, the same adsorbents are ideally used for both TLC and CC. Nevertheless, other solvent systems may also be used. Elution in CC may be carried out in one mobile phase, or its composition may be changed during chromatography (mobile phase gradient), thereby increasing the elution force. In this case, the TLC mobile phase should be changed accordingly.

The type and quality of the stationary phase greatly affect the separation of mixture components. Thus, the selection of an appropriate adsorbent is very important. However, most TLC analyses are carried out in a normal phase system where hydrophilic (polar) adsorbents are used. Reversed phase systems with lipophilic (non-polar) stationary phases are common and have little significance in the analysis of flavonoids. Currently, the most commonly used stationary phase for the analysis of flavonoids is silica gel. However, the use polyamide coated chromatography plates, in both normal and reversed phase systems, is not uncommon.

Detection of flavonoids on TLC plates is most often conducted under ultraviolet (UV) light at wavelengths of 254 or 366 nm. Luteolin derivatives also display fluorescence, which can be enhanced using the appropriate derivatization reagents, e.g., with the so-called NP/PEG reagent. The most frequent procedure consists of spraying the plate with 1% methanolic diphenylboric acid-β-ethylamino ester (natural product reagents, NP), followed by 5% ethanolic polyethylene glycol 4000 (PEG) solution. A densitometer may also be used for the qualitative analysis of the substance. The analysis is performed by comparing the retardation factor (Rf) and absorption spectrum of the test substance and the standard. Analytes can also be identified by extracting the separated substances from the plate. Then, the analysis is carried out using Fourier transform infrared mass spectrometry (MS), UV spectrometry, Raman spectrometry or other techniques]. Even though TLC separation of luteolin derivatives (Table 2) can be performed in different types of stationary phases such as a polyamide phase, it is most frequently performed on silica gel plates that are often coated with a fluorescent indicator (F254 plates) for preliminary detection. Such an approach has been used in case of analysis of luteolin 3'-O-glucoside, luteolin 6,8-C-dihexoside and luteolin 7-O-rutinoside in Phlomis sp. However, subsequent analysis with NP/PEG is the standard procedure for TLC analysis of luteolin and it is almost always performed regardless of additional types of detection such as detection of flavonoids in Ligustrum vulgare with aniline phthalate. Typically, the mobile phase for luteolin derivative separation consists of a mixture of aprotic organic solvents such as ethyl acetate (EtOAc) or acetone (Ace) and H2O with a significant amount of formic (FA) and/or acetic acid (AcOH) to avoid tailing of the separated zones.

Table 2 Thin layer chromatography in the analysis of luteolin derivatives.

Luteolin Derivative	Stationary Phase	Mobile Phase	Detection	Analyzed Species
Luteolin	Silica gel 60 f_{254}	Hx:EtOAc:AcOH (31:14:5 *v/v/v*); To:DI:AcOH (90:25:4 *v/v/v*)	FBS; UV, 254, 366 nm	Artemisia annua
	Silica gel 60 rp-18 f_{254}s	FA:H_2O:MeOH (5.5:47.25:47.25 *v/v/v*)		
Luteolin 7-rutoside	Silica gel	EtOAc:FA:AcOH:H2O (100:11:11:23 v/v/v/v)	NP/PEG, aniline phthalate; UV, 254,366 nm MS, NMR	Ligustrum vulgare
	Cellulose	30% AcOH		
Luteolin 7-rhamnoside	Polyamide	CHCl3:MeOH:MEK:AcAc (9:4:2:1 v/v/v/v); To:MeOH:MEK:BuOH (300:150:150:3 v/v/v/v)		

Luteolin Derivative	Stationary Phase	Mobile Phase	Detection	Analyzed Species
Luteolin 3'-glucoside	Silica gel 60 f254	MeOH:H2O (15:5 v/v); CHCl3:MeOH (15:5 v/v); 15% AcOH	NP/PEG; UV, 366 nm	Phlomis persica
Luteolin 6,8-dihexoside				
Luteolin 7-rutinoside				Ph. elliptica
Luteolin 7-glucoside	Silica gel	EtOAc:FA:H2O (18:1:1 v/v/v)	NP/PEG; UV, 366 nm	Carduus acanthoides
	Polyamide plates	EtOAc:FA:AcOH:H2O (100:10:10:13 v/v/v/v)		
Luteolin	Silica gel 60 f254	EtOAc:Ace:FA (8:1:1 v/v/v); EtOAc:H2O:AcOH:FA (10:3:1:1 v/v/v/v)	NP; UV, IR, NMR	Leontopodium alpium
Luteolin 7,4'-diglucoside				
Luteolin 6-hydroxy-7-glucoside				
Luteolin 4'-glucoside				
Luteolin 3'-glucoside				
Luteolin 7-glucoside				
Luteolin	Silica gel 60 f254	To:Et2O:AcOH (60:40:10 v/v/v); EtOAc:AcOH:FA:H2O (100:11:11:26 v/v/v/v)	NP/PEG; UV, 366 nm	Matricaria recutita, Achillea millefolium, Thymus vulgaris, Salvia officinalis
Luteolin 7-glucoside				

Abbreviations: AcAc, acetylacetone; Ace, acetone; AcOH, acetic acid; BuOH, butanol; $CHCl_3$, chloroform; DI, 1,4-dioxane; Et_2O, diethyl ether; EtOAc, ethyl acetate; FA, formic acid; FBS, Fast Blue B Salt; Hx, hexane; MeOH, methanol; MEK, methyl ethyl ketone; NP, 1% ethanolic diphenylboric acid-β-ethylamino ester - natural product reagents; PEG, 5% ethanolic polyethylene glycol 4000; To, toluene; UV, ultraviolet spectroscopy; MS, mass spectrometry; NMR, nuclear magnetic resonance; IR, infrared.

Thin layer chromatography is often used as a complementary method to other chromatographic techniques. For example, analysis of the butanol (BuOH) fraction of the methanol (MeOH) extract from the leaves of the common privet (Ligustrum vulgare) and allowed isolation of two luteolin derivatives from the plant. TLC was carried out, among other techniques, on polyamide plates, and the optimal mobile phase was found. TLC was used for analysis of individual fractions obtained in column chromatography with or without acid hydrolysis of compounds. MS detection and nuclear magnetic resonance (NMR) spectra were also used. Furthermore, TLC was used together with high performance liquid chromatography combined with mass spectrometry and pulsed amperometric detection (HPLC-PAD-MS) for analysis of luteolin and other phenolic compounds in Leontopodium alpinum. In addition to NP/PEG, UV, infrared (IR) and NMR analyses were used for identification of these compounds.

Compared to other chromatographic techniques, HPTLC results in reduced time and costs of analysis and provides much greater efficiency of separation. It is suitable even for the analysis of crude extracts similar to TLC, and a relatively small amount of solvent is used to analyze several samples, making this method environmentally friendly. In the analysis of luteolin derivatives (Table 3), HPTLC silica gel 60 is almost exclusively used as the stationary phase, while HPTLC NH_2 plates are rarely used, e.g., for separation of flavonoids in some Lamiaceae species such as Mentha piperita and Thymus sp.. In addition to NP/PEG, other detection systems may be employed for visualization of luteolin derivatives, such as bis-diazotized sulfanilamide or aqueous solutions of Al^{3+} ions for flavonoids in M. piperita , honey or Thymus sp.. Similar to TLC, mixtures of organic solvents, H_2O, and FA are most often used as the mobile phase.

Table 3 High performance thin layer chromatography in the analysis of luteolin derivatives.

Luteolin Derivative	Stationary Phase	Mobile Phase	Detection	Analyzed Species
Luteolin	HPTLC silica gel 60 F_{254}	Hx:EtOAc:AcOH (31:14:5 *v/v/v*); To:DI:AcOH (90:25:4 *v/v/v*)	FBS; UV, 254, 366 nm	Artemisia annua
	HPTLC diol $F_{254}S$	$CHCl_3$:Hx:EtOAc (34:4 *v/v/v*)	2% $AlCl_3$; UV, 366 nm	Oxytropis glabra
	HPTLC silica gel 60 RP-18W	BuOH:MeOH:H_2O (300:50:50 *v/v/v*); 15% AcOH	NPR/PEG; UV, 366 nm	
	HPTLC silica gel 60 F_{254}	EtOAc:MeOH:FA:H_2O (20:3:1:2 *v/v/v/v*)	MeOH:H_2SO_4 (95:5 *v/v*); UV, 254, 366 nm	Asparagus racemosus, Withania somnifera, Vitex negundo, Plumbago zylenica, Butea monosperma, Tephrosia purpurea
	HPTLC silica gel 60 F_{254}	To:EtOAc:FA (10:9:1 *v/v/v*)	UV, 254 nm	Cardiospermum halicacabum
	HPTLC silica gel 60 F_{254}	DCM:MeOH (70:30 *v/v*)	UV, 254 nm; NMR	Satureja montana
	HPTLC silica gel $G_{60}F_{254}$	To:EtOAc:FA (6:4:1 *v/v/v*)	UV, 349 nm	Hygrophila spinosa
	HPTLC silica gel 60 F_{254}	EtOAc:MeOH:H_2O:AcOH (3:1:1:1 *v/v/v/v*)	UV, 254, 366 nm	Foeniculum vulgare, Cuminum cyminum, Apium graveolens, Petroselinum crispum, Anethum graveolens, Ammi majus
	HPTLC silica gel 60 F_{254}	To:EtOAc:FA (3:3:0.8 *v/v/v*)	MeOH:H_2SO_4 (90:10 *v/v*); UV, 254 nm	Saraca asoca
	HPTLC silica gel 60 F_{254}	To:EtOAc:FA (6:4:0.3 *v/v/v*)	NP/PEG; UV, 366 nm	Premna mucronata
	HPTLC silica gel G_{60} F_{254}	To:EtOAc:FA (10:9:1 *v/v/v*)	UV, 254, 366 nm	Anisochilus carnosus
	HPTLC silica gel 60	*n*Hx:EtOAc:FA (30:20:1.5 *v/v/v*)	NP/PEG; UV, 349 nm	Satureja hortensis
	HPTLC silica gel 60 F_{254}	*n*Hx:EtOAc:AcOH (5:3:1 *v/v/v*)	UV	Propolis
	HPTLC silica gel 60 GF_{254}	To:EtOAc:FA:MeOH (3:3:0.8:0.2 *v/v/v/v*)	UV, 254, 365nm	Eclipta alba
	HPTLC silica gel 60 F_{254}	To:EtOAc:FA (5:3:1 *v/v/v*)	NP; UV, 366, 254 nm	Vitis vinifera
Luteolin 6-glucoside Luteolin 8-glucoside	HPTLC silica gel 60 F_{254}	EtOAc:FA:AcOH:H_2O (100:11:11:27 *v/v/v/v*)	NP/PEG; UV, 366 nm	Clinacanthus nutans
	HPTLC silica gel 60 F_{254}	EtOAc:FA:H_2O (82:9:9 *v/v/v*)	NP/PEG; UV, 366 nm	Passiflora alata, P. edulis
	Nano-DUASIL silica gel 60	THF:To:FA:H_2O (16:8:2:1 *v/v/v/v*)	UV, 350 nm	Phyllostachys pubescens

Luteolin Derivative	Stationary Phase	Mobile Phase	Detection	Analyzed Species
Luteolin glucoside	HPTLC silica gel 60 F_{254}	EtOAc:FA:AcOH:H_2O (100:11:11:26 *v/v/v/v*)	NP/PEG; UV, 366 nm	Equisetum arvense
	HPTLC silica gel 60 F_{254}	EtOAc:AcOH:FA:H_2O (10: 1.1:1.1:2.6 *v/v/v/v*)	UV, 254, 366 nm	Aerva javanica
Luteolin	NP-HPTLC silica gel	EtOAc:FA:AcOH:H_2O (100:11:11:27 *v/v/v/v*)	H_2O solution of 4% $Al_2(SO_4)_3$; UV, 365 nm	Apis mellifera, honey
		To:EtOAc:AcOH (50:45:5 *v/v/v*)		
	HPTLC silica gel 60 F_{254}	To:EtFo:FA (6:4:1 *v/v/v*)	NP; UV, 254, 366 nm	Potentilla grandiflora, P. recta, P. anserina, P. fruticose, P. rupestris, P. thuringiaca
Luteolin 7-glucoside	Modified HPTLC silica gel 60 F_{254} with CN, NH_2	To:EtFo:FA (7:5:1 *v/v/v*)		
	HPTLC diol F_{254}	EtOAc:MEK:DIPE:FA (3:10:4:1 *v/v/v/v*)		
Luteolin 7-glucoside	HPTLC silica gel 60	EtOAC:DCM:AcOH:FA:H_2O (100:25:10:10:11 *v/v*)	NP; UV, 366 nm	Lavandula stoechas
	HPTLC silica gel 60 F_{254}	EtOAc:AcOH:FA:H_2O (100:11:11:26 *v/v/v/v*)	NP/PEG; UV, 254, 366 nm	Stachys sylvatica, S. recta
Luteolin 7-rutinoside	HPTLC NH_2	Ace:AcOH (85:15 *v/v*)	MeOH:AlCl3 (98:2 v/v), bis-diazodized sulfanilamide; UV-VIS, 365 nm IR, MS, NMR	Mentha piperita
	HPTLC RP-18W	H_2O:MeOH (60:40 *v/v*)		
Luteolin 7-glucuronide	HPTLC silica gel 60	Et:Ace:H2O:FA (55:25:10:10 *v/v/v/v*)	NP/PEG; UV, 365 nm	Mentha piperita, Melissa officinalis, Salvia officinalis
Luteolin 7-glucoside	HPTLC NH_2	Ace:AcOH (85:15 *v/v*)		
Luteolin 7-rutinoside	HPTLC NH_2	Ace:AcOH (85:15 *v/v*)	MeOH:AlCl3 (98:2 v/v), NP/PEG; UV, 365 nm	Thymus vulgaris, Th. serpyllum, Majorana hortensis, Mentha piperita
Luteolin 7-glucoside				
Luteolin 7-glucuronide		Ace:FA (85:15 *v/v*)		
	HPTLC silica gel 60	DIPE:Ace:H_2O:FA (50:30:10:10 *v/v/v/v*)		
Luteolin	HPTLC silica gel 60 F_{254}	pre-develop: $CHCl_3$:MeOH (1:1 *v/v*); *n*Hx:EtOAc:FA (20:19:1 *v/v/v*)	NP: white light, 254, 366, 330 nm; PEG: 254, 297, 340, 366, 430 nm; paraffin–*n*Hx: 254, 320, 360, 366, 400 nm; ESI-MS/MS	Rosmarinus officinalis
Luteolin acetyl hexuronide				Salvia officinalis
Luteolin 7-rutinoside				
Luteolin acetyl hexoside				
Luteolin hexuronide				
Luteolin 3',7-diglucoside	HPTLC silica gel 60 F_{254}	EtOAc:MeOH: AcOH:FA:H_2O (30:1:2:1:3 *v/v/v/v/v*)	NP; UV, 366 nm	Colocasia esculenta
Luteolin 4'-glucoside				
Luteolin 7-glucoside				
Luteolin 8-glucoside				
Luteolin 6-glucoside				
Luteolin 8-glucoside	HPTLC RP C_{18}	FA:MeOH:H_2O (0.5:6.65:2.85 *v/v/v*)	ESI-MS/MS, 254, 366 nm	Cyclanthera pedata
	HPTLC silica gel F_{254}	To:EtOAc:FA (6:5:1 *v/v/v*)		

Abbreviations: Ace, acetone; AcOH, acetic acid; $AlCl_3$, aluminium trichloride; $Al2(SO_4)_3$, aluminium sulfate; BuOH, butanol; $CHCl_3$, chloroform; DCM, dichloromethane; DI,

1,4-dioxane; DIPE, diisopropyl ether; ESI-MS/MS, electrospray ionization combined with tandem mass spectrometry; EtFo, ethyl formate; EtOAc, ethyl acetate; Et, ether; FA, formic acid; H_2SO_4, sulfuric acid; Hx, hexane; MeOH, methanol; MEK, methyl ethyl ketone; ***n***Hx, n-hexane; NP, 1% methanolic diphenylboric acid-β-ethylamino ester - natural product reagents; neurotransmitters; PEG, 5% ethanolic polyethylene glycol 4000; THF, tetrahydrofuran; To, toluene; UV-VIS, ultraviolet-visible spectroscopy; ESI-MS, electrospray ionization combined with mass spectrometry.

High performance thin layer chromatography may also be used as a complimentary method to other chromatographic techniques. Scientist used the HPTLC technique in the analysis of the ethanol (EtOH) extract of Clinacanthus nutans leaves. The analysis revealed the presence of, among others, luteolin 8-C-glucoside (orientin) and luteolin 6-C-glucoside (isoorientin) in the raw material. Their detection was conducted by comparing Rf coefficients using derivatization reagents and 366 nm UV light. In this work, the characteristic fluorescent bands after derivatization provided important clues for the identification of the major flavone present in the samples, while high performance liquid chromatography combined with ultraviolet spectrometry or a diode array detector (HPLC-UV/DAD) technique was employed for the simultaneous detection and quantification of these compounds.

10.2.2 High Performance Liquid Chromatography in the Analysis of Luteolin Derivatives

Among the many chromatographic methods, adsorptive chromatography, in which the mobile phase is liquid and the stationary phase is solid, is of the greatest practical importance. For example, this method has a much wider application than GC because it allows the analysis of substances in the form of liquids and soluble solids. Furthermore, it is also suitable for analysis of thermolabile substances because it is usually performed at low temperatures, and such samples are not degraded. Due to the long analysis time, high mobile phase consumption and low efficiency, traditional CC is currently used mainly for preparative purposes. However, improved methods, such as HPLC and especially liquid chromatography combined with mass spectrometry (LC-MS), are increasingly used for analysis of natural compounds including luteolin derivatives.

High performance liquid chromatography has been performed since 1960. It is a dynamically developing method with a wide range of uses and has been proven to be very useful in phytochemical analysis. The principle of operation consists of pumping the mobile phase from the tank (or tanks) through the stationary phase-filled column. Eluents are previously filtered and degassed. Some systems are additionally equipped with thermostats that regulate the temperature of the column. If the chromatographic system is properly selected and applied, then the individual components are separated and detected. Strengthened signals are transmitted to the computer, where the obtained data are registered and properly processed. The separation is based on competition of the molecules of the eluent and the substance being analyzed for the space on the adsorbent surface in the column.

When choosing a column, one should be guided by the size of the sample, time of analysis and expected effect of the separation. Although columns with different diameters are available, those with a diameter of 4.6 mm are by far the most common. However, due to better detection of the components separated in smaller diameter columns, these columns are increasingly being used for separation and analysis. In addition, improved separation can be achieved using a column filled with smaller particles. In addition to 5 μm particles, which are most common, particles of 3 μm in diameter or even smaller can be used. For example (Table 4), this approach was chosen when separating luteolin 2''-O-feruloylhexosyl-6-C-hexoside and luteolin 6-C-glucoside from Arenaria montana and isoorientin in Achillea millefolium , as well as separating luteolin 6-C-hexosyl-8-C-pentoside, luteolin 2''-O-deoxyhexosyl-6-C-glucoside and luteolin 6-C-glucoside in Cymbopogon citratus . Particles less than 2 μm in diameter were used for ultra-performance liquid chromatography (UPLC) analysis of luteolin derivatives and other flavonoids in Lactuca sativa [72,73] and date palm (Phoenix dactylifera).

Abbreviations: Ace, acetone; AcOH, acetic acid; ACN, acetonitrile; AFNH4, ammonium formate; BuOH, butanol; DAD, diode array detector; DAD-ESI-MS, diode array detector combined with electrospray ionization and mass spectrometry; DAD-ESI-MS/MS, diode array detector combined with electrospray ionization and tandem mass spectrometry; DAD-ESI-QTOF-MS, diode array detector combined with electrospray ionization and quadrupole – time of flight mass spectrometry; DAD-MS, diode array detector combined with mass spectrometry; DAD-MS/MS, diode array detector combined with tandem mass spectrometry; DAD-QTOF-MS, diode array detector combined with quadrupole – time of flight mass spectrometry; ESI-IT-MS/MS, electrospray ionization combined with ion trap and tandem mass spectrometry; ESI-MS, electrospray ionization combined with mass spectrometry; ESI-MS/MS, electrospray ionization combined with tandem mass spectrometry; ESI-QTOF-MS, electrospray ionization combined with quadrupole – time of flight mass spectrometry; FA, formic acid; IMS-QTOF-MS, ion-mobility spectrometry combined with quadrupole – time of flight mass spectrometry; MeOH, methanol; MS/MS, tandem mass spectrometry; PA, phosphoric acid; PAD, pulsed amperometric detection; PDA-ESI-MS/MS, pulsed amperometric detection combined with electrospray ionization and tandem mass spectrometry; PDA-ESI-MS, pulsed amperometric detection combined with electrospray ionization and mass spectrometry; PDA-MS, pulsed amperometric detection combined with mass spectrometry; TFA, trifluoroacetic acid; UV/PAD-MS, UV/pulsed amperometric detection combined with mass spectrometry.

Table 4 High performance liquid chromatography/ultra performance liquid chromatography in the analysis of luteolin derivatives.

Luteolin Derivative	Stationary Phase/ Column	Mobile Phase	Conditions	Detection	Analyzed Species
Luteolin	ODS	0.2% PA:H2O (A), MeOH (B)	Injection volume: 20 µl; flow rate: 1.0 Ml/min	DAD, 360 nm	Honey
	ODS C18	0.2% PA:H2O (58:41 v/v)	Injection volume: 50 µl; flow rate: 1.0 Ml/min	UV, 350 nm	Chrysanthemum morifolium
	Kinetex C18	ACN (A), 0.05% TFA:H2O (B)	Flow rate: 0.8 Ml/min	UV, 254 nm	Enhalus acoroides
	ODS Hypersil C18	MeOH:ACN (1.25:1 v/v) (A), 0.5% AcOH:H2O (B)	Flow rate: 0.8 Ml/min	MS	Perilla frutescens
	Zorbax SB-C18	PA:H2O pH 4.0 (A), ACN (B)	Injection volume: 50 µl; flow rate: 0.6 Ml/min; 25 °c	DAD, 330 nm	Vernonia condensata
	ODS C18	MeOH (A), 0.05% TFA:H2O (B)	Injection volume: 20 µl; flow rate: 1.0 Ml/min	UV, 280 nm	Corchorus olitorius
	RP	0.5% PA:H2O (A), ACN (B)	Injection volume: 20 µl; flow rate: 0.8 Ml/min; at 35 ⊠	Triple-TOF-MS	Veronicastrum latifolium
	RP C18	1% FA:H2O (A), 40% solvent A:ACN (B)	Flow rate: 0.5 Ml/min; 25 °c	DAD-ESI-MS/MS	Olea europaea, olive oil
	Discovery HS C18	2% AcOH (A), ACN (B)	Injection volume: 20 µl; flow rate: 0.8 Ml/min	DAD 280, 320 nm	Agastache foeniculum
	Ultrasphere 5 C18	H2O (A), ACN (B)	Injection volume: 10 µl; flow rate: 1 ml/min; 25 °c	ESI-MS; 350 nm	Rosa rugosa
Luteolin 6-glucoside	Vydac RP C18	0.05% TFA:H2O (A), 0.038% TFA:ACN (v/v) (B)	Injection volume: 10 µl flow rate: 1 ml/min; 36 °c	UV, 342 nm	Ficaria verna
Luteolin 8-glucoside	RP C18	0.5% FA:H2O (A), ACN (B)	Injection volume: 20 µl; flow rate: 0.5 Ml/min; 23 °c	DAD, 254, 340 nm	Jatropha gossypiifolia, J. mollissima

Luteolin Derivative	Stationary Phase/ Column	Mobile Phase	Conditions	Detection	Analyzed Species
Luteolin 7-glucoside	Hypersil gold C18	0.1% FA:H2O (A), 0.1% FA:MeOH (B)	Injection volume: 10 µl; flow rate: 0.35 Ml/min; 38 °c	MS	Rhoeo discolor
	Thermo C18	0.3% FA:H2O (A), ACN (B)	Injection volume:10 µl; flow rate: 1.0 Ml/min; 30 °c	DAD-Q-Orbitrap-MS	Epimedium brevicornum, Anacyclus pyrethrum, Lycium barbarum, Cuscuta australis
Luteolin	Eclipse XDB C18	0.025% AcOH:H2O (A), 5% Ace:ACN (B)	30 °C	ESI-MS/MS	Olea europaea, olive oil
Luteolin 7-glucoside	Eclipse XDB C18	1% PA:H2O (A), ACN (B)	Injection volume: 10 µl; flow rate: 1 ml/min; 25 °c	DAD-ESI-QTOF-MS/MS	Verbascum ovalifolium
Luteolin 7-rutinoside	Eclipse XDB C18	ACN (A), 0.2% FA (B)	Flow rate: 0.3 Ml/min; 30 °c	DAD, 230, 254, 290, 334 nm; MS/MS; NMR	Ligustrum vulgare
Luteolin 7-rhamnoside					
Luteolin	Gemini C18	0.5% AcOH:H2O (A), ACN (B)	25 °C	ESI-QTOF-MS	Olea europaea
Luteolin 7,4-diglucoside					
Luteolin 7-rutinoside					
Luteolin 4-glucoside					
Luteolin 3-glucoside					
Luteolin 7-glucoside					
Luteolin 6-glucoside	Spherisorb S5 ODS-2	5% FA:H2O (A), MeOH (B)	Flow rate: 1 ml/min	DAD, 280 nm	Cymbopogon citratus
Luteolin 7-rhamnoside					
Luteolin 7-glucoside					
Luteolin 6-glucosyl-8-arabinoside					
Luteolin 2”-feruloylhexosyl-6-hexoside	Spherisorb S3 ODS-2 C18	0.1% FA:H2O (A), ACN (B)	Flow rate: 0.5 Ml/min; 35 °c	DAD-MS; 280, 370 nm	Arenaria montana
Luteolin 6-glucoside					

Luteolin Derivative	Stationary Phase/ Column	Mobile Phase	Conditions	Detection	Analyzed Species
Luteolin	Kinetex 100 C18	1% FA:H2O (A), 1% FA:ACN (B)	Flow rate: 1 ml/min; 30 °c	ESI-QTOF-MS;	Rosmarinus officinalis
Luteolin 7-rutinoside					
Luteolin 3-glucuronide					
Luteolin 3'-(2"-acetyl)-glucuronide	Superschera 100 RP C18			DAD, 324 nm	
Luteolin 6-hydroxy-7-glucoside					
Luteolin hexosyl-hexoside-malylester	AquasilW C18	TFA solution (pH 2.8) (A), ACN (B)	Flow rate: 1.0 Ml/min; 15 °c	UV/DAD, 340 nm	Dianthus versicolor
Luteolin 6-glucosyl-7-galactoside					
Luteolin 6-glucosyl-7-rutinoside					
Luteolin 6-glucosyl-7-rhamnosyl-galactoside					
Luteolin 7-apiofuranosyl (1→2)-glucopyranoside	Eclipse Plus C18	0.1% FA:MeOH (A), 0.1% FA:H2O (B)	Injection volume: 10 μl; isocratic mixture flow rate: 0.3 Ml/min	QTOF-MS/MS	Apium graveolens
Luteolin 7-glucopyranoside					
Luteolin 7-[apiofuranosyl (1→2)-(6"-malony)]-glucopyranoside					
Luteolin 7-(6-rhamnosyl)hexoside	Kinetex C18	1% FA (A), ACN (B)	Injection volume: 8 μl; flow rate: 0.8 Ml/min	DAD-ESI-MS; 340 nm	Phoenix dactylifera
Luteolin 7-(2-rhamnosyl)hexoside					
Luteolin 7-(2-hexosyl[6-sulfate]) hexoside					
Luteolin 7-hexosyl(6-sulfate)					
Luteolin	Gemini C18	0.1% FA:H2O (A), 0.1% FA:ACN (B)	0-60 Min, 10–60% b; flow rate: 1 ml/min	DAD-ESI-MS/MS; 210, 270, 310 and 350 nm	Achillea moschata
Luteolin 7-glucoside					
Luteolin 7-glucoside	Synergy Polar RP 80Å, LiChroCART 4-4 with guard column LiChrospher 100 C18	0.9% FA:0.1% AcOH:1.5% BuOH:H2O (A), 30% MeOH:0.9% FA:0.1% AcOH:ACN (B), MeOH (C)	Injection volume: 5 μl; flow rate: 1.0 Ml/min; 45 °c	PAD; MS; UV; IR; NMR	Leontopodium alpium
Luteolin 7,4'-diglucoside					
Luteolin 6-hydroxy-7-glucoside					
Luteolin 4'-glucoside					
Luteolin 3'-glucoside					

Luteolin Derivative	Stationary Phase/ Column	Mobile Phase	Conditions	Detection	Analyzed Species
Luteolin 7-diglucuronide	Eclipse Plus C18	1% FA:H2O:ACN (A), ACN (B)	Injection volume: 20 µl; flow rate: 0.5 Ml/min	DAD; UV-VIS, 190-450 nm; ESI-TOF-MS; ESI-IT-MS/MS	Lippia citrodora
Luteolin 5-rutinoside	Luna C18	0.1% FA:H2O (A), ACN (B)	Injection volume: 5 µl; flow rate: 1.0 Ml/min	DAD, 200-400 nm; ESI-MS	Nerium indicum
Luteolin 7-rutinoside					
Luteolin diglucoside	Luna C18	1% FA:H2O (A), ACN (B)	Injection volume: 20 µl; flow rate: 1 ml/min	PDA-ESI-MS/MS	Taraxacum officinale
Luteolin 7-rutinoside					
Luteolin 7-glucoside					
Luteolin	Gemini RP C18	2% AcOH:H2O (A), ACN (B)	---	ESI-MS; 257, 278 and 340 nm	Pistacia atlantica
Luteolin 2”-Galloyl-4’-glucoside					
Luteolin 4’-glucoside					
Luteolin 6-methoxy-8-arabinosyl-7-glucoside	Symmetry Shield Waters RP18	0.2% FA:H2O (A), ACN (B)	Flow rate: 1.2 Ml/min	UV/PAD-MS; UV; 200–600 nm	Saccharum officinarum
Luteolin 8-glucoside	Symmetry Shield RP18			UV/PAD	
Luteolin 8-rhamnosyl-7-rhamnoside					
Luteolin 8-glucoside	Kinetex PFP	0.8% AcOH:H2O (A), ACN (B)	Injection volume: 10 µl; flow rate: 0.7 Ml/min; 40 °c	UV/DAD, 330 nm	Clinacanthus nutans
Luteolin	Eclipse XDB C18	MeOH (A), 0.2% FA:H2O (B)	Flow rate: 0.8 Ml/min; 30 °c	DAD-MS/MS; UV; 254, 360 nm	Securigera securidaca
Luteolin 7-glucoside					
Luteolin 7-glucuronyl-3-glucoside					
Luteolin 6-glucoside					
Luteolin 6-glucosyl-2”-rhamnoside					
Luteolin 6-glucosyl-4’-glucoside					
Luteolin 3’-glucoside					

Luteolin Derivative	Stationary Phase/ Column	Mobile Phase	Conditions	Detection	Analyzed Species
Luteolin hexosyl-rhamnoside	Spherisorb S3 ODS-2 C8	0.1% FA:H2O (A), ACN (B)	Flow rate: 0.5 Ml/min; 35 °c	DAD-MS; 280, 370 nm	Coriandrum sativum
Luteolin 7-rutinoside					
Luteolin 7-glucoside					
Luteolin	XDB C18	1% FA:H2O (A), ACN (B)	Injection volume: 100 µl; flow rate: 4 ml/min	NMR; MS	Casimiroa edulis
Luteolin 6-glucosyl-8-arabinoside					
Luteolin 7-glucoside					
Luteolin 6-arabinosyl-8-glucoside					
Luteolin 3-glucopyranoside	PLRP-S 100Å	0.1% FA:H2O (A), ACN (B)	Injection volume: 100 µl; flow rate: 1.5 Ml/min	MS	Thymus alternans
Luteolin 7-glucopyranoside					
Luteolin 7-rutinoside					
Luteolin methoxy-hexoside					
Luteolin 4'-glucopyranoside	XDB C18	1% FA:H2O (A), ACN (B),	Injection volume: 100 µl; flow rate: 4 ml/min		
Luteolin 6-hydroxy-7-glucoside	Gemini C18	0.1% FA:H2O (A), ACN (B)	Flow rate: 0.8 Ml/min	UV-VIS, 200-400 nm; MS; DAD; NMR	Athrixia phylicoides
Luteolin	Syncronis C18	0.1% AcOH:H2O (A), ACN (B)	Injection volume: 5 µl; flow rate: 0.3 Ml/min	DAD-MS/MS	Honey
Luteolin 7-rhamnoside					
Luteolin	HSS T3	0.1% FA:H2O (A), 0.1% FA:ACN (B)	Injection volume: 3.1 Ml; flow rate: 0.15 Ml/min	PDA-MS	Phoenix dactylifera
Luteolin rhamnosyl hexoside					
Luteolin rhamnosyl dihexoside					
Luteolin diglucuronide	S3 ODS-2 C18	0.1% FA:H2O (A), ACN (B)	Flow rate: 0.5 Ml/min; 35 °c.	DAD-ESI-MS; 370, 330 and 280 nm	Thymus pallescens, Saccocalyx satureioides,
Luteolin 7-glucuronide					
Luteolin 7-rutinoside					

Luteolin Derivative	Stationary Phase/ Column	Mobile Phase	Conditions	Detection	Analyzed Species
Luteolin 7-glucoside					Ptychotis verticillata
Luteolin 6-glucoside					Coleostephus myconis
Luteolin	Kinetex C18	ACN (A), 0.1% FA:H2O (B)	Injection volume: 10 µl; flow rate: 0.4 Ml/min	ESI-MS	Lathyrus pratensis
Luteolin rutinoside					
Luteolin hexoside					
Luteolin 6-hexoside					L. aureus
Luteolin 7-glucoside	Luna C18	5% FA:H2O (A), MeOH (B)	Injection volume: 100 µl; flow rate: 1 ml/min; 35 °c	DAD, 200, 600 nm	Thymus pulegioides
Luteolin hexuronide		0.1% FA:H2O (A), MeOH (B)	Injection volume: 10 µl; flow rate: 0.5 Ml/min; 40 °c	ESI-MS/MS	
Luteolin	Eclipse Plus C18	0.5% AcOH:H2O (A), ACN (B)	Injection volume: 5 µl; flow rate: 0.5 Ml/min	DAD-QTOF-MS; 325-371 nm	Ficus carica
Luteolin hexosyl-pentoside					
Luteolin 6-glucoside					
Luteolin 8-glucoside					
Luteolin 7-glucoside					
Luteolin (3-hydroxy-3-methylglutaroyl)-X”-deoxyhexosyl-hexoside	ODS-2 C18 with ODS-2 C18 guard cartridge	1% FA:H2O (A), MeOH (B)	Flow rate: 0.2 Ml/min; 25 °c	PDA-ESI-MS	Urtica membranacea
Luteolin 6-hexoside					
Luteolin 6-rutinoside					
Luteolin dihexoside					

Luteolin Derivative	Stationary Phase/ Column	Mobile Phase	Conditions	Detection	Analyzed Species
Luteolin	Syncronis C18	0.01% AcOH:H2O (A), ACN (B)	Inaction volume: 5 µl; flow rate: 0.25 Ml/min	MS	Capsicum annuum
Luteolin 6,8-dihexoside					
Luteolin 6-hexosyl-8-pentoside					
Luteolin 6-pentosyl-8-hexoside					
Luteolin 6-hexoside					
Luteolin 8-hexoside					
Luteolin 7-(2”-pentosyl-4”-hexosyl)hexoside					
Luteolin 7-(2”-pentosyl)hexoside					
Luteolin 7-glucoside					
Luteolin 7-[2”-(5’”-sinapoyl) pentosyl]hexoside					
Luteolin 7-(2”-pentosyl-4”-hexosyl-6”-malonyl)hexoside					
Luteolin 7-(2”-pentosyl-6”-malonyl) hexoside					
Luteolin 7-[2”-(5’”-sinapoyl) pentosyl-6”-malonyl]hexoside					
Luteolin	Purospher star C18	10% FA:H2O (A), ACN (B)	25 °C	DAD-ESI-MS; 320-280 nm	Citrus aurantifolia
Luteolin 6,8-diglucoside					
Luteolin 8-glucoside					
Luteolin 7-rutinoside					
Luteolin	Zorbax SB-C18	AFNH4:H2O: ACN:FA (A), AFNH4:H2O: ACN:FA (B)	Injection volume: 5µl; flow rate: 1.0 Ml/min	ESI-MS; 325 nm	Caucalis platycarpos
Luteolin 7-glucoside					
Luteolin 7-rutinoside					

Luteolin Derivative	Stationary Phase/ Column	Mobile Phase	Conditions	Detection	Analyzed Species
Luteolin 7-glucoside	Beta-Basic C18	5% FA:ACN (A); 5% FA:H2O (B)	Injection volume: 20 µl; flow rate: 0.9 Ml/min	UV, 280 nm	Melissa officinalis
Luteolin 7-glucuronide					Mentha piperita, Salvia officinalis
Luteolin 7-rutinoside					
Luteolin Luteolin 7-glucuronide Luteolin 7-rhamnosyl-hexoside Luteolin 7-rutinoside Luteolin 7-glucoside Luteolin 7-glucuronide Luteolin 7-rutinoside Luteolin diglucoside	UPLC BEH C18 and a Acquity UPLC BEH C18 VanGuardTM pre-column	0.1% AcOH:H2O (A), 0.1% AcOH:MeOH (B)	Injection volume: 5 µl; flow rate: 0.5 Ml/min	DAD-ESI-QTOF-MS; 370 nm	Lactuca sativa
	HSS T3	0.1% FA:H2O (A), 0.1% FA:ACN (B)	Injection volume: 3 µl; flow rate: 0.4 Ml/min	IMS-QTOF-MS	
Luteolin hexoside	Hypersil Gold C18	0.1% FA:H2O (A), ACN (B)	---	DAD-ESI-MS	Salvia elegans, S. greggii, S. officinalis
Luteolin hydroxy-glucuronide					
Luteolin 7-glucoside					
Luteolin 7-glucuronide					
Luteolin malonyl-hexoside					
Luteolin glucuronide					
Luteolin glucoside					
Luteolin rutinoside			Flow rate: 0.2 Ml/min		Thymus barona,T. pseudolanuginosus, T. caespititius
Luteolin 7-rhamnosyl(1-6) galactoside	Eclipse XDB-C18	0.1% FA:H2O (A), 0.1% FA:ACN (B)	Flow rate: 0.5 Ml/min; 25 °c	MS	Filago germanica
Luteolin 4'-glucuronide	C18µ-Bondapak RP18	MeOH:H2O (42:58 v/v)	Flow rate: 2.0 Ml/min	NMR	Galactites elegans

Luteolin Derivative	Stationary Phase/ Column	Mobile Phase	Conditions	Detection	Analyzed Species
Luteolin	Kinetex 100 A C18	1% FA:H2O (A), ACN (B)	Injection volume: 5 µl; flow rate: 0.8 Ml/min; 25 °c	DAD, 340 nm	Allophylus africanus
Luteolin 3'-7-diglucoside					
Luteolin 7-glucoside					
Luteolin 6-(2-rhamnosyl)-hexoside				DAD-ESI-MS/MS; 280, 340 nm	
Luteolin (pentosyl)-hexoside					
Luteolin hexoside					
Luteolin 6-glucoside	ODS2 C8	0.1% FA:H2O (A), ACN (B)	Flow rate: 0.5 Ml/min; 35 °c	DAD-ESI-MS; 280, 370 nm	Achillea millefolium
Luteolin 6-hexosyl-8-pentoside	S3 ODS-2 C18	0.1% FA:H2O (A), ACN (B)	Flow rate: 0.5 Ml/min; 35 °c	DAD-MS; 280, 370 nm	Cymbopogon citratus
Luteolin 2"-deoxyhexosyl-6-glucoside					
Luteolin 6-glucoside					
Luteolin 6-pentosyl-8-pentoside					
Luteolin-7-rhamnoside					
Luteolin 7-glucoside					
Luteolin 2"-deoxyhexosyl-pentoside					
Luteolin 6-pentoside					
Luteolin 2"-deoxyosyl-6-(6-deoxy-pento-hexosuloside					
Luteolin	C18	0.005 % FA:H2O (A), MeOH (B)	Injection volume: 10 µl; flow rate: 0.5 Ml/min; 30 °c	MS-Orbitrap	Chrysanthemum trifurcatum
Luteolin 8-glucoside					
Luteolin 4'-glucoside					
Luteolin 7-glucoside					
Luteolin 6-glucoside					

Luteolin Derivative	Stationary Phase/ Column	Mobile Phase	Conditions	Detection	Analyzed Species
Luteolin 6-glucoside	HSS T3	0.1% FA:H2O (A), 0.1% FA:ACN (B)	Injection volume: 3.1 Ml; flow rate: 0.15 Ml/min	PDA-MS	Passiflora edulis
Luteolin 8-glucoside					
Luteolin 6-deoxyhexosyl-8-pentoside					
Luteolin 6-fucoside					
Luteolin 8-deoxyhexoside					
Luteolin 6,8-diglucoside					
Luteolin 6,8-diglucoside	Kinetex C18	0.05% FA:H2O (A), 0.05% FA:ACN (B)	Flow rate: 0.4 Ml/min; 40 °c	MS/MS	Eragrostis tef
Luteolin 8-glucosyl-7-glucoside isomer					
Luteolin 6-glucosyl-7-glucoside isomer					
Luteolin 8-glucosyl-7-rhamnoside					
Luteolin 8-glucoside					
Luteolin 7-glucoside					
Luteolin 7-rhamnoside					
Luteolin 7-(6”-syringly)glucosyl-6-glucoside					
Luteolin 7-(2”-syringyl)arabinosyl-6-glucoside					
Luteolin 8-(6”-diacetyl)glucoside					
Luteolin	Nucleosil 100-3.5 C18	FA:H2O (A), FA:ACN (B)	Injection volume: 20 µl	MS	Arum hygrophilum
Luteolin 6-glucoside					
Luteolin 6-[6”-glocosy-caffeoyl-glucopyranosyl(”→2)-glucopyranoside	Eclipse Plus C18	0.2% FA:H2O (A), 0.2% FA:ACN (B)	30 °C	PDA-ESI-MS/MS	Triticum aestivum
Luteolin 6-glucopyranoside				UV, 350 nm	

Luteolin Derivative	Stationary Phase/ Column	Mobile Phase	Conditions	Detection	Analyzed Species
Luteolin 7-glucoside	Syncronis C18	0.1% FA:H2O (A), ACN (B)	Injection volume: 5 µl; flow rate: 0.25 Ml/min	MS/MS	Iris pumila, I. variegata, I. humilis
Luteolin 6-glucoside					
Luteolin 7-(2”-p-coumaroyl)-rhamnoside					
Luteolin	Luna Omega Polar C18 with Polar C18 Security Guard cartridge	0.1% FA:H2O (A), ACN (B)	Flow rate: 0.4 Ml/min	ESI-MS/MS; 320, 350 nm	Parentucellia latifolia
Luteolin hexoside					
Luteolin	Intersil ODS	ACN:H2O:FA (10:89:1 v/v/v) (A), ACN:H2O:FA (89:10:1 v/v/v) (B)	Injection volume: 10 µl; flow rate: 0.5 Ml/min; 40 ⊠	DAD-MS/ MS; 360 nm	Verbascum eskisehirensis
Luteolin glucoside					
Luteolin glucuronide					
Luteolin pentosyl-glucoside					
Luteolin 4’-glucoside	Kinetex RP C18	PA:H2O pH 3 (A), PA:ACN pH 3 (B)	Injection volume: 1 µl; flow rate: 1.2 Ml/min; 35 ⊠	UV 300 nm	Matricaria recutita
Luteolin 7-O-glucuronide	C18	0.02% TFA:H2O (A), MeOH:ACN (3:7 v/v) (B)	Injection volume: 2.5 Ml; flow rate: 0.9 Ml/min; 45 °c	DAD-ESI-MS 240, 254, 325 nm	Lippia alba
Luteolin glucoside	YMC-Triart C18	0.1% FA:H2O (A), 0.1% FA:ACN B)	Injection volume: 10 µl; flow rate: 0.8 Ml/min	DAD-ESI-MA 265, 280, 330, and 360 nm	Chrysanthemum morifolium
Luteolin 7-O-β-glucoside					
Luteolin glucuronide					
Luteolin malonylglucoside					
Luteolin					
Luteolin 7-diglucuronide	Eclipse XDB C18	0.03% PA:H2O (A), solvent A:ACN (1:9 v/v) (B)	Injection volume: 20 µl; flow rate: 0.8 Ml/min; 25 °c	DAD, 210, 250, 320, 350 and 370 nm	Thymus pannonicus
Luteolin 7-O-glucuronide					

Stationary phases with different polarity may be used in HPLC. In the normal phase system, polar column fillings are used, and most often the filling is silica gel. However, silica gel can adsorb water, which leads to the loss of the original separating properties of the column and thus to impaired reproducibility of the obtained results. Therefore, the gel is often modified with the aim of enabling better separation of mixture components. This is performed mainly by bonding alkyl chains (or alkyl chains bearing other functional groups) to functional groups on the gel surface. Silica gels optimized in this way are referred to as the associated phase. In the so-called reversed phase (RP) system, non-polar associated phases are used. Such systems are especially useful in the analysis of insoluble or poorly water-soluble compounds, as well as in the analysis of polar compounds, provided that a mobile phase with high water content is used. Typically, an octadecylsilane (ODS) phase, composed of 18 carbon atoms (RP-18), is employed. Such fillers can have different properties depending on the silica gel type and/or production method. According to the available literature, the analysis of luteolin derivatives (Table 4) was performed exclusively with RP systems, and the RP-18 system was the most frequently used system, as exemplified in the separation of six luteolin derivatives in Securigera securidaca or as many as ten derivatives in Capsicum annuum . Stationary phases composed of 8 carbon atoms (RP-8) were rarely used. Examples include the separation of flavonoids in Coriandrum sativum and Achillea millefolium .

The selection of an appropriate mobile phase is extremely important for chromatographic separation. The type of analyte, mixture composition, stationary phase and detector used should be taken into account. Mixtures of up to three components are most commonly used. In the reverse system, mixtures of MeOH/H_2O or acetonitrile (ACN)/H_2O are routinely used. As the amount of organic solvent increases, the retention time for non-polar substances decreases, while the addition of H2O extends their retention times. As a rule of thumb, systems containing ACN/H_2O eluents are more efficient than those containing MeOH/H_2O. In the normal phase system, however, the base solvent is a non-polar eluent, and its polarity is appropriately modified by the addition of another solvent of higher polarity, e.g., chloroform ($CHCl_3$). According to the available literature, virtually all mobile phases used for analysis of luteolin derivatives consisted either of ACN/H_2O, e.g., or, less often, MeOH/H_2O mixtures. Additionally, a modifier such as FA, e.g.,, AcOH, e.g., or trifluoroacetic acid (TFA) was added to avoid peak tailing. Analyses performed without acidic modifiers are rare.

In the chromatography process, the elution can be carried out in two ways. The first method is isocratic elution, which involves running with the same composition of mobile phase during the analysis. This means that a constant elution force is maintained throughout the entire separation period. In this case, even a slight change in the composition of the mobile phase may affect the results of the analysis. Isocratic elution is only rarely used for analysis of flavonoids. In the case of luteolin derivatives, this method was used only for analysis of luteolin 7-rhamnosyl- (1-6)-galactoside in Filago germanica , luteolin

7-rhamnosyl(1-6)galactoside in Galactites elegans , and various luteolin derivatives in Apium graveolens . If the composition of the eluent changes during the division of the mixture, then the gradient elution can be described. The gradient can be linear or specifically programmed. With a change in the composition of the eluent, its elution force increases. For this reason, this method is used particularly for the separation of mixtures composed of substances of different polarity. The vast majority of papers describing the separation of luteolin derivatives in plant mixtures use a gradient approach. Examples include the separation of luteolin 8-C-glucoside, luteolin 6-C-glucoside, luteolin 4'-O-glucoside and luteolin 7-O-glucoside from Chrysanthemum trifurcatum , thirteen luteolin derivatives from Cymbopogon citratus and many others.

By far, the most common method for detection of luteolin derivatives is using diode array detectors (DADs) (also called photodiode array detectors, PDA), which were used in as many as 46 instances (e.g., for the analysis of luteolin derivatives and other flavonoids in Cymbopogon citratus , Dianthus versicolor , Clinacanthus nutans , Thymus alternans and others). More often than not, however, DADs were combined with other detectors for additional structure determination or confirmation, with or without prior isolation. Examples of DADs combined with other detection methods for analysis of luteolin derivatives in plants include the use of tandem mass spectrometry (MS/MS) and NMR for Ligustrum vulgare , diode array detectors combined with electrospray ionization and mass spectrometry (DAD-ESI-MS) for Phoenix dactylifera and Nerium indicum , electrospray ionization combined with tandem mass spectrometry (ESI-MS/MS) for Achillea moschata , electrospray ionization combined with time of flight mass spectrometry (ESI-TOF-MS) and electrospray ionization combined with ion trap and tandem mass spectrometry (ESI-IT-MS/MS) for Aloysia citrodora and many others.

For example, the HPLC-PAD-MS technique was used in the analysis of aerial flowering parts of the Edelweiss alpine region (Leontopodium alpinum). This method allowed the basic separation of almost all components of the L. alpinum extract prepared by exhaustive dichloromethane (DCM) followed by MeOH extraction. In total, 14 compounds have been isolated from the extract, including several luteolin derivatives. The authors used a gradient as the mobile phase as follows: H2O:0.9% FA:0.1% AcOH:1.5% BuOH (A) and ACN:30% MeOH:0.9% FA:0.1% AcOH (B) and MeOH (C). The structure of the isolated components was additionally confirmed using NMR spectroscopy].

10.2.3. Liquid Chromatography in the Analysis of Luteolin Derivatives

LC combined with tandem mass spectrometry (LC-MS/MS) (<u>Table 5</u>) is especially useful in the analysis of multicomponent mixtures, such as herbal extracts because it does not require a large amount of the sample or previous separation. To further reduce the influence of other factors on the analysis, more advanced techniques, involving the combination of more than one detection method, e.g., LC combined with NMR and MS (LC-NMR-MS), are increasingly used.

Table 5. Liquid chromatography in the analysis of luteolin derivatives.

Luteolin Derivative	Stationary Phase	Mobile Phase	Conditions	Detection	Analyzed Species
Luteolin	Zorbax SB C18 column with Security-Guard C18	MeOH (A), 0.5% AcOH:H_2O (B)	Flow rate: 1.0 Ml/min; injection volume: 20 µl	Ms/ms	Abri herba, A. mollis
	Inertsil ODS-3	0.1% FA:H_2O (A), ACN (B)	Flow rate: 0.5 Ml/min; injection volume: 10 µl	Ms/ms	Castanea mollissima
	RP C18	0.1% FA:H_2O (A), 0.1% FA:ACN (B)	Injection volume: 5 µl; flow rate: 0.3 Ml/min; 40 °c	Esi-ms/ms	Centaurea cyanus
Luteolin 7-glucoside	Eclipse XDB C18	1% FA:H_2O (A), MeOH (B)	Injection volume: 5 ml; flow rate: 0.6 Ml/min; 45 °c	Ms/ms	Plantago atrata, P. coronopus, P. holosteum, P. lanceolata, P. reniformis, P. schwarzenbergiana
Luteolin	Zorbax Plus C18	0.1% FA:H_2O (A), 0.1% FA:ACN (B)	Injection volume: 1µl; flow rate: 0.4 Ml/min	Ms	Matricaria recutita, Achillea millefolium, Thymus vulgaris, Salvia officinalis
	Eclipse XDB C18	0.05% FA:H_2O (A), MeOH (B)	Flow rate: 1 ml/min	Ms/ms	Vitis vinifera
Luteolin 7-apiosyl-glucoside	Symmetry C18	0.1% FA:H_2O (A), 0.1% FA:ACN (B)	Flow rate: 1.0 Ml/min; 25 °c	Dad-esi-ms; dad, 350, 310, 270 nm; uv, 190-650 nm	Apium graveolens
Luteolin 7-glucoside					
Luteolin 7-malonyl-apiosyl-glucoside					
Luteolin 7-6'-malonyl-apiosyl-glucoside					
Luteolin 7-6'-malonyl-glucoside					
Luteolin 3'-glucoside	Zorbax SB C18	0.1% FA:H_2O (A); 0.1% FA:ACN (B)	Injection volume: 5 µl; flow rate: 0.3 Ml/min; 25 °c	Ms/ms; tqms; esi	Phlomis persica
Luteolin 6-glucoside					
Luteolin 6,8-dihexoside					
Luteolin 7-rutinoside					Ph. eliptica
Luteolin 7-rutinoside	XTerra MS C18	ACN (A), 0.05% AcOH:H_2O (B)	Injection volume: 20 µl; flow rate: 1.0 Ml/min	Pda-ms; nmr	Sechium edule
Luteolin 7-glucopyranoside					
Luteolin	Octadecyl silica gel ODS	H_2O:MeOH:FA (89:10:1 v/v/v) (A), MeOH:H_2O:FA (89:10:1 v/v/v) (B)	Flow rate: 1 ml/min; 40 °c	Ms/ms	Nepeta cilicica
Luteolin glucuronide					
Luteolin glucoside					

Luteolin Derivative	Stationary Phase	Mobile Phase	Conditions	Detection	Analyzed Species
Luteolin	Capcell Park C18	0.5% FA:H2O (A), 0.5% FA:ACN (B)	Flow rate: 0.5 Ml/min; 25 °c	Ms/ms	Humulus japonicus
Luteolin dihexoside					
Luteolin 7-dihexoside					
Luteolin 7-rutinoside					
Luteolin glucoside					
Luteolin 7-acetylglucoside					
Luteolin 6-glucoside	Intersil ODS	ACN:H2O:FA (10:89:1 v/v/v) (A), ACN:H2O:FA (89:10:1 v/v/v) (B)	Flow rate: 0.7 Ml/min; 40 °c	Ms/ms	Achillea sivasica
Luteolin glucoside					
Luteolin methoxy-2”-pentosyl-6-hexoside	ODS2 C8	0.1% FA:H2O (A), ACN (B)	Flow rate: 0.5 Ml/min; 35 °c	Dad-esi-ms; 280, 370 nm	Geranium molle
Luteolin 2”-rhamnosyl-6”-hexosyl-glucoside					
Luteolin 7-glucosyl-8-glucoside					
Luteolin 6-glucoside					
Luteolin 6-hexosyl-8-pentoside					
Luteolin acetylhexoside	Spherisorb S3 ODS2 C18	0.1% FA:H2O (A), ACN (B)	Flow rate: 0.5 Ml/min; 35 °c	Dad-esi-ms/ms	Achillea millefolium
Luteolin 6-glucoside					Coleostephus myconis
Luteolin 3-glucuronide					Rosmarinus officinalis
Luteolin glucuronide					
Luteolin hexosyl-pentoside	Zorbax RP C18	0.1% FA:H2O (A), 0.1% FA:ACN (B)	Injection volume: 5 µl; flow rate: 0.4 Ml/min; 35 °c	Qtof	Ficus carica
Luteolin 7-glucoside					
Luteolin 7-hexoside					
Luteolin 6-hexoside					
Luteolin 8-glucoside					
Luteolin 6-hexosyl-8-acetyl-hexoside	ODS2	H2O:ACN:FA	Injection volume: 20 µl; flow rate: 0.8 Ml/min	Ms/ms	Oxalis pes-caprae
Luteolin 8-glucosyl-7,3'-dimethoxyl-2”-O-glycoside					
Luteolin 8-glucosyl-7,3'-dimethoxyl-6-desoxyhexoside					
Luteolin 6-glucoside	Hydro RP	H2O (A), MeOH (B), 5% AcOH:MeOH (C)	Flow rate: 1 ml/min	Ms/ms	Passiflora morifolia
Luteolin 8-glucoside					

Luteolin Derivative	Stationary Phase	Mobile Phase	Conditions	Detection	Analyzed Species
Luteolin Luteolin 6-glucoside Luteolin 8-glucoside Luteolin 7-glucoside	Luna C18	0.1% FA:H2O (A), 0.1% FA:ACN (B)	Flow rate: 0.4 Ml/min	Dad, 280, 320, 365 nm; ms; ms/ms	Theobroma cacao
Luteolin 6-glucosie Luteolin 7-glucoside	XBridge C18	0.3% FA:H2O (A), MeOH (B)	Injection volume: 1 µl; flow rate: 1 ml/min; 40 °c	Ms/ms; esi-ms	Lophatherum gracile
Luteolin Luteolin glucuronyl-hexoside Luteolin 7-pentosyl-hexoside Luteolin 7-rutinoside Luteolin 7-glucoside Luteolin 7-glucuronide Luteolin glucoside Luteolin 7-malonyl-6"-glucoside Luteolin 7-acetyl-6"-glucoside Luteolin 7-dihexoside	Symmetry C18	0.1% FA:H2O (A), 0.1% FA:ACN (B)	Flow rate: 1.0 Ml/min; 25 °c	Dad-esi-ms; 350, 310, 270 and 520 nm	Chrysanthemum morifolium
Luteolin Luteolin7-glucuronide	Acquity BEH C18	0.1% FA:H2O (A), 0.1% FA:ACN (B)	Flow rate: 0.3 Ml/ min; 40 °c	Qtof-ms/ms	Ageratum conyoides
Luteolin hexoside	Zorbax SB C18	0.5% FA:H2O (A), ACN (B)	Injection volume: 5 µl; flow rate: 0.4 Ml/min	Ms/ms	Matricaria recutita
Luteolin hexoside Luteolin hexosyl-deoxy-hexose	Hypersil gold C18	1% FA:H2O (A), 0.1% FA:ACN (B)	Flow rate: 1 ml/min	Ms/ms	Cecropia obtusa
Luteolin dihexoside Luteolin pentosyl-hexoside Luteolin 7-glucoside Luteolin malonyl-hexoside	Spherisorb S3 ODS-2 C18	0.1% FA:H2O (A), ACN (B)	Flow rate: 0.5 Ml/min; 35 °c.	Dad-esi-ms/ms	Cotula cinerea
Luteolin Luteolin 8-glucoside Luteolin 5-glucopyranoside	Zorbax SB C18	0.1% FA:AFNH4 (A), 0.1% FA:ACN (B)	Injection volume: 1 ml; flow rate: 0.2 Ml/min	Esi-ms/ms	Ocimum sanctum

Abbreviations: AcOH, acetic acid; ACN, acetonitrile; AFNH4, ammonium formate; DAD-ESI-MS, diode array detector combined with electrospray ionization and mass spectrometry; DAD, diode array detector; DAD-ESI-MS/MS, diode array detector combined with electrospray ionization and tandem mass spectrometry; ESI, electrospray ionization; ESI-MS, electrospray ionization combined with mass spectrometry; FA, formic acid; MeOH, methanol; MS/MS, tandem mass spectrometry; PDA-MS, pulsed amperometric detection combined with mass spectrometry; QTOF, quadrupole time-of-flight; TQMS, triple quadrupole mass spectrometer.

Research on ethanol extract from Lophatherum gracile stems and leaves was performed using LC coupled with MS/MS. Gradient elution was performed using 0.3% FA and MeOH. Analysis of the species revealed the presence of, among others, luteolin 7-O-β-D-glucoside, and luteolin 6-C-glucoside.

Another raw material containing luteolin and its derivatives is the cocoa seed (Theobroma cacao). In the analysis of the H2O:MeOH extract, the following mobile phase was used: H2O:0.1% FA and ACN:0.1% FA. The elution was carried out in a linear gradient, whereas LC combined with electrospray ionization and tandem mass spectrometry (LC-ESI-MS/MS) coupling allowed the identification of the tested compounds.

Lin and Harnly performed a water-methanol analysis of the flower extract of Chrysanthemum morifolium and distinguished many compounds, including numerous derivatives of luteolin. In this case, the mobile phase was a mixture of 0.1% FA:H2O and 0.1% FA:ACN, in varying proportions. The qualitative determination of the analyzed substances was based on a comparison of retention times as well as mass and UV/Vis spectra.

10.2.4. Gas Chromatography in the Analysis of Luteolin and its Derivatives

Gas chromatography (GC) is a technique used to analyze volatile as well as non-volatile compounds after their derivatization. GC is characterized by chromatographic distribution of either a gas mobile phase on a solid adsorbent (gas-solid chromatography) or a liquid on an inert support (gas-liquid chromatography). GC can be hyphenated with various detection techniques such as GC combined with mass spectrometry (GC-MS), GC combined with tandem mass spectrometry (GC-MS/MS) or GC combined with time of flight mass spectrometry (GC-TOF-MS), thus greatly increasing the versatility, sensitivity and accuracy of the method. The analyzed substances should be thermally stable, and their boiling (or sublimation) temperature should not exceed 350-400°C. To achieve this, non-volatile substances are often derivatived. Polar functional groups are transformed into their less polar counterparts, thus increasing the volatility of the prepared compounds. The most common examples include substitution with a trimethylsilyl (TM) group, organic radicals or compounds such as trimethylchlorosilane (TMCS), hexamethyldisilazane (HMDS) or N,O-bis(trimethylsilyl)-trifluoroacetamide (BSTFA). Such an approach greatly increases the number of possible analytes. For example, phenolic groups of flavonoids are often transformed into their less polar trimethylsilyl counterparts allowing for rapid and effective separation of complex mixtures.

In adsorption GC, a gas that is chemically inert to the stationary phase as well as the components being analyzed is used as the mobile phase. Most often hydrogen, nitrogen or argon is used. Helium is being used less often due to its higher cost than other gases and the implementation of the principles of chemical safety. The mobile phase must be properly selected for compatibility with the detector used. However, the carrier gas itself does not have a significant influence on the separation effects of the analyzed mixtures. In the process of separation, the method of application of the sample to the chromatography column is very important. The sample should always have as small of a volume and the shortest dosing time as possible. This ensures better separation and narrower bands.

Gas chromatography uses open-ended columns, i.e., capillary columns and packed columns. Open-ended (OT-open tubular) columns are characterized by much higher efficiency than packed columns; therefore, they are chosen much more often. Capillary columns are particularly useful in the separation of substances with significantly different boiling points. Ideally, the column should have a similar polarity to the analyzed components. However, due to the higher efficiency and greater durability of stationary phases with low polarity, so capillary columns are recommended for chromatographic analyses.

Another factor that influences the efficiency of the separation of the analyzed sample components, as well as the time of analysis, is temperature. The separation temperature should be selected depending on the stationary phase used and the boiling point of the analytes.

The analysis of luteolin derivatives has been carried out in accordance with the general procedure used for analysis of other flavonoids (Table 6). MS was used for detection of volatile derivatives of luteolin, while the use of a flame ionization detector (FID) was described only in two papers. Most commonly, helium was used as a carrier gas, but the use of nitrogen was also recorded. The derivatization of luteolin, which is necessary for chromatographic separation, was mostly achieved with BSTFA/TCMS. For the analysis of lipophilic luteolin derivatives, such as 7,3',4'-trimethyl-luteolin in Arnica alpina, no derivatization was necessary.

Table 6 Gas chromatography in the analysis of luteolin and its derivatives.

Luteolin Derivative	Column	Derivatization	Conditions	Detection	Analyzed Species
Luteolin	HP-5-MS	1% TMCS:BSTFA	Carrier gas: He; injector temperature: 250 °C A: (column temperature) 5 min, 170°C; 3 °C/min, 170–255 °C; 1 min, 255 °C; 2 °C/min, 255–310 °C; flow rate: 0.5 mL/min; analysis time: 70 min; B: (column temperature) 5 min, 160°C; 3°C/min, 160–188°C; 1 min, 188 °C; 15 °C/min, 188–241 °C; 1 min, 241 °C; 2 °C/min, 241–282 °C; 5 °C/min, 282–310 °C; 5 min, 310 °C; flow rate: 1.0 mL/min; analysis time: 50 min	APCI-TOF-MS	Fruits of various olives species (Olea L.)
	Capillary column Supelco SPBM-5	Two- and three-phase transfer catalysis (PTC), methyl iodide	Carrier gas: He; injector temperature: 260 °C; detector temperature: 280 °C; furnace temperature: 5 min, 50 °C; 5 °C/min, 50–150 °C; 10 °C/min, 150–210 °C; analysis time: 45 min	MS	Mentha spicata, Hypericum perforatum
	Non-polar RSL 200 BP	0.2 M trimethylaniline hydroxide (TMAH):H2O-free MeOH Py:HMDS:TMCS	Carrier gas: N2; linear speed of the carrier 17.5 cm/s; 0–2 min, 280 °C and 235 °C; 1 °C/min, 280–290 °C; flow rate: 30 mL/min	FID	Different samples form AFRC Institute of Plant Science Research and John Innes Institute, Norwich, U.K.
	BPX5	Py:BSTFA:TMCS) (50:50:1 v/v/v)	Carrier gas: He; injector temperature: 310 °C; 1 min, 100 °C; 30 °C/min, 100–210 °C; 2 °C/min, 210–240 °C; 4 °C/min, 240–270 °C; 5 °C/min, 270–310 °C; 5 min, 310 °C; flow rate: 1.5 mL/min	QMS	Propolis, Chrysanthemum sp, Theobroma cacao (bitter chocolate)

Luteolin Derivative	Column	Derivatization	Conditions	Detection	Analyzed Species
	Capillary column Low-bleed CP-Sil 8 CB-MS	TMCS (100 µL), BSTFA (200 µL) HMDS:TMCS:Py (3:1:9, v/v/v)	Carrier gas: He; 2 °C/min, 70–135 °C; 10 min, 135 °C; 4 °C/min, 135–220 °C; 10 min, 220 °C; 3.5 °C/min, 220–270 °C; 20 min, 270 °C; injector temperature: 280 °C; detector temperature: 290 °C; flow rate: 1.9 mL/min	MS	Teucrium polium
	Quartz capillary column	Py:BSTFA (1:1 v/v)	Carrier gas: He;injector temperature 220 °C; detector temperature: 270 °C; 2.3 °C/min 200–270 °C; 30 min, 270 °C	MS	Aspalathus linearis
Luteolin 7,3',4'-trimethyl	Capillary column Permabond OV-1	---	Crrier gas: N2, He; injector temperature: 300 °C; column temperature: 270 °C (isothermal); flow rate: 1.3 mL/min	FID	Arnica alpina
	Capillary column OV-1	---		MS	

Abbreviations:Py,pyridine,BSTFA,N,O-bis(trimethylsilyl)trifluoroacetamide;TMS,trimethylsily;HMDS,hexamethyldisilazane; TMCS, trimethylchlorosilane; APCI-TOF-MS, atmospheric pressure chemical ionization combined with quadrupole – time of flight mass spectrometry; FID, flame ionization detector; QMS, quadrupole mass spectrometry.

10.2.5. Counter-Current Chromatography in the Analysis of Luteolin Derivatives

Counter-current chromatography (CCC) is a variation of liquid chromatography in which both the stationary and mobile phases are liquid. The separation of the constituents of the mixture is carried out in a system of immiscible liquids that are in equilibrium with each other. The method is simple and rapid, offering the possibility of introducing the raw sample to the column without need for previous clean-up. Counter-current chromatography is mainly used for purification of natural compounds, while its use as an analytical technique is far less common. In CCC, various dividing techniques can be used, thus distinguishing centrifugal partition chromatography (CPC) and rapid- or high-speed CCC (HSCCC), often referred to as hydrodynamic chromatography. High-speed CCC is particularly often encountered in studies involving the separation of flavonoids. Table 7 presents the conditions for the separation of mixtures containing luteolin and its derivatives using rapid counter-current chromatography. Separation of luteolin derivatives from mixtures of different phytochemicals is usually carried out with mixtures of EtOAc with one of alcohol (BuOH, EtOH or AcOH) and H_2O. Upon separation, structure determination by NMR is often necessary. However, instead of NMR, high resolution mass spectrometry (HRMS) can also be employed for structure confirmation, as evidenced by the use of this technique for differentiation of various luteolin derivatives in Lippia origanoides .

Table 7 Counter-current chromatography, as a preparative technique in the separation of luteolin derivatives.

Luteolin Derivative	Solvent System	Conditions	Detection	Analyzed Species
Luteolin 6-glucoside	EtOAc:BuOH:H2O (2:1:3 v/v/v)	Rotation speed: 800 rpm; flow rate (lower phase): 2.4 Ml/min	UV-VIS, 254 nm; NMR; MS	Patrinia villosa
Luteolin 7-glucoside	EtOAc:EtOH:AcOH:H2O (4:1:0.25:5 v/v/v/v)	Rotation speed: 800 rpm; flow rate (lower phase): 1.5 Ml/min	UV, 254 nm; NMR; MS	Paeonia suffruticosa
Luteolin 6,8-dihexoside	Hx:EtOH:H2O (4:3:1 v/v/v)	Rotation speed: 850 rpm; flow rate (lower phase): 2 ml/min	TLC; HPLC-UV-HRMS	Lippia origanoides
Luteolin 8-glucoside				
Luteolin 6-glucoside				
Luteolin 7-glucoside				

Abbreviations: AcOH, acetic acid; BuOH, butanol; EtOAc, ethyl acetate; EtOH, ethanol; HRMS, high resolution mass spectrometry; Hx, hexane, rpm, revolutions per minute.

10.3. PERSPECTIVES

- The presented comparison of chromatographic methods currently used to determine luteolin and its derivatives provides a systematic summary of the available knowledge. Without a doubt, chromatographic analysis may be successfully employed as an efficient method for both qualitative assessment (fingerprinting) and quantitative determination of luteolin derivatives. In tedious and time-consuming determinations of multi-ingredient plant extracts, including those that contain the most prevalent luteolin derivatives, combinations of large-scale chromatographic techniques with serially aligned detection modalities such as LC-MS/MS or LC/NMR/MS were found to be particularly useful. Such beneficial coupling of chromatography with other analytic techniques expands analytical capabilities while additionally improving the accuracy, sensitivity, and precision of assays. Despite the dominant position of LC in the analysis of natural compounds and the dynamic development of novel chromatographic methods, the TLC/HPTLC has not lost its important place in the phytochemical analysis of luteolin derivatives. The technique is relatively simple and inexpensive while facilitating rapid qualitative and quantitative analysis of test compounds. In addition, it facilitates large quantities of diluted samples being deposited in the stationary phase, allowing for a wider choice of mobile phase carriers. The technique is subject to continuous improvements and its range of applicability is expanding.

- Effective chromatographic analysis in the determination of luteolin derivatives requires appropriately selected chromatographic separation conditions. The appropriate choice of stationary phase sorbent may significantly improve test conditions. Most TLC analyses of luteolin derivatives are carried out in normal phase systems featuring a polar (hydrophilic) sorbent phase. In HPLC, separation conditions can be chosen more arbitrarily as the technique allows for the use of stationary phases of varying polarity which is particularly advantageous in the analysis of compounds that are either insoluble or poorly soluble in water. In order to additionally improve the sensitivity of HPLC analyses, one should focus on the column parameters responsible for appropriate separation. Quite often, the fairly successful analyses are used as starting points for further modifications, including the development of preparative-scale analyses. Of all chromatographic techniques, GC is the least applicable in the analysis of plant extracts, including those that contain luteolin derivatives. Despite its high sensitivity and efficiency when coupled with various detection techniques (MS, MS/MS, TOF-MS), the chromatographic separation process involves high temperatures and derivatization of analytes. Due to this important aspect, GC-MS is less frequently used in the analysis of polyphenolic compounds. The complexity of plant matrices is unquestionably a significant problem in the chromatographic analysis of plant extracts. It has a negative impact on the efficacy of analyses and prevents complete identification of all components of the plant extract. However, chromatography remains the primary and most effective analytical technique available in the current state of the art for the determination of compounds of natural origin.

11

PRINCIPLES OF ION EXCHANGE

11.1 INTRODUCTION: AN OVERVIEW

This chapter provides a general introduction to the theoretical principles that underlie every ion exchange separation. An understanding of these principles will enable the separation power of ion exchange chromatography (IEX) to be fully appreciated.

Net surface charge and pH

IEX separates molecules on the basis of differences in their *net surface charge.* Molecules vary considerably in their charge properties and will exhibit different degrees of interaction with charged chromatography media according to differences in their overall charge, charge density, and surface charge distribution. The charged groups within a molecule that contribute to the net surface charge possess different *pKa* values (acid ionization constant) depending on their structure and chemical microenvironment.

Since all molecules with ionizable groups can be titrated, their net surface charge is highly pH dependent. In the case of proteins, which are built up of many different amino acids containing weak acidic and basic groups, net surface charge will change gradually as the pH of the environment changes, that is, proteins are *amphoteric.* Each protein has its own unique *net charge versus pH relationship* which can be visualized as a *titration curve.* This curve reflects how the overall net charge of the protein changes according to the pH of the surroundings.

Figure 1 illustrates several theoretical protein titration curves (these curves can be generated using a combination of isoelectric focusing and electrophoresis, but with modern solutions for rapid method development, actual titration curves are rarely used).

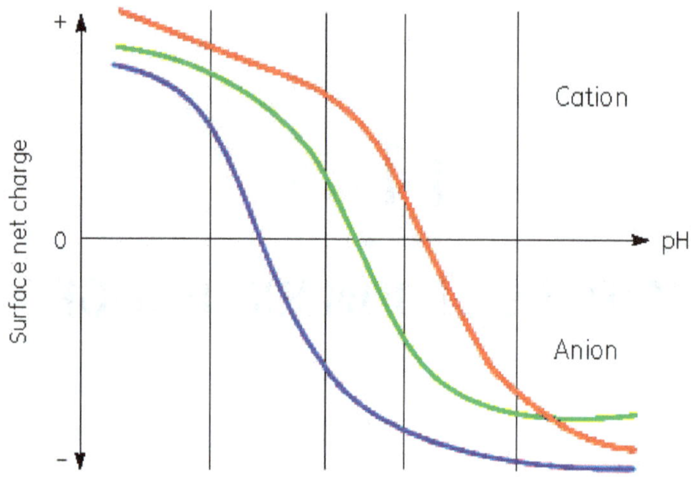

Fig. 1: Theoretical protein titration curves, showing how net surface charge varies with pH.

IEX takes advantage of the fact that the relationship between net surface charge and pH is unique for a specific protein. In an IEX separation, reversible interactions between charged molecules and oppositely charged IEX media are controlled in order to favor binding or elution of specific molecules and achieve separation. A protein that has no net charge at a pH equivalent to its isoelectric point (pI) will not interact with a charged medium. However, at a pH above its pl, a protein will bind to a positively charged medium or anion exchanger and, at a pH below its pI, a protein will bind to a negatively charged medium or cation exchanger. In addition to the ion-exchange interaction, other types of binding can occur, but these effects are very small and mainly due to van der Waals forces and nonpolar interactions.

11.2 STEPS IN AN IEX SEPARATION

An IEX medium comprises a matrix of spherical particles substituted with ionic groups that are negatively or positively charged. The matrix is usually porous to give a high internal surface area. The medium is packed into a column to form a packed bed. The bed is then equilibrated with buffer which fills the pores of the matrix and the space in between the particles.

11.2.1 Equilibration

The first step is the equilibration of the stationary phase to the desired start conditions. When equilibrium is reached, all stationary phase charged groups are bound with exchangeable counterions, such as chloride or sodium. The pH and ionic strength of the start buffer are selected to ensure that, when sample is loaded, proteins of interest bind to the medium and as many impurities as possible do not bind.

11.2.2 Sample Application and Wash

The second step is sample application and wash. The goal in this step is to bind the target molecule(s) and wash out all unbound material. The sample buffer should have the same pH and ionic strength as the start buffer in order to bind all charged target proteins. Oppositely charged proteins bind to ionic groups of the IEX medium, becoming concentrated on the column. Uncharged proteins, or those with the same charge as the ionic group, pass through the column at the same speed as the flow of buffer, eluting during or just after sample application, depending on the total volume of sample loaded.

11.3 ELUTION

When all the sample has been loaded and the column washed with start buffer so that all nonbinding proteins have passed through the column, conditions are altered in order to elute the bound proteins. Most frequently, proteins are eluted by increasing the ionic strength (salt concentration) of the buffer or, occasionally, by changing the pH. As ionic strength increases the salt ions (typically Na+ or Cl-) compete with the bound components for charges on the surface of the medium and one or more of the bound species begin to elute and move down the column. The proteins with the lowest net charge at the selected pH will be the first ones eluted from the column as ionic strength increases. Similarly, the proteins with the highest charge at a certain pH will be most strongly retained and will be eluted last. The higher the net charge of the protein, the higher the ionic strength that is needed for elution. By controlling changes in ionic strength using different forms of gradient, proteins are eluted differently in a purified, concentrated form.

11.4 REGENERATION

A final wash with high ionic strength buffer regenerates the column and removes any molecules still bound. This ensures that the full capacity of the stationary phase is available for the next run. The column is then re-equilibrated in start buffer before starting the next run. The above describes a typical IEX separation. Alternatively, conditions can be chosen to maximize the binding of contaminants to allow the target protein(s) to first pass through the column to be collected.

11.5 RESOLUTION

The resolution of an IEX separation is a combination of the degree of separation between the peaks eluted from the column (the selectivity of the medium), the ability of the column to produce narrow, symmetrical peaks (efficiency) and, of course, the amount (mass) of sample applied. These factors are influenced by practical issues such as matrix properties, binding and elution conditions, column packing, and flow rates which are covered in detail in Chapter 2, *Ion exchange in practice*.

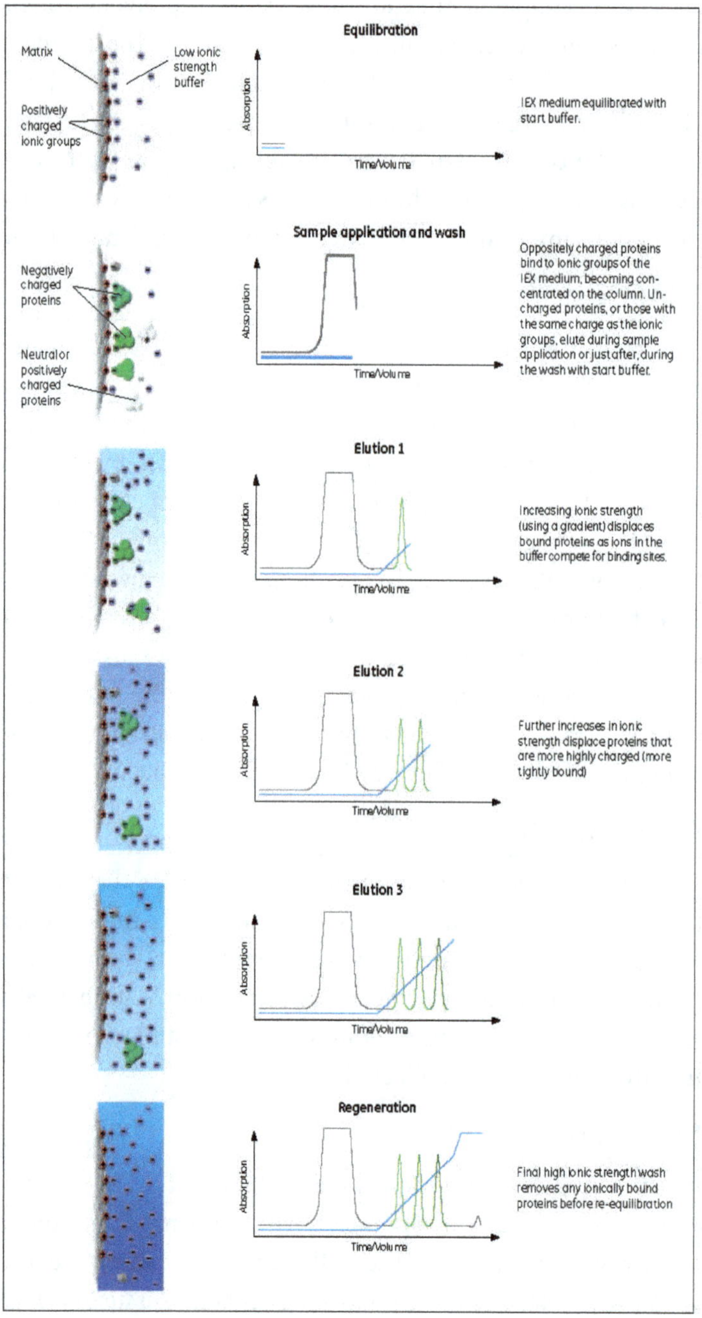

Fig 1.2. Principles of an anion exchange separation.

11000421 AC 13

Fig. 2: Principles of an anion exchange separation.

Resolution (Rs) is defined as the distance between peak maxima compared with the average base width of the two peaks. Rs can be determined from a chromatogram, as shown in Figure 3.

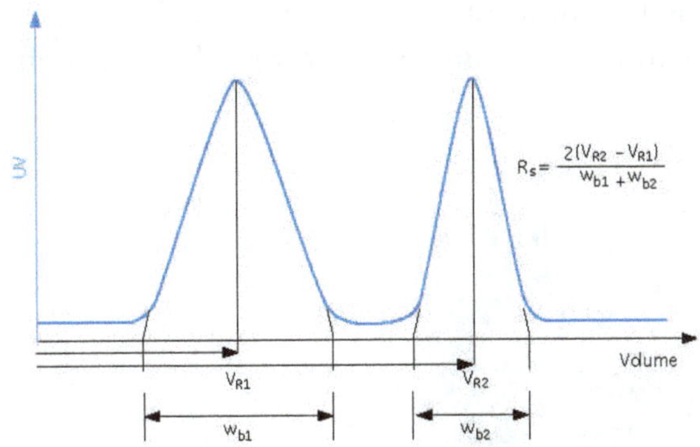

Fig. 3: Determination of the resolution (Rs) between two peaks.

Elution volumes and peak widths are measured with the same units to give a dimensionless resolution value. Rs gives a measure of the relative separation between two peaks and can be used to determine if further optimization of the chromatographic procedure is necessary. If Rs = 1.0 (Fig 1.4) then 98% purity has been achieved at 98% of peak recovery, provided the peaks are symmetrical and approximately equal in size. Baseline resolution requires that Rs ≥ 1.5. At this value, peak purity is 100%.

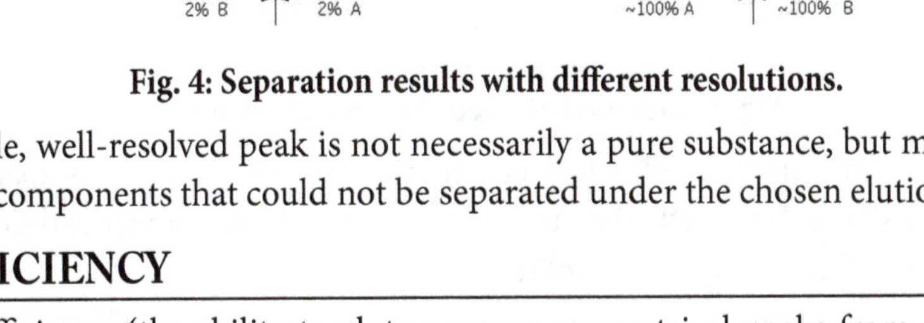

Fig. 4: Separation results with different resolutions.

A single, well-resolved peak is not necessarily a pure substance, but might represent a series of components that could not be separated under the chosen elution conditions.

11.6 EFFICIENCY

Column efficiency (the ability to elute narrow, symmetrical peaks from a packed bed) relates to the *zone broadening* which occurs on the column and is frequently stated in terms of the number of theoretical plates (see Appendix 3 for determination of column efficiency). One of the main causes of zone broadening is longitudinal diffusion of the solute molecules, that is, proteins, peptides, or oligonucleotides. Zone broadening can be minimized if the distances available for diffusion are minimized. In all situations, a well-packed column will contribute significantly to resolution. Columns that are packed unevenly, too tightly, too loosely, or that contain air bubbles will lead to channeling (uneven passage of buffer through the column), zone broadening and hence loss of resolution. Figure 1.5 illustrates the parameters that contribute to good column efficiency. Obviously particle size is a significant factor in resolution and, in general, the smallest particles will produce the narrowest peaks under the correct elution conditions and in a well-packed column.

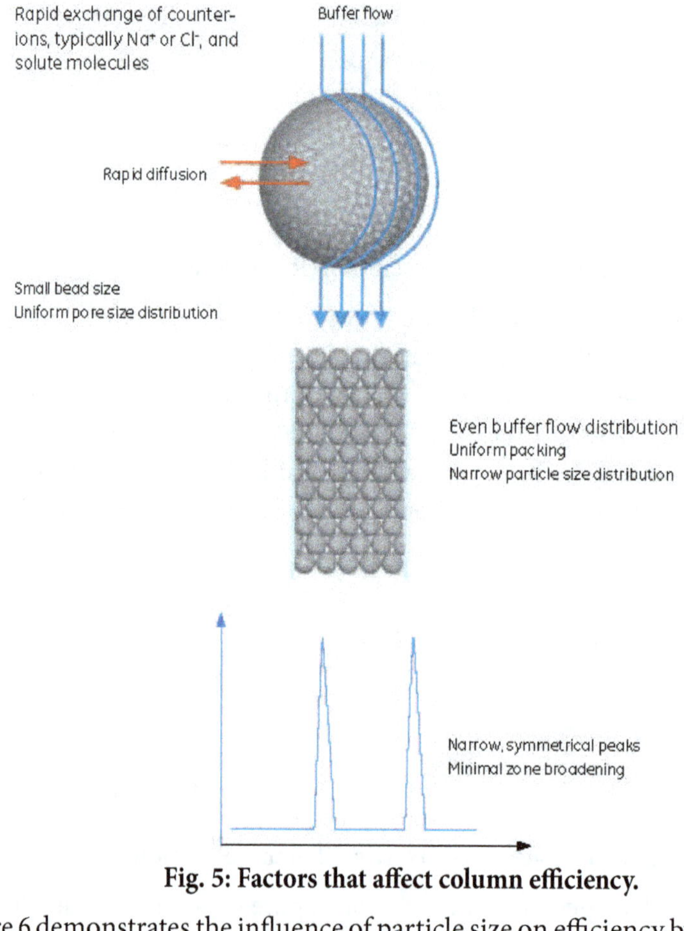

Fig. 5: Factors that affect column efficiency.

Figure 6 demonstrates the influence of particle size on efficiency by comparing several different IEX media under exactly the same running conditions. Note that different media selectivities also influence the final resolution.

Although resolution in terms of efficiency can be improved by decreasing the particle size of the matrix, using a smaller particle size often creates an increase in back pressure so that flow rates need to be decreased, lengthening the run time. Hence the need to match the medium with the requirements for the purification (speed, resolution, recovery, and capacity).

Figure 1.6 demonstrates the influence of particle size on efficiency by comparing several different IEX media under exactly the same running conditions. Note that different media selectivities also influence the final resolution.

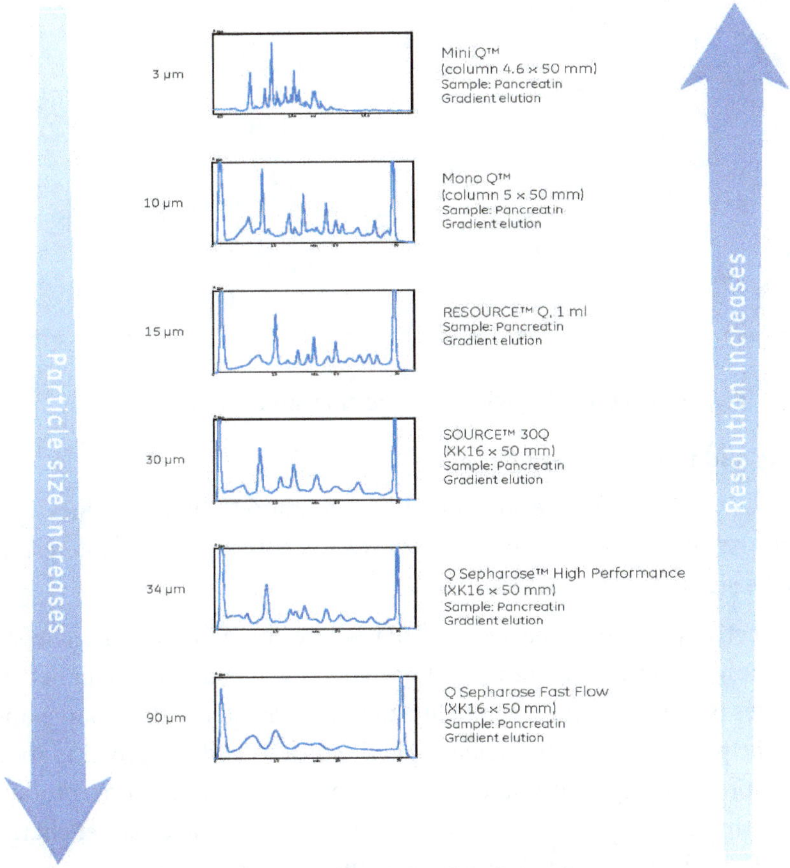

Fig 1.6. Examples of the influence of particle size and selectivity on final resolution.

16 11000421 AC

Fig. 6: Examples of the influence of particle size and selectivity on final resolution.

The viscosity of highly concentrated samples might reduce resolution if large sample volumes are loaded onto columns packed with small particles. Samples may be diluted or, alternatively, a larger particle size should be used.

11.7 SELECTIVITY

Good selectivity (the degree of separation between peaks) is a more important factor than high efficiency in determining resolution (Fig 7) and depends not only on the nature and number of the functional groups on the matrix, but also on the experimental conditions, such as pH (influencing the protein charge), ionic strength, and elution conditions. It is the ease and predictability with which these experimental conditions can be manipulated, when using a suitably designed chromatography medium, that gives IEX the potential of extremely high resolution.

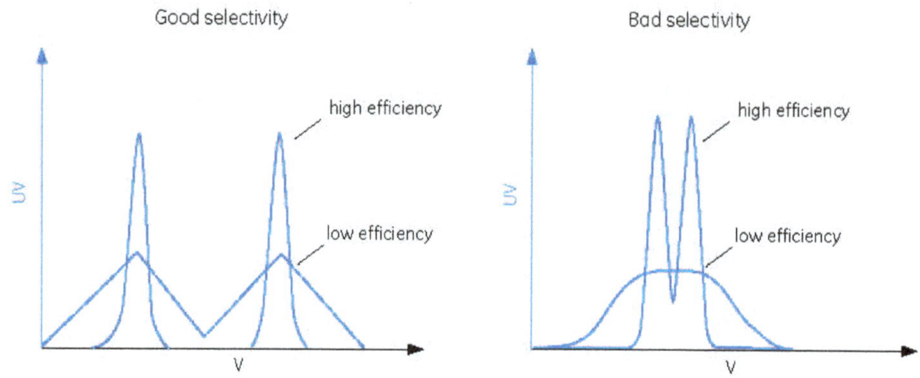

Fig. 7: Effect of selectivity and efficiency on resolution.

11.7.1 Selectivity and pH

Good selectivity is achieved by performing IEX separations at pH values carefully selected to maximize the differences in net charge of the components of interest. Figure 1.8 emphasizes the significance of pH.

Optimum selectivity can be expected at a pH where there is maximum separation between the titration curves for the individual proteins (i.e., the difference in net charges between the species is greatest) and when using an ion exchanger with a charge opposite to the charge of the proteins at the particular pH. The order in which proteins are eluted cannot always be predicted with absolute certainty since a titration curve (produced in practice by measuring electrophoretic mobility in a gel) reflects the total net charge on a protein and IEX depends on the net charge on the surface of the protein.

11.7.2 Selectivity and Elution

The figures below illustrate the most common forms of IEX separation in which proteins are eluted by increasing the ionic strength of a buffer (typically with NaCl) using *linear gradient or step elution.* The UV absorbance and conductivity traces show the elution of protein peaks and the changes in salt concentration, respectively, during elution.

Buffer volumes used during sample application, elution, washing and re-equilibration are expressed in *column volumes (CV)*, for example 5 CV = 5 ml for a column with a 1 ml bed volume. Using CV to describe a separation profile facilitates method development and transfer of methods to columns of different dimensions when scaling-up. *Gradient elution* (Fig 1.9) is often used when starting with an unknown sample (as many components as possible are bound to the column and eluted differentially to see a total protein profile) and for high-resolution separation or analysis.

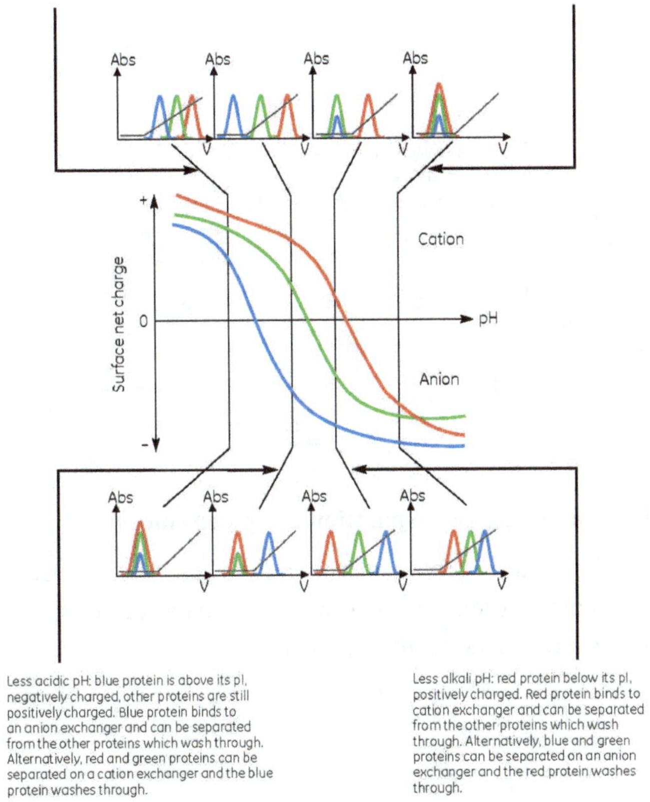

Fig. 8: Effect of pH on protein binding and elution patterns.

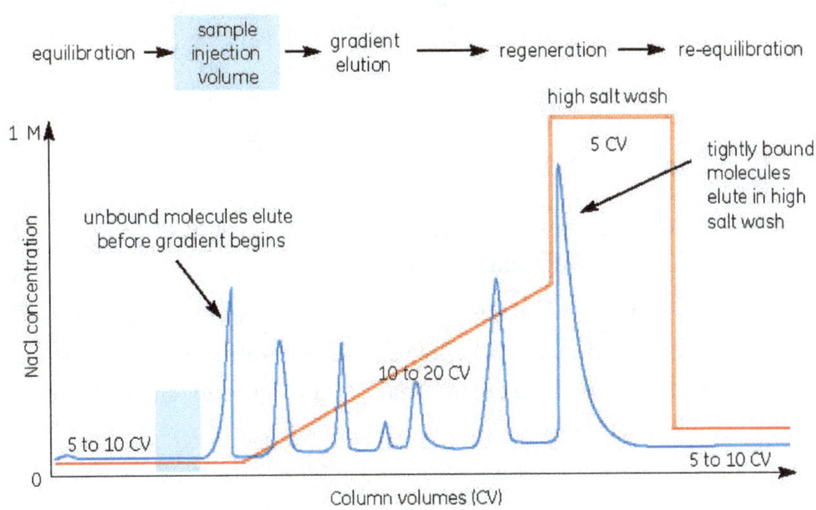

Fig. 9: Typical high-resolution IEX separation using linear gradient elution.

Step elution is used in several ways. When an IEX separation has been optimized using gradient elution, changing to a step elution speeds up separation times and reduces buffer consumptionwhile retaining the required purity level (Fig 10).

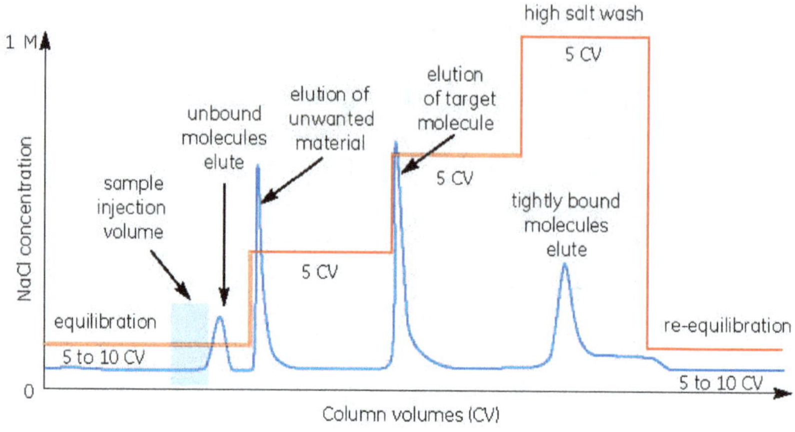

Fig. 10: Typical IEX separation using step elution.

Step elution can also be used for *group separation* in order to concentrate the proteins of interest and rapidly remove them from unwanted substances (Fig 11). The target protein(s) is eluted in an enriched, concentrated form.

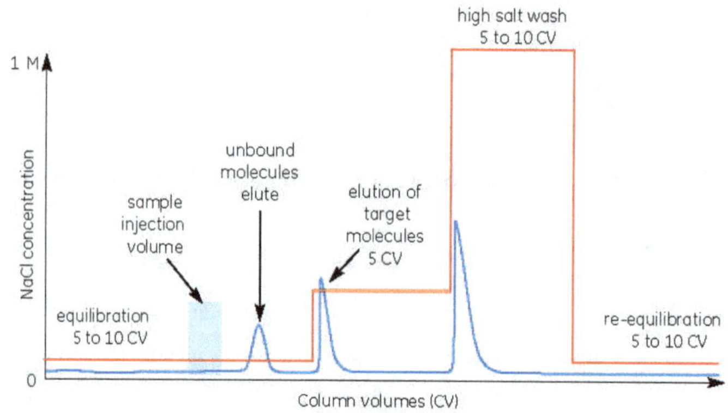

Fig. 11: Typical IEX separation using a step elution to separate groups of proteins with very different charge properties.

Occasionally, step elution is used to *remove contaminants* by choosing conditions that maximize binding of the contaminants and allow the target protein(s) to pass through the column (Fig 12). Care must be taken to ensure that the binding capacity of the column is sufficient to bind all contaminants.

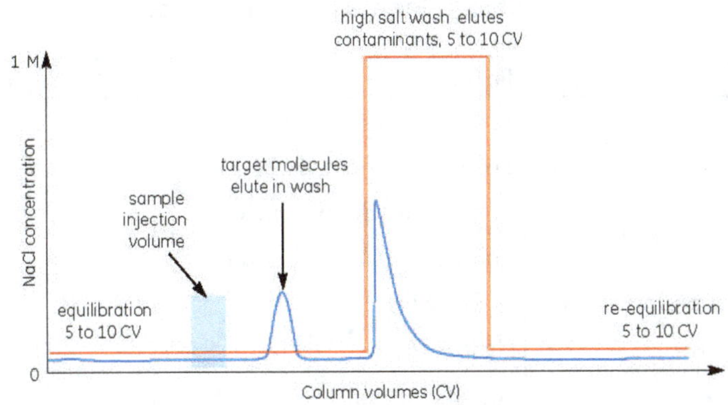

Fig. 12: Contaminant removal: target protein(s) elute in the wash, contaminants bind to the column.

11.8 COMPONENTS OF IEX MEDIA

Chromatography media for ion exchange are made from porous or nonporous matrices, chosen for their physical stability, their chemical resistance to stringent cleaning conditions, and their low level of nonspecific interaction. The matrices are substituted with functional groups that determine the charge of the medium.

11.8.1 Matrix

- High porosity offers a large surface area covered by charged groups and so ensures a high binding capacity. High porosity is also an advantage when separating large biomolecules. Nonporous matrices are preferable for extremely high-resolution separations when diffusion effects must be avoided.
- An inert matrix minimizes nonspecific interactions with sample components.
- High physical stability ensures that the volume of the packed medium remains constant despite extreme changes in salt concentration or pH thus improving reproducibility and avoiding the need to repack columns.
- High physical stability and uniformity of particle size facilitate high flow rates, particularly during cleaning or re-equilibration steps, to improve throughput and productivity.
- High chemical stability ensures that the matrix can be cleaned using stringent cleaning solutions if required.
- Modern IEX media use either polymer or agarose-based matrices to fulfil not only the requirements for high binding capacity, chemical and physical stability, but to generate media with suitable particle sizes for a range of applications (Table 1).

Table.1. Ion Exchange matrices

	Form	Mean particle size (μm)
MiniBeads™	Polystyrene/divinyl benzene	3
MonoBeads™	Polystyrene/divinyl benzene	10
Source 15	Polystyrene/diviny benzene	15
Source 30	Polystyrene/diviny benzene	30
Sepharose High performance	Agarose 6%	34
Sepharose fast flow	Agarose 6%	90
Spharose XL	Agarose 4%	90
	Agarose 6%, dextran chains coupled to agarose	90
Sepharose Big Beads	Agarose 6%	200
Capto™ Impres	High-flow agarose	40
Capto Impact	High-flow agarose	50
Capto	High-flow agarose	90

Mini Beads is a matrix made from polystyrene, with divinyl benzene as cross-linker, to produce highly spherical (monodispersed), very small (3 μm), nonporous particles that facilitate micropreparative or analytical separations when *extremely high resolution* is more important than high binding capacity or high flow rates.

Mono-beads and SOURCE are matrices made from polystyrene with divinyl benzene to produce highly spherical (monodispersed), small (10, 15, or 30 μm), porous particles (Fig 13) that facilitate *high resolution separations at high flow rates.*

Fig. 13: Electron micrograph of MonoBeads showing spherical, monodispersed particles

Sepharose media are based on chains of agarose, arranged in bundles and with different degrees of cross-linking (Fig 14), to give a range of rigid, macroporous matrices with good capacity and low nonspecific adsorption. The most suitable matrix can be selected according to the degree of resolution, binding capacity, and flow rates desired for the separation. For example, gradient elution on Sepharose High Performance (34 μm) will give a high-resolution separation whereas the larger particles of Sepharose Fast Flow (90 μm) or Sepharose Big Beads (200 μm) would be most suited to high-capacity step elution at high flow rate.

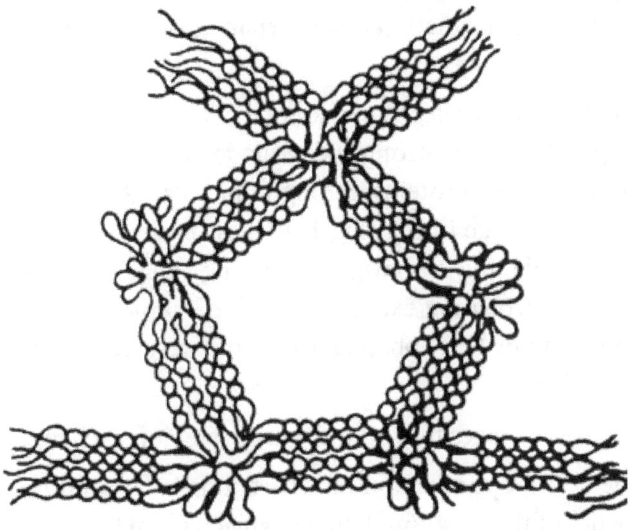

Fig. 14: Structure of cross-linked agarose media (Sepharose).

Capto media are based on a chemically modified, high-flow agarose matrix. This matrix provides particle rigidity without compromising pore size, outstanding pressure/flow properties, and high chemical stability to support CIP procedures. Capto media are suitable for scaling up and for use in large-scale bioprocess purifications. The basic characteristics of Capto (90 μm), Capto ImpAct (50 μm), and Capto ImpRes (40 μm) IEX media are summarized in Chapter 3.

11.8.2 Functional Groups

The functional groups substituted onto a chromatographic matrix (Table 2) determine the charge of an IEX medium, that is, a positively charged anion exchanger or a negatively charged cation exchanger.

Table 2. Functional groups used on ion exchangers

Anion exchangers		Functional group
Quaternary ammonium (Q)	Strong	$-CH_2-N^+-(Ch_3)_3$
Diethylaminoethy (DEAE)*	Weak	$-CH_2-CH_2-N^+-(CH_2-Ch_3)_2$
Diethylaminopropy (ANX)*	Weak	$-CH_2CHOH-CH_2-N^+-(CH_2-Ch_3)_2$

Cation exhangers		Functional group
Sulfopropy (SP)	Strong	$-CH_2-CH_2-CH_2-SO_3$
Methyl sulfonate (S)	Strong	$-CH_2-SO_3$
Carboxymethyl(CM)	Weak	$-CH_2-COO$

* The active end of the charged group is the same for DEAE and ANX. The difference between them is in the length of the carbon chain of the charged group. DEAE has a diethylaminoethyl group bound to the agarose. ANX has a diethylaminopropyl group attached which prevents the formation of quaternary groups, giving a different selectivity compared to DEAE.

The terms strong and weak refer to the extent that the ionization state of the functional groups varies with pH. The terms strong and weak do not refer to the strength with which the functional groups bind to proteins. Strong ion exchangers show no variation in ion exchange capacity with change in pH (Fig 15). These exchangers do not take up or lose protons with changing pH and so have no buffering capacity, remaining fully charged over a broad pH range. Strong ion exchangers include Q (anionic), S, and SP (cationic). Approximately 5 ml of Q or SP Sepharose Fast Flow were equilibrated in 1 M KCl and titrated with 100 mM NaOH.

There are several advantages to working with strong ion exchangers:

- Development and optimization of separations is fast and easy since the charge characteristics of the medium do not change with pH.
- The mechanism of interaction is simple since there are no intermediate forms of charge interaction.
- Sample loading (binding) capacity is maintained at high or low pH since there is no loss of charge from the ion exchanger

The majority of proteins have pI within the range 5.5 to 7.5 and can be separated on either strong or weak ion exchangers. An advantage of a weak ion exchanger, such as DEAE (anionic), ANX (anionic), and CM (cationic) is that they can offer a different selectivity compared to strong ion exchangers. A disadvantage is that because weak ion exchangers can take up or lose protons with changing pH, their ion exchange capacity varies with pH (Fig 16).

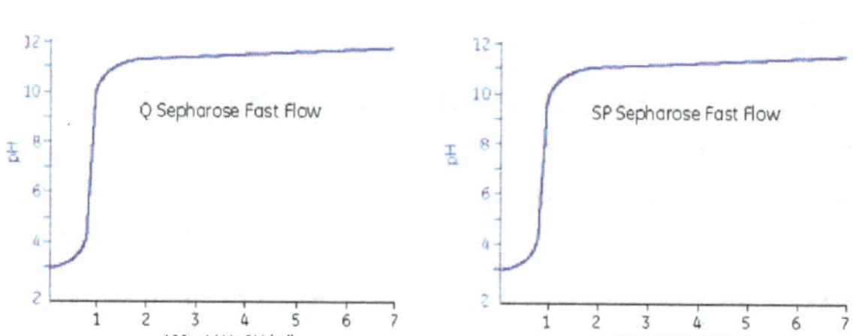

Fig. 16: Titration curves show how the ion exchange capacity of weak ion exchangers varies with pH.

Try a weak ion exchanger such as DEAE, CM, or ANX Sepharose Fast Flow, if a strong ion exchanger (substituted with Q, S, or SP) does not give the required selectivity.

11.9 BINDING CAPACITY AND RECOVERY

The capacity of an IEX medium is a quantitative measure of its ability to take up counterions (proteins or other charged molecules). The *total ionic capacity* is the number of charged functional groups/ml medium, a fixed parameter of each medium. Of more practical relevance is the actual amount of protein that can bind to an IEX medium, under defined experimental conditions. The capacity of a chromatography medium can be measured in two ways: static binding capacity (SBC) and dynamic binding capacity (DBC). SBC is the maximum amount of protein that can bind at given conditions. SBC is often obtained during excess load of sample.

DBC is measured during given conditions, including flow rate, and is the amount of protein that binds before a significant breakthrough of the target protein appears. Figures for binding capacity in this handbook refer to the dynamic binding capacity.

The static and dynamic binding capacities depend upon the properties of the protein, the IEX medium, and the experimental conditions. The capacity of an IEX medium will vary according to the molecular size of the specific protein (which affects its ability to enter all the pores of the matrix) and its charge/pH relationship (the protein must carry the correct net charge at a sufficient surface density at the chosen pH). With earlier IEX media, larger biomolecules had limited access to the functional groups, significantly reducing the binding capacity.

Nowadays, ion exchange matrices such as MonoBeads, Capto, SOURCE, and Sepharose media all have exclusion limits for globular proteins in excess of 1 × 106 and are therefore suitable for the majority of biomolecule separations. Binding capacities will still vary according to the molecular size of the biomolecules. For example, a matrix with a high degree of small pores will exhibit a higher binding capacity for smaller molecules. Experimental conditions such as pH, ionic strength, counterion, flow rate, and temperature should all be considered when comparing binding capacities of different IEX media. Modern IEX media show very low levels of nonspecific adsorption so that sample recovery under suitable separation conditions is very high, typically between 90% and 100%.

11.10 CHROMATOFOCUSING

Chromatofocusing (CF) is a purification method separating proteins on the basis of differences in their isoelectric points (pI). The matrix is usually a weak anion exchanger in which the functional groups are amines. The eluent is a buffer containing a large number of buffering substances which together give a uniform buffering capacity over a broad pH range. A pH gradient is generated on the column as buffer and medium interacts. Proteins with different pI values migrate at different rates as the pH gradient develops, continually binding and dissociating while being focused into narrow bands and finally eluted.

CF is a powerful method and can resolve very small differences in pI (down to 0.02 pH units) and thus separates very similar proteins. However, the capacity of the method is low and should preferably only be used for partially pure samples. CF can be considered if IEX or other methods do not give satisfactory purification.

12

APPLICATION OF THIN LAYER CHROMATOGRAPHY FOR QUALITATIVE ANALYSIS OF GUNPOWDER

12.1 INTRODUCTION

Thin layer chromatography can be used to monitor the progress of a reaction, identify compounds present in a given mixture, and determine the purity of a substance. A number of enhancements can be made to the original method to automate the different steps, to increase the resolution achieved with TLC and to allow more accurate quantitative analysis. This method is referred to as "high-performance TLC". Our goal is to examine and to compare the quality of the results obtained by thin-layer chromatography, high-performance thin layer chromatography and vacuum stability test in the analysis of samples of gunpowders. Thin layer chromatography as a qualitative method has shown good quality of results that can be used for a good life prediction of the ammunition.

One of the major problems concerned with the manufacture and use of propellants is the measurement of their chemical stability and the prediction of their safe storage life.1 At normal storage temperature of 21–30°C, propellants should have a safe shelf-life of 140 years, and a minimum of 35 years. It is a characteristic property of nitrocellulose-based propellants that the nitric ester constituents e.g. nitroglycerin, nitrocellulose, undergo a slow decomposition even at ambient temperatures. A solution of nitrocellulose in methanol (concentration 0.40-0.45%) showed depolymerization after exposure to light to the wave-length 302m/i, whereas no depolymerisation was detectable after exposure to the wave-lengths 334 and 365µm. Solutions of the same nitrocellulose sensitized by adding ca. 1% β-naphthylamine showed depolymerization at all the wave-lengths. The degradation products formed, unless removed, can cause a reduction in chemical stability which can lead to cook-off/self-ignition due to the exothermic nature of the reactions involved. In addition they can lead to a loss of calorific value, changes in ballistic properties and cracking in large diameter charges. Small amounts of stabilizing compounds are included in propellant formulations, either singly or as mixtures, in order to react with the degradation products, thus reducing the probability of the adverse effects. The single-temperature ageing procedure has shown to be a cost-effective method to assure that a propellant is sufficiently stable to be stored for 10 years at ambient storage conditions (25°C) – this at least for propellants consisting of well-known and generally used propellant components. For propellants of new formulations, in particular with stabilizers and other relevant components which were not used before in propellant formulations, more elaborate methods, such as the multi-temperature ageing procedure, have to be used.

12.2 STABILIZERS IN GUNPOWDERS

Until recently, the stabilizers considered a group of aromatic compounds with the ability to react quickly with nitric oxides that occur during the thermal decomposition of nitrocellulose, thereby preventing their autocatalytic effect on further decomposition more recently, this group of stabilizers is treated as "secondary stabilizers" because they reduce the effects of decomposition that is already occurring. The term "primary stabilizers" refers to a group of compounds that prevent or slow down the thermal decomposition of nitrocellulose. Typical stabilizers for propellants are diphenylamine DPA, N-nitrosodiphenylamine NODPA and the urea derivatives ethylcentralite CI and methylcentralite CII. The list of the most common organic stabilizers is shown in Table 1.

Table 1 List of most common organic stabilizers used in production of ammunition

Name	Chemical name	Formula
Akardite I	1,1-Diphenylurea	$C_{13}H_{12}N_2O$
Akardite II	3-methyl-1,1-diphenylurea	$C_{14}H_{13}N_3O_2$
Akardite III	1-Ethyl-1,3-diphenylurea	$C_{15}H_{16}N_2O$
Centralite I	1,3-Diethyl-1,3-diphenylurea	$C_{17}H_{20}N_2O$
Centralite II	1,3-Dimethyl-1,3-diphenylurea	$C_{15}H_{16}N_2O$
DPA	Diphenylamine	C12H11N or C_6H5NHC_6H_5

12.2.1 Nitrocellulose and Mechanisms of its Degradation

The gunpowders for classical ammunition, whether single-phase, double or triple, as well as a large number of rocket fuels as the basic component of the composition contain nitrocellulose. Nitrocellulose has a similar aspect to cotton, is white and has a fibrous texture. It is produced from cellulose. Nitrocellulose (R-ONO_2) has good energy and technological properties that enable the production of various types of gunpowders of different ballistic properties, temperature and burning speed, etc. The appearance of the nitrocellulose molecule is shown in Figure 1. The undesirable characteristic of nitrocellulose is chemical degradation with the release of gaseous products. One of simplified mechanisms of nitrocellulose degradation is shown bellow.

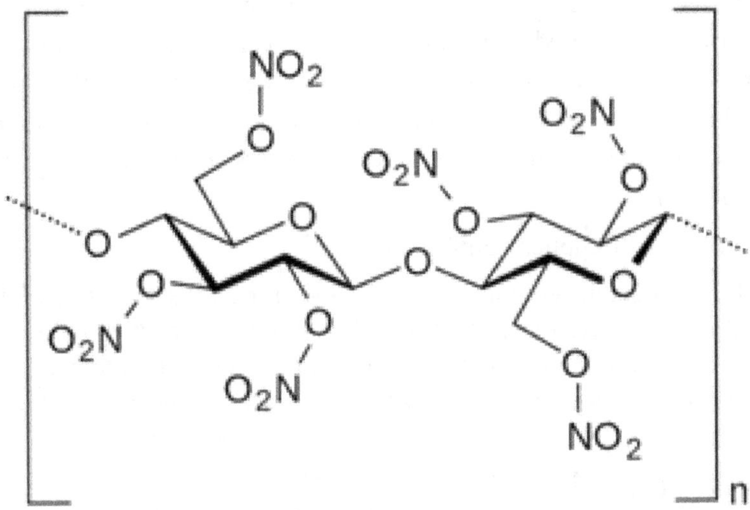

Fig. 1: Nitrocellulose, R-ONO$_2$.

12.2.2 Hydrolysis

Long-term storage of nitrate ester compounds, they react with moisture inside or moisture around the packaging, resulting in nitric acid (HNO_3) and nitrogen oxides such as NO_2 and NO_3. These reactions of nitric oxide and nitric acid are exothermic reactions that increase the ambient temperature during decomposition:

$RONO_2 + H_2O \rightarrow ROH + H^+ + NO^{-3}$

$HNO_3 \Leftrightarrow H^+ + NO^{-3}$

$R'ONO_2 + H+ \longrightarrow R'OH + NO^{+2}$

$NO^{+2} + NO^{-3} \rightarrow N_2O_5$

$N_2O_5 + H_2O \rightarrow 2HNO_3 \Leftrightarrow 2H^+ + 2NO^{-3}$

12.2.3 Definitions (AOP-48)

Effective stabilizer is a defined term which is used to assess the stability in the single temperature test. Daughter stabilizers are substances with stabilizing capabilities that were not included in the propellant formulation but were produced from the initial stabilizers during propellant manufacture or ageing. Most stabilizer depletion products fall into this class. Certain stabilizers can appear both as initial stabilizer (incorporated in basic formulation) and as daughter stabilizer (produced from initial stabilizer), this even in the same propellant. Typical examples for such stabilizers are 2NDPA, 4NDPA, pNMA and pNEA – they can be used as initial stabilizers but are also formed from DPA, AKA-II, MC or EC. INITIAL STABILIZERS ("parent stabilizers") are stabilizers that were incorporated in the propellant formulation during manufacture. The content of "effective stabilizer" is calculated from the contents of all initial stabilizers (diphenylamine, 2-nitro-diphenylamine, ethyl centralite, methyl centralite, akardite-II, p-nitro-N-methylaniline, resorcinol-except if they are used as surface moderants) and the content of N-nitrosodiphenylamineas follows. For propellants without diphenylamine as well as for propellants with diphenylamine and other stabilizers. Effective stabilizer=Σ (contents of initial stabilizers).

For propellants with diphenylamine only:

a. Effective stabilizer=content of diphenylamine + 0.85 N-nitrosodiphenylamine. Percentage effective stabilizer is the amount of effective stabilizer found, expressed as a percentage by weight of the propellant sample. Initial level is the percentage of effective stabilizer found in the propellant sample prior to ageing.

12.3 THIN LAYER CHROMATOGRAPHY

Thin layer chromatography (TLC) and high performance thin layer chromatography (HPTLC) – now also called planar chromatography – are, like all chromatographic techniques, based on a multistage distribution process. This process involves: a suitable adsorbent (the stationary phase), solvents or solvent mixtures (the mobile phase or eluent), and the sample molecules. For thin layer chromatography the adsorbent is coated as a thin layer onto a suitable support (e.g. glass plate, polyester or aluminium sheet). On this layer the substance mixture is separated by elution with a suitable solvent. The principle of TLC is known for more than 100 years now). TLC is a chromatography technique used to separate non-volatile mixtures. Thin-layer chromatography can be used to monitor the progress of a reaction, identify compounds present in a given mixture, and determine the purity of a substance. Specific examples of these applications include: analyzing ceramides and fatty acids, detection of pesticides or insecticides in food and water, analyzing the dye composition of fibers in forensics, assaying the radiochemical purity of radiopharmaceuticals, or identification of medicinal plants and their constituents. Thin-layer chromatography is performed on a sheet of glass, plastic, or aluminium foil, which is coated with a thin layer of adsorbent material, usually silica gel, aluminium oxide (alumina), or cellulose. This layer of adsorbent is known as the stationary phase.

After the sample has been applied on the plate, a solvent or solvent mixture (known as the mobile phase) is drawn up the plate via capillary action. Because different analytes ascend the TLC plate at different rates, separation is achieved. The mobile phase has different properties from the stationary phase. For example, with silica gel, a very polar substance, non-polar mobile phases such as heptane are used. The mobile phase may be a mixture, allowing chemists to fine-tune the bulk properties of the mobile phase. After the experiment, the spots are visualized. Often this can be done simply by projecting ultraviolet light onto the sheet; the sheets are treated with a phosphor, and dark spots appear on the sheet where compounds absorb the light impinging on a certain area. The plate is shown in (Figure 2-5) Chemical processes can also be used to visualize spots; anisaldehyde, for example, forms colored adducts with many compounds, and sulfuric acid will char most organic compounds, leaving a dark spot on the sheet. To quantify the results, the distance traveled by the substance being considered is divided by the total distance traveled by the mobile phase. (The mobile phase must not be allowed to reach the end of the stationary phase.) This ratio is called the retardation factor (Rf). If the solvent front is 6 cm then the R_f value for the pigment at 3cm would simply be 0.5 showed in Figure 2,Figure 3.

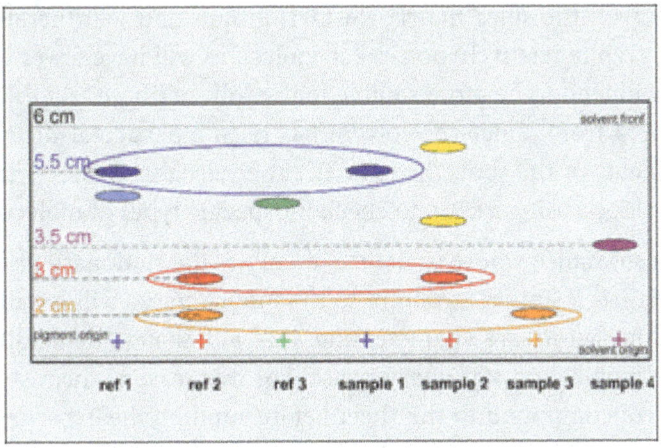

Fig. 2: TLC plate with developed samples.

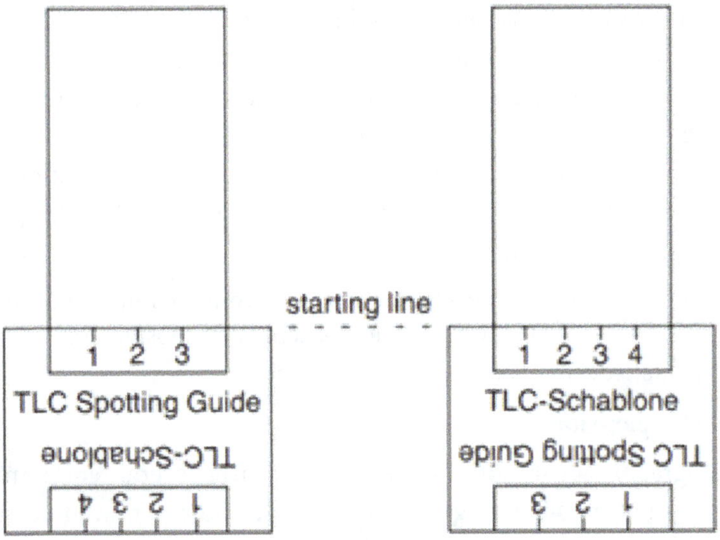

Fig. 3: Application of a sample on a micro precoated sheet with the aid of a capillary and a TLC spotting guide.

R_f=distance travelled by the component/distance travelled by the solvent. In the thin layer chromatography analysis, a glass plate is coated by adsorbent materials such as silica. Silica is the most commonly used adsorbent material for TLC analysis. Structurally silica gel consists of a matrix of Si-OH groups which can interact with molecules via hydrogen bonding and adsorption. A few micro liters of a dilute solution is put onto the silica surface of the plate using a micro capillary. This is known as "spotting". The plate is then placed in a jar containing a solvent (generally mixtures of ethylacetate and hexanes). As time goes, the solvent gradually rises up the plate due to capillary action carrying the components of the sample with it. Different molecules are carried up the plate to different distances due to variable interactions with the adsorbent material. For example, when silica is used, polar molecules with groups such as hydroxy (OH) or amine (NH_2) will tend to form hydrogen bonds with the silica matrix (Si-OH) groups and will therefore not move as fast up the plate. While relatively non-polar molecules will have fewer interactions with the matrix and will tend to be more soluble in the solvent phase and therefore rise faster up with the solvent front. Once the solvent has risen a particular distance sufficient to separate components of the spot, the plate is either visualized directly using ultraviolet light, or it is developed using a stain to check for specific types of molecules.

In general, a substance whose structure resembles the stationary phase will have low R_f, while one that has a similar structure to the mobile phase will have high retardation factor. Retardation factors are characteristic, but will change depending on the exact condition of the mobile and stationary phase. For this reason, chemists usually apply a sample of a known compound to the sheet before running the experiment. The success of thin layer chromatography as a highly efficient micro analytical separation method is based on a large number of advantageous properties: high sample throughput in a short time suitable for screening tests pilot procedure for HPLC after separation the analytical information can be stored for a longer period of time (the TLC ready-to-use layer acts as storage medium for data) separated substances can be subjected to subsequent analytical procedures (e.g. IR, MS) at a later date rapid and cost-efficient optimisation of the separation due to easy change of mobile and stationary phase. For a better understanding of a thin layer chromatographic separation we describe here the basic steps:

a. Sample preparation.
b. Sample application development of a chromatogram, separation techniques evaluation in TLC – visualisation of separated substances.
c. Qualitative and quantitative determinations.

12.3.1 Sample Preparation

For a chromatographic separation the sample must meet several requirements to obtain good results. It is not possible do go into detail here. However, eventually several steps for sample pretreatment may be necessary. These include sampling, mechanical crushing of a sample, extraction steps, filtration and sometimes enrichment of interesting components or clean-up, i.e. removal of undesired impurities. In our experiment, we prepared the samples of gunpowders into pieces of diameter up to 2mm and extracted into dichloroethane.

12.3.2 Sample application

The aim of a chromatographic separation determines how the sample should be applied to the TLC plate or sheet. The most frequent technique still is application with a glass capillary as spot or short streak. Application as streak will yield better results especially for instrumental quantification. For both types of application some manual skill is required to obtain reproducible results. Substance zones which are too large from the beginning will cause poor separations since during chromatography they will become even larger and more diffuse.

12.3.3 Our TLC Sets for Example Use this Very Simple Way of Application.

The mixture to be separated and the reference solution are applied to the micro pre coated sheets as spots by means of glass or plastic capillaries. Only use each capillary once to avoid contamination of the following samples. The capillaries fill themselves quickly when dipped into organic sample solutions, with aqueous solutions filling will be much slower. Before emptying the capillary roll the submerged end horizontally on filter paper. If the capillary is filled to the upper end, however, you may apply the "clean" upper end to the adsorbent layer. Place the capillary on the layer vertically and carefully, vertically so that the capillary empties itself and carefully to avoid damage to the layer. Damaged layers result in unevenly formed spots. To keep spots as small and compact as possible, it is advisable to apply a solution in several portions with intermediate drying (blow with cold or hot air). This is especially important for aqueous sample solutions. The following figures demonstrate the clean and easy application of samples with the above-mentioned spotting guide. It is recommended to apply 0.1-1% solution in the amount of 1-25µl. The sample zone on the starting line should be 1-3mm in diameter, 1-2cm apart from the edge of the plate. For analytical work, the sample is applied in the amount of 1µg to 1ng. A valuable aid for manual application especially of large volumes of very dilute samples is the concentrating zone (e.g. SIL G/UV254), which consists of a chromatographically inactive adsorbent. After application allow the solvent of the samples to evaporate completely (about 10minutes) or blow with cold or hot air. Development of a chromatogram should never start before the solvent of the applied samples is evaporated completely.

12.3.4 Developing a Chromatogram Separation Techniques

The most frequently used separation technique is ascending TLC in the customary trough chamber (standard method, linear development). Usually it is applied as single development. However, multiple development, with or without change of eluent (step technique) can improve separation results. For 2-dimensional development only 1 spot of the sample is applied in one edge of a plate. After chromatography in the first direction the plate is dried, turned by 90° and developed in the 2nd dimension with another eluent. Thus complicated mixtures give 2-dimensional chromatograms taking advantage of the different separating properties of two eluents. A generally applicable standardised optimisation method is described by H Keuker et al., It is important to pay attention to the atmosphere in the developing chamber. If reproducible migration distances are required, saturation of the chamber atmosphere with eluent vapour is necessary. For this purpose the developing chamber is lined with well absorbing chromatography paper (e.g. MN 260) and charged with a correspondingly larger volume of eluent. Figure 4 shows:

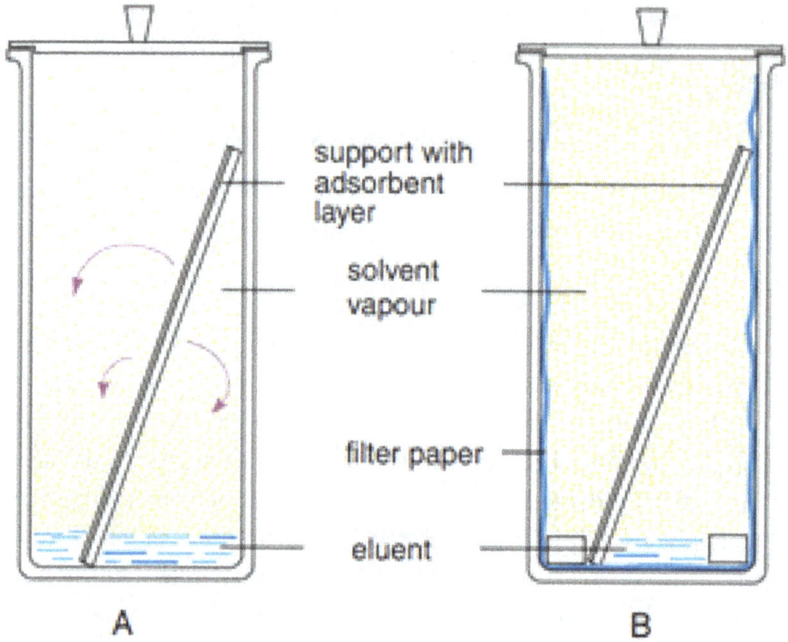

Fig. 4: Developing chambers and saturation.[14]

a. Chamber with normal saturation. The arrows stand for the evaporation of the eluent from the layer and the dots symbolise the vapour density.
b. Chamber lined with filter paper, saturated with eluent vapour.
c. A better control of chromatographic conditions, especially of the vapour phase, is achieved in flat chambers (horizontal chambers) or a trough with sandwich configuration. In these techniques the vapour phase is excluded by a covering plate.

12.3.5 Qualitative Detection

Qualitative evaluation is generally made directly on the TLC plate via the characteristic Rf values of substances, i.e. the ratio of distance start – substance zone to distance start – solvent front and specific chemical reactions.

We made a qualitative determination in our experiments according to the Figure 5 & Table 2-5 & Table 6.

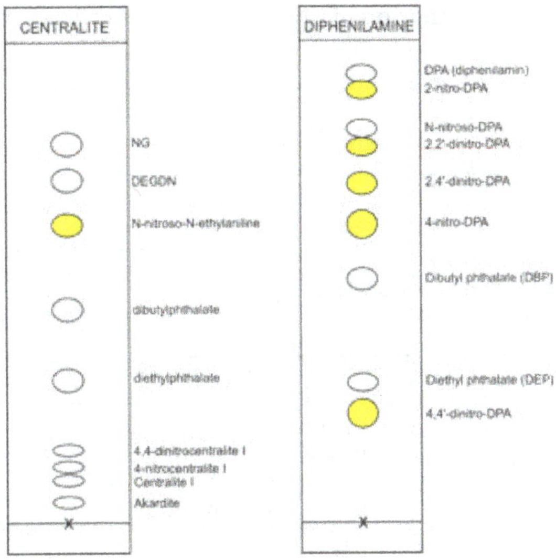

Fig. 5: Chromatogram of standard mixtures, CENTRALIT (left) and DIPHENILAMIN (right) in the appropriate solutions.

Table 2 An example of coloring that can occur when determining the type of stabilizer

Diphenylamine	Centralite	DPA i CEN	Akardit
Blue	Red	Purple	Brown

Table 3 Appearance of color as evidence of nitroglycerin

+	-	-	-

Table 4 Definitions of DPA stabilizer image

A	Diphenylamine spot is much stronger than nitrodiphenylamine; barely recognizable 4-nitro-diphenylamine and 2-nitro-diphenylamine.
B	Diphenylamine spot and nitrodiphenylamine are clearly dispersed, 4-nitr-diphenylamine and 2-nitro-diphenylamine clearly recognizable.
C	Diphenylamine spot is less intense than nitrodiphenylamine, 4-nitro-diphenylamine and 2-nitro-diphenylamine clearly recognizable.
D	Diphenylamine spot has virtually disappeared; Nitrodiphenylamine clearly recognizable, 4-nitro-diphenylamine and 2 nitro-diphenylamine are clearly recognizable, no dinitroproducts are not recognizable.
E	Nitrodifenylamine recognizable, dinitroproducts are recognizable.
F	Nitrodifenylamines are not recognizable, dinitroproducts are recognizable.

Table 5 Estimation of diphenylamine stabilizer state

	A	B	C	D	E	F
Initial state						
After aging						

CATEGORY	Description	HPTLC	VST	Safe storage time at 25°C
1	Gunpowder and rocket fuel good for storing	At least 50% of initial stabilizer concentration	0-1 ml/g	4 years
2	Gunpowder and rocket fuel good for storing, priority in the use of ammunition labored with this gunpowder.	50% of initial stabilizer concentration - 0.2% of stabilizer concentration	1-2 ml/g	2 years
3	Gunpowder and rocket fuel are unstable, ammunition must be removed from the storage place, processed or destroyed.	Less than 0.2% of stabilizer concentration	> 2ml/g	Unsafe

12.3.6 Visualisation of Separated Substances

First of all it is necessary to recognise the position of a substance spot. Only in very few cases the sample is a dye which can be seen with the naked eye. Much more often for unspecific visualisation substances can be viewed under UV light, since many substances show a UV absorption. If a fluorescent indicator is added to the layer, all substances absorbing in the respective region of wave length cause a quenching of the fluorescence, i.e. they appear as dark spots on the fluorescent layer. Customary fluorescent indicators are excited at 254nm or (less frequently) at 366nm with a mercury lamp showed in Table 7. The identification of separated substances is possible via the Rf value compared to the pure compound, which is often applied simultaneously on the same plate. For a number of compounds their native fluorescence can be used for visualisation, which is excited by UV light (mostly long-wave UV) (e.g. aflatoxins). This allows not only determination of the Rf value, but often enables a further qualitative assignment. Such compounds often occur in the extract of gunpowders. If these methods do not allow localisation or characterisation of substances, especially post-chromatographic detection methods can be applied chemical reactions on the TLC plate. Quite unspecific reactions are iodine adsorption and the charring technique (spraying with sulphuric acid and heat treatment). More reliable results are possible with specific reagents for spraying or dipping, which form colored or fluorescent compounds with the substances to be detected.

Table 7 Fluorescent indicators

Composition	Absorption maximum [nm]	Colour of fluorescence
Manganese-activated zinc silicate	254	Green
Inorganic fluorescent pigment	366	Blue

12.3.7 Qualitative Assessment Of Tlc Plates with Gunpowder Samples

Because of the potential dangers of spontaneous combustion of NC propellants in Warehouse ordnance, which could be with disastrous consequences, a series of procedures was developed in order to control their chemical stability.16 The assessment process is based on comparing the developed sample with the standard.

Experiment

Used equipment

i. Block for sample heating.
ii. Thermometer -10 to 1500C.
iii. TLC plates.
iv. Plan in the form of a plate of plastic.
v. UV lamp 254mm.
vi. Pipes with threaded plugs 100 ml (3pcs).
vii. Set a blade.
viii. Hubkolben pipettes 1000µ
ix. 1000 µl spikes (100pcs).
x. Eppendorf cap 2ml (100pcs).
xi. Standard bottle with thread 4 ml (100pieces).
xii. Plastic bottles with a dropper.
xiii. Safety glasses.
xiv. Laboratory gloves.

12.3.8 Aging in Metal Block Thermostats

The kinetic behaviours at thermal accelerated ageing temperature is one of the most important aspects of propellants. The thermal decomposition of NC is a complex process which is believed to be catalyzed by the decomposition products, especially by reactive nitrogen oxides. The aging of the gunpowders in the laboratory involves the artificial warming of the sample of gunpowders in a thermocouple at a constant temperature. For aging, glass vials (different sizes) are used. Two samples of the same barrel are used in the experiment. To determine the safe storage stability at 25°C, the following experimental temperatures are used:

Test temperature 80°C:

i. Four days for category 1 classification.
ii. Two days for category 2 classification.

The temperature selection is based on the Table 8 & Table 9. In our experiment; we selected a test temperature of 80 °C, and a test period of 20 and 48 hours. This decision is based on the Table 8.

Table 8 Data of the required testing periods in order to determine the safe minimum storage stability at 25°C (a)

Storage time at 25°C [years]	Equivalent test duration at 70°C [hours]	Equivalent test duration at 70°C [days]	Equivalent test duration at 80°C [hours]	Equivalent test duration at 80°C [days]
	70	70	80	80
0.5	31	1.3	10.4	0.43
1	62.5	2.6	20.8	0.87
2	125	5.2	41.7	1.74
4	250	10.4	83	3.5
6	375	15.6	125	5.2
8	500	20.8	167	6.9
10	625	26	208	8.7

12.3.9 Determine the Type of Stabilizer

When assessing the type of stabilizer, take into account the precise identification of the stabilizer, whether it is centralit (ethyl centralit, EC or C1) or diphenylamine (DFA or DPA). If they are together C1 and DPA, then the main stabilizer is DPA. The purple color indicates the presence of centralite and diphenylamine (DPA is considered or estimated as a more active stabilizer) showed in Table 9. The experiment is performed on a small quantity of gunpowder (or it extract in dichloroethane). Two drops of the appropriate reagent are added to the gunpowder sample, whereupon the color appears according to the table above.

Table 9 Data of the required testing periods in order to determine the safe minimum storage stability at 25°C (b)

Test duration [days]	Equivalent storage life at 25°C for test temperature 80°C [years]	Equivalent storage life at 25°C for test temperature 70°C [years]
1	1.1	0.35
2	2.3	0.75
3	3.4	1.1
4	4.6	1.5
5	5.7	1.9
6	6.9	2.3
7	8	2.6
8	9.2	3
9	10.3	3.4

Test duration [days]	Equivalent storage life at 25°C for test temperature 80°C [years]	Equivalent storage life at 25°C for test temperature 70°C [years]
10	11.5	3.8
11	12.6	4.2
12	13.8	4.6

Extraction of the stabilizer and determination of the energy base

To determine the content of the stabilizer by the thin layer chromatography method, the sample is a gunpowder extract. In the experiment we extracted gunpowder samples in dichloroethane. The gunpowder can be mono-basic, double-basic or multi-basic. I A small amount of extract (the extract described in step two) succeeds in the trough (a few drops ≈ 50μl).

c. Add one drop of acetic acid solution.
d. Add about 1 mg Zn.
e. Add three drops of reagent which contain sulfanilic acid and 1-naphthylamine.

If as a result we get is intense purple dyeing, then it's a double-basic gunpowder. Nitroglycerin or some other nitrate ester, as the second energy base in the powder - mild violet color is not a proof of nitroglycerin. Double-base rocket propellants are nitrocellulose (NC)-based energetic materials with a tendency towards slow, but constant thermal decomposition during aging. The visualization is shown in the Table 3.

12.3.10 Development of Samples on TLC Plate

The development of samples on the TLC panel is shown in the Sample application paragraph. First, add the appropriate solution in a glass (chamber) to cover the bottom in a thickness of 2mm of liquid. The solution we pour into a glass bottle is selected as follows:

1. Stabilizer diphenylamine: solution "DPA"
2. Stabilizer centralit: solution "CEN"
3. Stabilizer mixture: diphenylamine and centralite: solution "DFA"
4. Akardite stabilizer: analysis with bottle solutions ("CEN" and "DPA") is required, because the result of the accreditation is the production of nitro-DPA and diNitro-DPA, nitro-akardite and akardite, which are determined by the "CEN" solution.

12.3.11 Evaluation of TLC Plates

After drying, in a darkened room, and under UV light, mark the visible points or points with their own color like in Figure 5.

12.3.12 Estimation of Gunpowder with Stabilizer Diphenylamine

Estimation of gunpowders with stabilizer diphenylamine is based on the information shown in the Table 4. It is essential to compare the TLC plate of the sample with the standard in the (Figure 5) (Figure 6) according to.

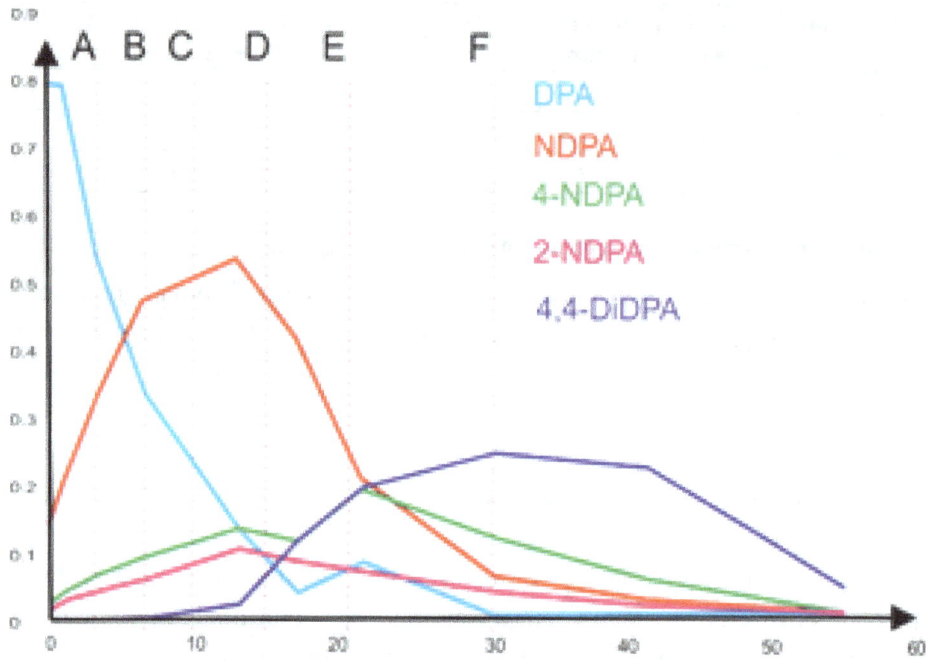

Fig. 6: Degradation process of diphenylamine.

Awarding category 1:

v. Image of the stabilizer in the initial state: A, B

vi. Image of the stabilizer after the equivalent of four years of aging at 25°C: not worse than C.

Awarding category 2

vii. Image of the stabilizer in the initial state: A, B, C, D

viii. Image of the stabilizer after aging equivalent of four years at 25°C: not worse than E.

Awarding category 3

i. Image of the stabilizer in the initial state: E, F

ii. Image of the stabilizer after aging equivalent of four years at 25°C: F.

Decisions can be made according to the following table.

Estimation of gunpowder with stabilizer centralite I

iii. Estimation of gunpowders with stabilizer centralite I is based on the information shown in the Table 10.

iv. It is essential to compare the TLC plate of the sample with the standard in the (Figure 5) (Figure 7).

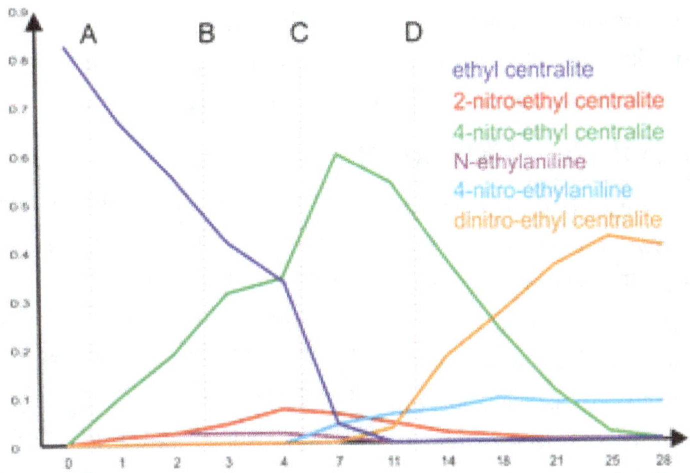

Fig. 7: Degradation process of centralite I (ethyl centralite).

Table 10 Definitions of a CEN I stabilizer image

A	Centralit clear, barely recognizable N-nitroso-ethylaniline and 4-nitro centralite.
B	Centralit is clearly recognizable, nitroso-ethylaniline and 4-nitro-centralite are clearly recognizable.
C	Nitrozo ethylaniline and 4-nitro-centralite are clearly recognizable, dintro-centralite is eventually recognizable, centralit is still recognizable.
D	Dintro-centralite recognizable, nitrozo-ethylaniline and 4-nitro-centralite recognizable, the centrality virtually disappeared.

Awarding category 1:

i. Image of the stabilizer in the initial state: A.
ii. Image of the stabilizer after the equivalent of four years of aging at 25°C: not worse than B.

Awarding category 2:

i. Image of the stabilizer in the initial state: A, B.
ii. Image of the stabilizer after aging equivalent of four years at 25°C: not worse than C.

Awarding category 3:

i. Image of the stabilizer in the initial state: C or worse.
ii. Image of the stabilizer after aging equivalent of four years at 25°C: D or worse.

Decisions can be made according to the following table.

Categorization of gunpowder according to standards AOP 48 and STANAG 4556

Each country has its own categorization, and the indicative categorization is given in NATO standards. Smaller caliber ammunition contains 0.5 to 1% of initial DPA stabilizer; bigger caliber ammunition contains 1.5 to 2% of initial DPA stabilizer.[20]

12.4 RESULTS OF THE EXPERIMENT

The results of the described experiment, in the titles above, are shown in the following Table 11. The same gunpowder samples were examined on TLC and two additional methods and we will make a comparison of the results. We analyzed 11 different samples of gunpowders of 6 types of cartridges (20mm, 40mm, 76mm, 100 mm, 105 mm, 122mm). The results were processed and passed the appropriate categories and in the table above we can compare the categories obtained by the TLC method with the categories adopted on the basis of the other two methods: HPTLC and VST. All gunpowder samples are single-base (without nitroglycerin), which is confirmed by TLC method and VST method. The main stabilizer according to TLC is determined chemically where the result of the reaction is dyeing, as explained in the title Determine the type of stabilizer. DPA was determined in all gunpowder samples, which was confirmed by the HPTLC method. Eight samples showed a category matching, these are samples no. 1, 2, 3, 6, 8, 9, 10 and 11. Samples 4, 5, and 7 are identified by category 2 according to TLC method, and 1st category by HPTLC and VST. The difference in grade may be due to the fact that TLC is qualitative, while HPTLC and VST are quantitative methods. The classical method depends **Table**

The results of the experiments on three different methods, TLC, HPTLC and VST, have shown very good agreement. Determining the base of the gunpowders showed complete agreement by TLC and VST methods. Determining the type of stabilizers in the gunpowders showed complete agreement by TLC and HPTLC methods. Determining the category of the gunpowders by TLC showed very good agreement with HPTLC and VST methods. 8 of the 11 samples showed the identical category. The remaining 3 samples are in the transition category according to TLC method and Tables 8 and 10, which implies the situation in which the products of the decay of the stabilizer appear, and in the second hand the stabilizer has about 1% or more. The method of interpreting the results leads to apparent disagreement of the categories for the three samples under the numbers 4,5 and 7 in Table 11. For these three samples, this means that they are closer to the 2nd than the 1st category. Obviously, the results of the TLC method are very useful and usable for the purpose of life prediction of ammunition. Thin layer chromatography for the purpose of qualitative analysis of gunpowder samples is a reliable method and largely depends on the expertise of individuals and from the interpretations of the results. Results of qualitative thin layer chromatography analysis provides very good guidance in purpose of life prediction of gunpowders. Thin layer chromatography method can be economically employed for routine use because the consumption of mobile phase is low and, hence, there are scarcely any disposal problems. Thin layer chromatography method does not require the use of machines or special devices, it is fully portable and easy to handle and considerably cheaper than most commercial methods.

11 Results of experiments on TLC, HPTLC and VST methods for 11 samples of gunpowder

No.	Name of ammunition	Gunpowder	Series of gunpowder	High preasure thin layer chomatography							
				Scanned stabilizer	Main stabilizer	Relative humidity [%]	Temoerature [°C]	Atmospheric pressure [hPa]	[%] DPA	[%] NNODPA (on 254 λ)	[%] NNODPA 550 λ
1	Bullet 100 mm	-	14/7 - 39/67 K	DFA, NNODFA	DPA	46	24	1019	1,63	0,50	
2	Bullet 100 mm	NC 18	MBL 7216	DFA, NNODFA	DPA	47	25	1025	1,11		
3	Bullet 122 mm	NC-43	MBL 8318	DFA, NNODFA	DPA	46	24	1019	0,49		
4	Bullet 40 mm	NCD-22	MBL 8828	DFA	DPA	68	25	1013	1,02		-
5	Bullet 76 mm	-	9/7 1/68 T	DFA, NNODFA	DPA	46	24	1019	1,73	0,34	
6	Bullet 20 mm	NCD-06	MBL 6439	DFA, NNODFA	DPA	44	23	1023	0,70		
7	Bullet 20 mm	NCD-07	MBL 6746	DFA, NNODFA	DPA	44	23	1023	1,39		
8	Bullet 122 mm	NC-122	MBL 8108	DFA, NNODFA	DPA	-	-	-	<0,2		-
9	Bullet 122 mm	NC-27	MBL 8442	DFA, NNODFA	DPA	-	-	-	<0,2		
10	Bullet 122 mm	NC-122	MBL 7901	DFA, NNODFA	DPA	-	-	-	<0,2		
11	Bullet 105 mm	B19T(1.6)	31-BE-78	DFA, NNODFA	DPA	46	24	1019	<0,2		

Vacum stability test		Thin layer chomatorgraphy						
NGL (+,-)	Δ VST gas [ml/g]	Fuel base	Type of stabilizer	Initial state of sample	State after aging 48 h at 80 °C	State after aging 96 h at 80 °C	Category	Category by hptlc and vst
-	0,871	1	DPA	A	B	C	1	1
-	1,639	1	DPA	A	C	D	2	2
-	2,003	1	DPA	B	E	E	2	2
-	0,620	1	DPA	C	C	D	2	1
-	0,590	1	DPA	A	B	D	2	1
-	1,537	1	DPA	B	D	E	2	2
-	0,629	1	DPA	B	C	E	2	1
-	2,513	1	DPA	C	E	F	3	3
-	0,990	1	DPA	C	F	F	3	3
-	2,246	1	DPA	C	E	F	3	3
-	3,437	1	DPA	E	F	F	3	3

on the eye of the observer, and this can be the cause of the different results. Sample number 4 showed a stabilizer ratio of 1.02%, which is a category 1 characteristic, while the VST method showed 0.62ml/g and it is a category 1 characteristic. We conclude that this is a transition group, as shown in (Table 5) (Table 12) by mixed green-yellow color. Sample number 5 is similar, however, the content of DPA is very high and the cause of disagreement in this case is probably human eye or inexperience. With this sample of gunpowder we can not say with certainty that it is the first category for HPTLC because we do not know how much the initial concentration of DPA was in the gunpowder. In case it was high, about 4%, it would still be category 2. In this case, we were guided by a theoretical mean concentration of DPA quoted in gunpowders, about 1.5-2 %.20 Sample number 7 is also in the transition group and it is not a mistake to say that it falls into category 2 of stability.

Table 12 Estimation of centralite I stabilizer state

	A	B	C	D
Initial state				
After aging				

Samples 4, 5, and 7 belong to the transition group according to the TLC qualitative method, and because the method is qualitative, the attitude of our laboratory is to take a tougher decision, and this is one reason for disagreement with the mathematical decision of the other two methods. Obviously, this is a good matching result, and also, for this sample the initial concentration of the DPA stabilizer is not known. Samples 8, 9, 10 and 11 clearly showed the third category, which was confirmed by the other two methods. Clarity in determining the third category is the most important in this experiment. The problem that occurs in qualitative analysis on TLC is the appearance of colouration of spots at sites characteristic of DPA, N-nitroso-DPA, 2,2' and 2,4'-dinitro-DPA at the same time according to Figure 5. In this case, it is confusing which grade to select from (Table 4) (Table 10). Analyzing Figure 6, we can conclude that this is the sample in life range between E and F. In practice, there is a high likelihood of errors in these cases. In such confusions, we advise group work, or to make a quantitative experiment on one of the appropriate methods.

13

ROLE OF STANDARD FLOW LIQUID CHROMATOGRAPHY FOR SHOTGUN PROTEOMICS IN BIOENERGY RESEARCH

Over the past 10 years, the bioenergy field has realized significant achievements that have encouraged many follow on efforts centered on biosynthetic production of fuel-like compounds. Key to the success of these efforts has been transformational developments in feedstock characterization and metabolic engineering of biofuel-producing microbes. Lagging far behind these advancements are analytical methods to characterize and quantify systems of interest to the bioenergy field. In particular, the utilization of proteomics, while valuable for identifying novel enzymes and diagnosing problems associated with biofuel-producing microbes, is limited by a lack of robustness and limited throughput. Nano-flow liquid chromatography coupled to high-mass accuracy, high-resolution mass spectrometers has become the dominant approach for the analysis of complex proteomic samples, yet such assays still require dedicated experts for data acquisition, analysis, and instrument upkeep. The recent adoption of standard flow chromatography (ca. 0.5 mL/min) for targeted proteomics has highlighted the robust nature and increased throughput of this approach for sample analysis. Consequently, we assessed the applicability of standard flow liquid chromatography for shotgun proteomics using samples from Escherichia coli and Arabidopsis thaliana, organisms commonly used as model systems for lignocellulosic biofuels research. Employing 120 min gradients with standard flow chromatography, we were able to routinely identify nearly 800 proteins from E. coli samples; while for samples from Arabidopsis, over 1,000 proteins could be reliably identified. An examination of identified peptides indicated that the method was suitable for reproducible applications in shotgun proteomics. Standard flow liquid chromatography for shotgun proteomics provides a robust approach for the analysis of complex samples. To the best of our knowledge, this study represents the first attempt to validate the standard flow approach for shotgun proteomics.

13.1 INTRODUCTION: AN OVERVIEW

Advances in biofuels research focusing on feedstock characterization and engineering as well as the genetic manipulation of microbes have progressed significantly in the last few years. Unfortunately, analytical capabilities required to efficiently monitor and assess these changes are lagging. Furthermore, many modern bioanalytical techniques are focused toward medical and health related research, which have significantly different priorities and requirements for success. Most biotechnology research challenges are not constrained by sensitivity or resolution of an assay, rather they depend on accurate identification and quantitation of target molecules for a large number of samples. Consequently, an important component for biotechnological research is sample throughput supported by a robust analytical platform. Recent advances in proteomics and metabolomics have focused on liquid chromatography-mass spectrometry (LC-MS) methods by increasing their sensitivity to aid discovery-based research efforts. This is most evident with the development of nano-LC couple to high-resolution mass spectrometers; yet, this technology is yet to mature into a robust platform capable of consistently analyzing hundreds of samples per week. Consequently, alternate technologies capable of answering the questions necessary for biotechnology progress are needed. Recently, we published a high throughput targeted proteomic toolkit based on standard flow chromatography coupled to mass spectrometry (MS) to help address these issues for Escherichia coli; however, a significant amount of methods development was necessary. This work prompted us to assess the utility of standard flow liquid chromatography (LC) for shotgun proteomic methods related to biotechnology.

The discipline of proteomics has been dominated by nano-flow LC coupled to MS since its early development over 20 years ago. The adoption of nano-flow LC for protein identification was driven by the substantial increases in sensitivity and detection capabilities of nano-flow (ca. 500 nL/min) over capillary (ca. 50 μL/min) and standard flow (ca. 0.5 mL/min) chromatography. Typically, shotgun proteomic studies utilize nano-flow chromatography methods due to limited amounts of sample and to obtain optimal ionization efficiency for sensitive detection of peptides but at the cost of ease of use and system robustness. These issues are particularly problematic for biotechnology research that depends heavily on high sample throughput. Recently, ultra-high performance LC coupled to triple quadrupole mass spectrometers has been shown to yield comparable sensitivity and better analytical metrics (coefficient of variation, dynamic range) than nano-flow LC for MRM-based analysis of biomarker proteins when the amount of sample is adjusted to the column size. Additionally, the well-established robust operation of standard flow chromatography makes this instrumentation attractive for applications that rely on high sample throughput, consistent results, and less system downtime. For applications where sample abundance is not limited, one of the greatest concerns with ultra-high pressure liquid chromatography-mass spectrometry (UHPLC-MS) workflows is the loss of sensitivity, relative to nano-LC-MS workflows, leading to datasets of insufficient depth to answer questions of interest.

The adoption of a UHPLC-MS workflow for sample delivery requires both efficient ionization and instrument speed to handle both the sample delivery rate and reduced elution times for peptides. The past decade has seen significant developments and advances in instrumentation associated with proteomics-based MS. Advances in reversed phase C18 chromatography columns yield greater separation efficiency and more stable retention times. The current generation of instrumentation is faster, more sensitive, and is better able to deal with the dynamic range inherent in biological samples. Moreover, the development and adoption of off-axis nebulizers for sample delivery when using electrospray ionization significantly reduces contamination of capillaries and skimmers. Collectively, these improvements have enabled the development of a UHPLC-MS workflow for MRM-based analysis of biomarker proteins. Consequently, we were interested in assessing the capacity of this workflow in shotgun proteomic experiments. The work described here details the results of an analysis of a prokaryote (E. coli) and an eukaryote (Arabidopsis thaliana) whole cell proteomes on an Agilent UHPLC-QTOFMS system, but would be generally applicable for any current generation of tandem mass spectrometer being utilized for shotgun proteomics.

13.2 METHODS

13.2.1 Protein Extraction

Protein was extracted using standard techniques with analytical reagents where suitable. For E. coli DH5α samples, cell lysis and protein precipitation was accomplished using a chloroform/methanol precipitation. A 100 μL aliquot of cells was transferred to a 1.7 mL tube, followed by the addition of 400 μL of methanol, 100 μL of chloroform, and 300 μL of water, with mixing by vortex after each addition. Following centrifugation at 21,000 × g for phase separation, the methanol and water layer was removed and 300 μL of methanol was added. The tube was briefly vortexed to dislodge the protein pellet, then centrifuged at 21,000 × g for 2 min. The chloroform and methanol layer was removed and the protein pellet was dried for 5 min in a vacuum concentrator. The protein pellet was re-suspended in 100 mM (NH4)HCO3 with 20% methanol, reduced with 5 mM TCEP [Tris(2-carboxyethyl)phosphine hydrochloride] for 30 min at room temperature, treated with 10 mM iodoacetamide (IAA) for 30 min in the dark at room temperature, and digested with trypsin (1:50 w/w) overnight at 37°C. Aliquots of 40 μg were taken for analysis by LC-MS/MS. For A. thaliana (L.) Heynh. (ecotype Landsberg erecta), protein was extracted from a previously described heterotrophic cell culture. A total of 1 g plant material (fresh weight) was used for the isolation of total protein. The plant material was harvested and frozen with liquid nitrogen in an Eppendorf tube with two small steel balls. The protein extraction was performed by the addition of 0.4 mL of fresh disruption buffer [125 mM Tris-HCl, 7% (w/v) SDS, and 10% β-mercaptoethanol], followed by vortex for 10 min at room temperature. The samples were centrifuged at 10,000 × g for 5 min at 4°C and the supernatant separated into two 2 mL tubes. Samples were further extracted in 800 μL methanol and mixed, then 200 μL chloroform added and mixed, and finally 500 μL of ddH2O added and vortexed (30 sec each time). Samples were centrifuged for 5 min at 10,000 × g at 4°C, the aqueous phase removed, and 500 μL of methanol added. Samples were vortexed for 30 s and centrifuged for 10 min at 9,000 × g at 4°C. The supernatant was discarded and the pellet air-dried. The dried pellet was suspended in 200 μL of re-suspension buffer [3M urea, 50 mM (NH4)HCO3, and 5 mM dithiothreitol, pH 8], and incubated with IAA at a final concentration 10 mM for 30 min in the dark. Prior to analysis by MS, 40 μg of extracted protein was digested with trypsin (1:10 w/w) overnight at 37°C. Peptides were desalted using C18 Micro SpinColumns (Harvard Apparatus) as previously outlined . Eluted peptides were re-suspended in 2% acetonitrile, 0.1% formic acid prior to analysis by LC-MS/MS.

13.2.2 Standard Flow Mass Spectrometry

All samples were analyzed on an Agilent 6550 iFunnel Q-TOF mass spectrometer (Agilent Technologies) coupled to an Agilent 1290 UHPLC system. Peptide samples were loaded onto a Sigma–Aldrich Ascentis Peptides ES-C18 column (2.1 mm × 100 mm, 2.7 μm particle size, operated at 60°C) via an Infinity Autosampler (Agilent Technologies) with Buffer A (2% acetonitrile, 0.1% formic acid) flowing at 0.400 mL/min. Peptides were eluted into the mass spectrometer via a gradient with initial starting condition of 5% buffer B (98% acetonitrile, 0.1% formic acid). For analysis of all samples, buffer B was increased to 35% over 120 min. Buffer B was then increased to 50% over 5 min, then up to 90% over 1 min, and held for 7 min at a flow rate of 0.6 mL/min, followed by a ramp back down to 5% B over 1 min where it was held for 6 min to re-equilibrate the column to original conditions. Peptides were introduced to the mass spectrometer from the LC by using a Jet Stream source (Agilent Technologies) operating in positive-ion mode (3,500 V). Source parameters employed gas temp (250°C), drying gas (14 L/min), nebulizer (35 psig), sheath gas temp (250°C), sheath gas flow (11 L/min), VCap (3,500 V), fragmentor (180 V), OCT 1 RF Vpp (750 V). The data were acquired with Agilent MassHunter Workstation Software, LC/MS Data Acquisition B.05.00 (Build 5.0.5042.2) operating in Auto MS/MS mode whereby the 20 most intense ions (charge states, 2–5) within 300–1,400 m/z mass range above a threshold of 1,500 counts were selected for MS/MS analysis. MS/MS spectra (100–1,700 m/z) were collected with the quadrupole set to "Medium" resolution and were acquired until 45,000 total counts were collected or for a maximum accumulation time of 333 ms. Former parent ions were excluded for 0.1 min following MS/MS acquisition.

13.2.3 LC-MS/MS Data Analysis and Integration

The acquired data were exported as .mgf files using the Export as MGF function of the MassHunter Workstation Software, Qualitative Analysis (Version B.05.00 Build 5.0.519.13 Service Pack 1, Agilent Technologies) using the following settings: peak Filters (MS/MS), the absolute height (≥ 20 counts), relative height (≥ 0.100% of largest peak), maximum number of peaks (300) by height; for charge state (MS/MS), the peak spacing tolerance (0.0025 m/z plus 7.0 ppm), isotope model (peptides), charge state limit assigned to (5) maximum. Resultant data files were interrogated with the Mascot search engine version

2.3.02 (Matrix Science) with a peptide tolerance of ±50 ppm and MS/MS tolerance of ±0.1 Da; fixed modifications Carbamidomethyl (C); variable modifications Oxidation (M); up to one missed cleavage for trypsin; peptide charge 2+, 3+, and 4+; and the instrument type was set to ESI-QUAD-TOF. Searches were performed against either an E. coli (strain K12) dataset obtained from UniProt or the latest release of the A. thaliana dataset comprising TAIR10 obtained from The Arabidopsis Information Resource . Both databases incorporated proteins comprising the common Repository of Adventitious Proteins (cRAP v2012.01.01 from The Global Proteome Machine). The E. coli database comprised 4,429 sequences (1,398,775 residues) while the Arabidopsis database comprised 35,508 sequences (14,522,421 residues). Protein and peptide matches identified after interrogation of MS/MS data by Mascot were filtered and validated using Scaffold v4.3.0 (Proteome Software Inc.). Peptide identifications were accepted if they could be established at >95.0% probability by the Peptide Prophet algorithm with Scaffold delta-mass correction. Protein identifications were accepted if they could be established at >95.0% probability and contained at least 1 identified peptide (at 95% and greater). Protein probabilities were assigned by the Protein Prophet algorithm. This resulted in false discovery rates of 0.9 (E. coli) and 0.3% (Arabidopsis) for protein and 0.37 (E. coli) and 1.66% (Arabidopsis) for peptides. Proteins that contained similar peptides and could not be differentiated based on MS/MS analysis alone were grouped to satisfy the principles of parsimony.

13.3 STATISTICAL DATA ANALYSIS

Statistical analysis was performed using the statistical toolbox in MATLAB v2009b (Mathworks). The following analysis was performed on data from both Arabidopsis and E. coli. After data filtering using Scaffold v4.3.0 (outlined above), peptide data were exported as .csv files and only peptides derived from the same protein identified across all replicates were considered. The Mascot ions score and total ion current values were only used for the best matching peptide (based on ion score) in each replicate. The total spectrum count for each peptide from each replicate was calculated manually. The coefficient of variation (CV: SD/mean) of total spectrum counts, Mascot ions score, and total ion current for each of the common peptides across the replicates was calculated to determine the variation across the technical replicates. The histograms of the CV were plotted to determine distribution and general extreme value fit algorithm was used to determine fit and mean of the distribution. For principal component analysis (PCA), the total spectrum count, Mascot ions score, and total ion current for all the common peptides of each technical replicate were mean centered. PCA was done by eigenvalue decomposition of the data covariance, resulting in a set of linearly uncorrelated variables. The principal component scores were then plotted to identify groupings.

13.4 RESULTS AND DISCUSSION

The initial setup and tuning of the standard flow LC-MS/MS parameters was undertaken using 20 fmol aliquots of trypsin digested bovine serum albumin (BSA). The system settings were deemed adequate once identification of BSA was comparable (i.e., unique peptides, total peptides, and coverage) to that achieved using previously benchmarked nano-flow LC-MS/MS approaches on a variety of instruments over the past 5 years.

13.4.1 Application of Standard Flow LC-MS/MS with Prokaryotic Samples

The recent progress of bioenergy research has relied heavily on transformational developments in feedstock characterization and metabolic engineering. Yet, omics methods to characterize and quantify systems of interest have mainly been adapted from the health and clinical fields that have very different research needs (i.e., high sensitivity, deep proteome coverage). We report the application of a standard flow LC-MS/MS approach that is suitable for large numbers of shotgun proteomic experiments where sample abundance is not limiting. The setup is capable of undertaking a rapid analysis of low complexity samples as well as handling highly complex samples by employing extended analysis times. Although its application requires instrumentation with the ability to deal with increased flow rates and shorter peptide elution times, it is apparent that the current generation of tandem mass spectrometers is capable of handling these parameters. While traditional nano-LC-MS/MS approaches are likely to continue to dominate shotgun analyses as they produce a greater number of protein identifications and have the ability to deal with increased sample complexity and dynamic range, the robust nature and simplicity of standard flow coupled to MS makes this approach an attractive alternative for applications where sample throughput and reproducibility are important factors.

Initial experiments were performed on E. coli, a well-characterized organism with minimal proteomic complexity. The E.coli genome of the widely utilized laboratory strain K-12 was completed nearly 20 years ago. The genome has undergone multiple revisions and is estimated to codes for over 4,300 proteins. Shake flasks of E. coli DH5α cultures were grown aerobically at 37°C on Luria Broth (LB) medium supplemented with 1% glucose. Total protein was extracted and digested overnight with trypsin at 37°C. The analysis procedure for E. coli samples was developed to take advantage of the speed and robustness of a standard flow analysis and as such, a method was developed that incorporated a 120 min gradient. A total of four biological replicates each equivalent to 40 μg of peptide were analyzed by a standard flow UHPLC-QTOFMS operating with typical shotgun proteomics data acquisition parameters. The samples yielded on average 31,938 ± 989 (SE) MS/MS spectra (Table 1). Each of these MS/MS datasets was used to query the E. coli (K12) protein database from UniProt using the Mascot search engine (Matrix Science) with resultant protein matches filtered using Scaffold (Proteome Software). In total, 802 proteins were identified from the four replicates with an average of 786 ± 4.5 (SE) per sample (Table 1 and Table S1 in Supplementary Material). This represents about 20% of the protein coding capacity of E.coli. On average, 14,639 ± 484 (SE) MS/MS spectra were successfully matched to a peptide, corresponding to around 46% of the total queries for each sample (Table 1). This would be regarded as a high conversion rate indicating that the acquired MS/MS was of sufficient quality to confidently assign nearly 50% of the queries using the approach. The consistency of the standard flow approach was highlighted by the total number of proteins assigned in each sample. A total of 768 proteins were identified in three out of four replicates, while 746 proteins (93%) were identified collectively in all four samples. This compares well to the total number of proteins identified in all samples (802). Few proteins were identified uniquely in a single replicate except for replicate Ecoli-1, where 30 unique proteins were identified (Figure 1A). Consequently, we sought to examine whether this setup was adequately dealing with the sample complexity given the flow rate and an approximate peptide elution time of around 6 s (Figure S1 in Supplementary Material). Nearly 75% of the matched MS/MS spectra were redundant, with an average of 4,009 unique peptides per sample (Table 1). Only 2,825 unique spectra (54%) were shared between all four samples; however, the new unique peptides (2,393) accounted for only 56 new proteins (Figure 1A). The approach consistently identifies the major proteins in a sample and a majority of unique spectra assigned between replicates are derived from previously identified proteins. This indicates that the QTOFMS has the capacity to handle the standard flow UHPLC setup in shotgun mode even at these higher flow rates and reduced peptide elution times.

Table 1. Values obtained from E. coli (Ec) and Arabidopsis (At) samples analyzed using the standard flow technique.

	Ec 1	Ec 2	Ec 3	Ec 4	At 1	At 2	At 3
Queries	31,452	29,348	33,638	33,638	26,135	37,388	28,203
Spectra	14,677	13,265	15,173	15,173	6,038	9,712	6,846
Sectra matched (%)	47	45	46	46	23	26	24
Unique peptides	4,028	3,729	4,123	4,123	3,050	6,643	3,553
Total proteins	788	772	791	791	1,214	1,359	1,291

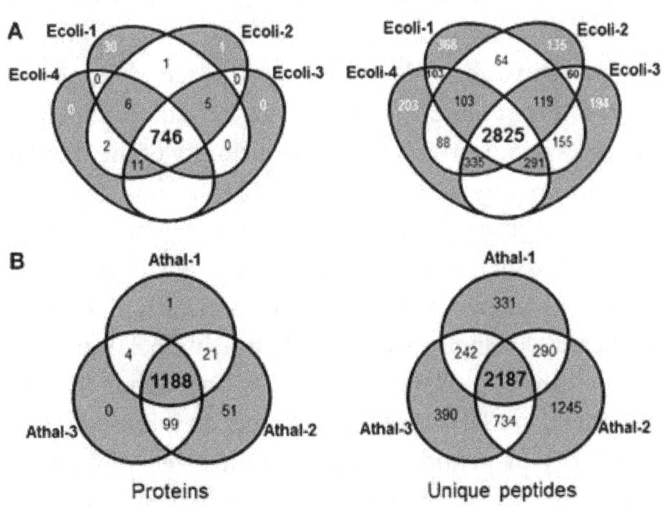

Fig. 1: Venn diagrams highlighting the shared matched proteins and unique peptides between technical replicates from (A) E. coli and (B) Arabidopsis total protein samples analyzed with the standard flow LC-MS/MS approach.

13.4.2 Application of Standard Flow LC-MS/MS to a Complex Eukaryotic Sample

Next, we were interested in assessing the suitability of the standard flow proteomics platform on a complex eukaryotic proteome. The genome of the reference plant Arabidopsis was completed over a decade ago and now represents one of the most highly curated and annotated genomes in biology. As a consequence, Arabidopsis is being utilized as a proving ground for plant synthetic biology approaches, many of which have focused on biomass manipulation for efficient biofuels production. The most recent proteome release (TAIR10) from The Arabidopsis Information Resource comprises over 27,000 loci and over 35,000 distinct protein products. Protein extractions were performed from a 7-day old Arabidopsis cell cultures that have previously been extensively employed for proteomic assays. A total protein sample was analyzed in triplicate by UHPLC-QTOFMS using the standard flow proteomics platform. We analyzed the equivalent of ca. 40 μg of digested total protein over a 120 min gradient with MS conditions identical to those used with the E. coli samples. An average of 30,575 ± 3,458 (SE) MS/MS spectra were collected over the 2-hour run from the three replicates (Table 1). These numbers are similar to those obtained using the E.coli samples and may reflect the upper capacity of the method given the increased peptide complexity that would be expected from a eukaryotic sample.

These Arabidopsis datasets were each used to interrogate the most recent Arabidopsis protein dataset using the Mascot search engine and matches filtered and integrated using Scaffold. This resulted in an average of 7,532 ± 1,115 (SE) matched spectra from the three replicates and corresponds to about 25% of the total spectra obtained (Table 1). This corresponds to about half the conversion rate observed for the E.coli samples and likely reflects the quality of the MS/MS spectra given the overall ion intensities are likely on average considerably lower due to the increased size of the proteome. The average number of proteins identified over the three samples was 1,288 ± 73 (SE) with a total of 1,364 unique proteins identified in all three replicates (Table S2 in Supplementary Material). The number of proteins consistently identified by all three replicates was 1,188 proteins or 87% of the total number (Figure 1B). The minor variation in identifications between these technical replicates demonstrates the reproducibility of the standard flow approach for shotgun proteomics as the method was capable of consistently identify the same proteins in each of the replicates. In an attempt to understand whether the approach is adequately dealing with the increased complexity of this eukaryotic sample, we examined the proportion of unique matched peptides identified in each replicate. The 1,364 unique proteins identified in these samples were matched using 5,419 unique peptides, while the 1,188 proteins identified in all three replicates only required 2,187 (40.4% of the total) unique peptides (Figure 1B). These results indicate that again, as found with the E. coli samples, the majority of peptides exclusive to a given replicate are derived from proteins that have already yielded a high scoring peptide match. It is conceivable that the complexity of the eukaryotic sample contributes to many more co-eluting peptides during the 120 min gradient than the *E. coli* sample, resulting in unique peptides selected for MS/

MS by the data-dependent acquisition method in each replicate (Figure 1B). Although similar results with regard to unique peptides were also observed with the E. coli samples, on average, an *E.coli* protein was identified by 6.8 unique peptides (768 proteins from 5,218 unique peptides) compared to the Arabidopsis proteins at 4.0 unique peptides each (1,364 proteins from 5,419 unique peptides). Nearly twice as many proteins were identified in the Arabidopsis samples (1,364 proteins compared to 768 proteins), which would be expected given the differences in coding capacity between these species. However, considerably more identified proteins would have been expected given proteome sizes, indicating that the standard flow approach has a more limited capacity to deal with the eukaryotic sample due to issues, such as increased sample complexity at any given point in time, lower overall ion intensity, and dynamic range limitations (as discussed above) of this QTOFMS system.

13.4.3 Assessing the Performance of Standard Flow LC-MS/MS

Last, we investigated the reproducibility of the standard flow UHPLC-QTOFMS by comparing parameters of the identified peptides between the replicates for both E. coli and Arabidopsis samples. The peptide ions score obtained from Mascot after data interrogation can be indicative of the quality of the fragmentation spectra, total ion current obtained from the MS/MS spectra can provide information about the intensity of an eluted peptide while total spectral count can provide an indication of peptide intensity and spectral complexity. These values can be used as a proxy to assess sample limitations and reproducibility by the LC-MS system. PCA was employed to ascertain whether there were any differences between ion scores, total ion current, and spectral count for peptides identified in all the replicates for either E. coli or Arabidopsis samples (Figure 2). The analysis demonstrated that none of the replicates for either E. coli or Arabidopsis could be separated by principal component scores for these attributes, indicating that majority of the differences for either data set can be largely attributed to biological heterogeneity of the sample (Figure 2).

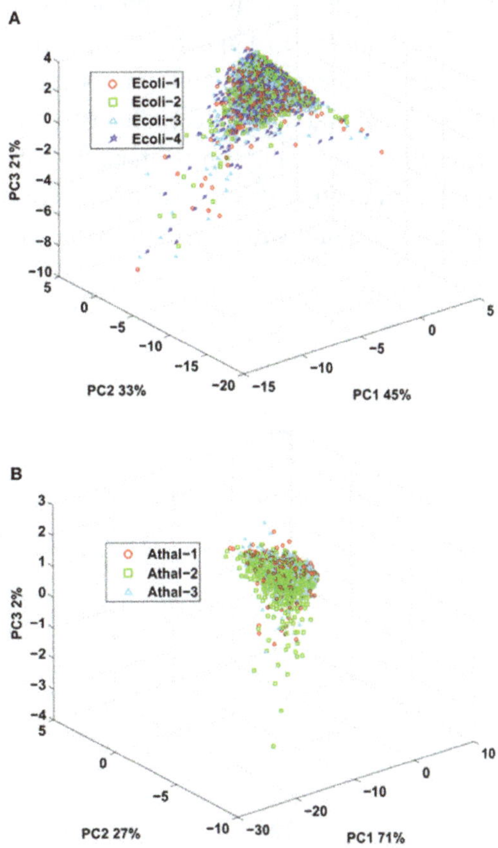

Fig. 2: Principal component analysis of spectral counts, Mascot ions score, and total ion current (derived from MS/MS) from common peptides identified over all replicates from (A) E. coli and (B) Arabidopsis.

The similarities in ion score, total ion current, and spectral counts can also be observed when analyzing the distributions of CV for the each identified peptides across the replicates (Figure S2 in Supplementary Material). The variations in Mascot ion score and total ion current (derived from MS/MS spectra) for the peptides identified over the replicates were similar for both E.coli and Arabidopsis. This indicates that from sample to sample, identical ions performed similarly with regard to the intensity of matched peptides (total ion current) as measured by the mass spectrometer. This is further supported by the small variation in Mascot ion scores with ion intensity having a relationship to the quality of the MS/MS spectra and subsequent spectral matching procedures. The variation in total spectrum count was more pronounced between the identified peptides of E. coli (0.67) and Arabidopsis (0.32), with samples of lower complexity (i.e., E.coli) having a larger variation in spectral counts for a peptide across replicates. This could be due to the higher repeat sampling rate that likely occurred during analysis of E.coli samples due to the reduced number of distinct ions/peptides in the sample. This conclusion is supported when looking at the average number of spectral counts for a given identified peptide; for E.coli, it was 3.74 spectra per peptide while for Arabidopsis it was 2.70. The lower average number of spectral counts for a peptide from Arabidopsis is likely indicative of the increased sample complexity. Collectively, these data indicate that between the replicates, the quality and intensities of peptides identified across the replicates was similar, indicating that the standard flow approach was not significantly impeding the performance of the mass spectrometer to acquire tandem mass spectra

14

SUPERCRITICAL FLUID CHROMATOGRAPHY

14.1 SUPERCRITICAL FLUID

Supercritical fluid is a state of matter that is intermediate between a gas and liquid in its properties. This state formed when a gas or liquid solvent is subjected to temperature and pressure condition exceeding a particular critical point. The temperature and pressure at which this pint Occurs are known as the Critical temperature and Critical pressure and are Characteristic of the solvent. Beyond this point, the solvent Will be neither a gas or liquid, but will possess properties of both phases . Whether this supercritical fluid acts more like a gas or liquid will depend on the pressure and temperature.

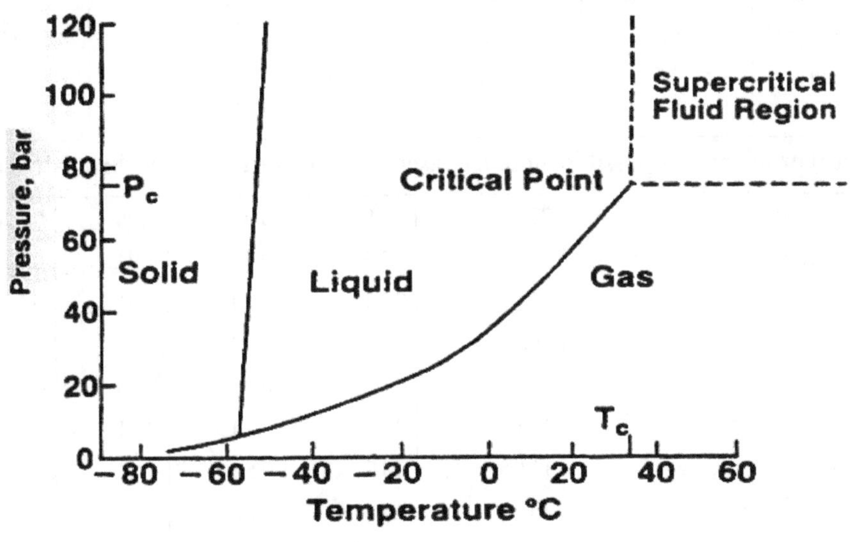

Figure 7.1. Phase diagram for carbon dioxide

]

Fig. 1: phase diagram for carbon dioxide

Table 1.Change in density of supercritical fluid carbon dioxide with pressure and temperature

Temperature (°C)	Pressure (atm)	Density (g/ml)
40	72	0.22
	400	0.96
60	72	0.17
	400	0.90
80	72	0.14
	400	0.82
100	72	0.13
	400	0.76
120	72	0.12
	400	0.70
140	72	0.11
	400	0.64

Table 2 Representative properties of typical chromatographic mobile phases

Mobile phase	Temperature (°C)	Pressure (atm)	Density (g/ml)	Diffusivity (cm^2/s)	Viscosity (cP)
Helium	200	1.5	$2x10^{-4}$	0.1-1	0.02
Carbon dioxide					
Low density	100	80	0.15	10^{-3}	0.02
High density	35	200	0.8	10^{-4}	0.1
Water	20		1.0	10^{-5}	1.0

Physical properties of possible mobile phases for super critical fluid chromatography

Carbon dioxide	31.3	72.9	0.47	0.96	0
Nitrous oxide	36.5	72.5	0.45	0.94	0.51
Sulfur hexafluoride	45.5	37.1	0.74	1.61	0
Xenon	16.6	58.4	1.10	2.30	0
Butane	152.0	37.5	0.23	0.50	0
Pentane	196.6	33.3	0.23	0.51	0
Dichloro-difluoromethane	111.8	40.7	0.56	1.12	0.17
Trifluoromethane	25.9	46.9	0.52	-----	
Ammonia	132.5	112.5	0.24	0.40	1.47
Water	374.4	226.8	0.34		1.65
Methanol	240.5	78.9	0.27		

1. Supercritical Fluid Extraction There are several advantages of supercritical fluid extraction (SFE):
2. *SFE is generally fast*. The rate of mass transfer between a sample matrix and an extraction fluid is determined by the rate of diffusion of a species in the fluid and the viscosity of the fluid—the greater the diffusion rate and the lower the viscosity, the greater will be the rate of mass transfer

The solvent strength of a supercritical fluid can be varied by changes in the pressure and to a less extent in the temperature.

Many supercritical fluids are gases at ambient condition. Some supercritical fluid are cheap, inert, and nontoxic.

1. Supercritical fluid chromatography (SFC) *SFC is* a chromatographic technique in which the mobile phase is a supercritical fluid. *The use of a supercritical* fluid mobile phase in chromatography was first proposed in 1958 by J. Lovelock. The first actual report use of this in a chromatographic system was in 1962 by Klesper et al, who used it to separate thermally-labile porphyrins. SFC is of importance because it permits the separation and determination of *a group of compounds that are not conveniently handled by either GC or LC.* These compounds (1) are either non-volatile or thermally labile so that the

GC are in-applicable, and (2) contain no functional groups that make impossible detection by spectroscopic or electrochemical techniques employed in LC.

2. Theory of SFC Since supercritical fluids have properties between those of gases and liquid, their use as a mobile phase offers several advantages. Typical physical properties of liquid, gases and supercritical fluids are shown below: One of advantages is that supercritical fluid have lower densities and viscosities than liquids. This results in larger diffusion coefficients for solutes is SFC than LC. This results in Better efficiencies and higher optimum linear Velocities in SFC than LC. The plate height of a SFC System is given by the van Deemter equation.

$$\mathbf{H = A + B/u + Cu}$$

3. SFs have higher densities than gas, so that mobile phase has a greater chance of interacting with the solute than that in GC (i.e., carrier gas). This makes the mobile phase important in determining the retention of solutes on the system and give more flexibility in optimizing the separation. For example, retention of solutes in SFC can be changed by using a different column (i.e. different stationary phases) as in GC, or by changing the mobile phase strength as in LC.

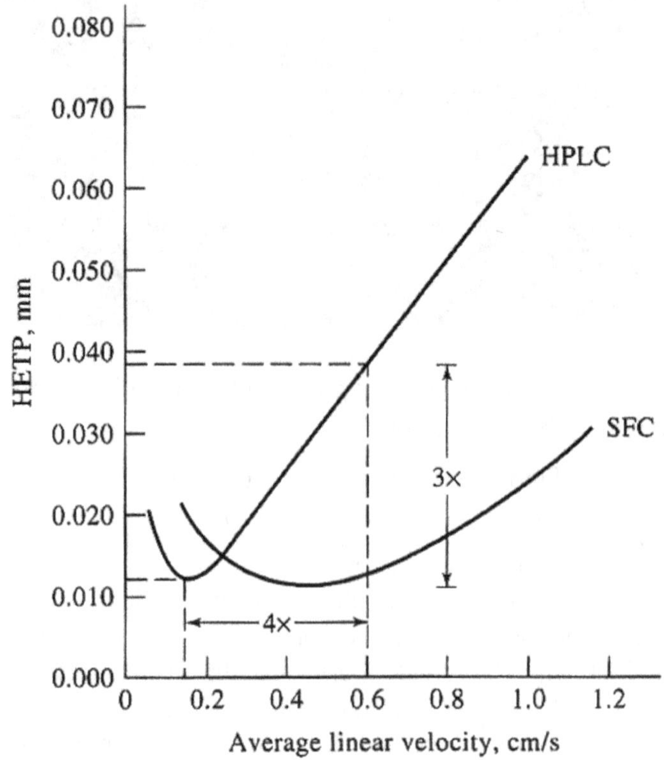

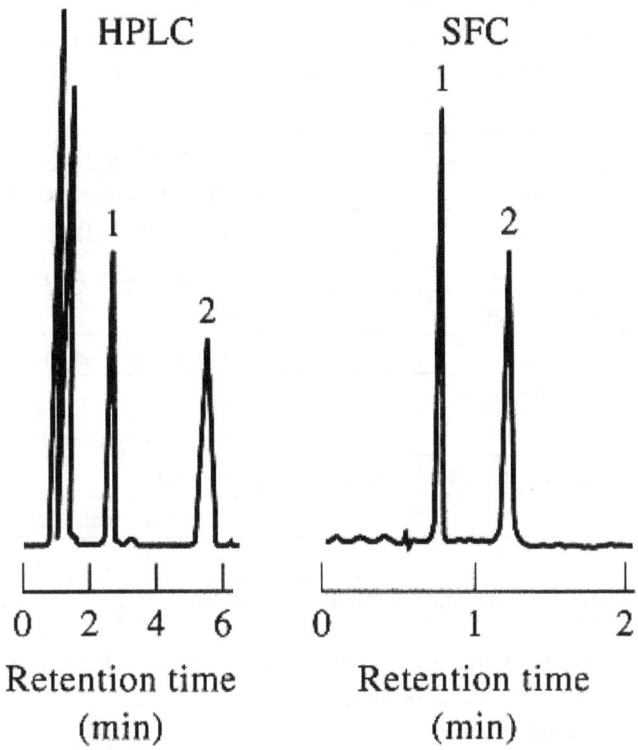

Fig. : Comparison of chromatograms obtained by conventional partition chromatography (HPLC) and super critical fluid chromatography (SFC). Column: 20 cm x 4.6 mm packed with 10 p.m reversed-phase bonded. Analyses: (1) biphenyl; (2) terphenyl. For HPLC: mobile phase, 65/35% CH3OH/H20; flow rate, 4 mL/min; linear velocity, 0.55 cm/s; sample size, 10 p.L. For SFC: mobile phase, CO2; flow rate, 5.4 mL/min; linear velocity, 0.76/s; sample size, 3 ILL. (From D. R. Gere, T. 1. Stark, and T N. Tlyeeten, Application Note 800-4. Hewlett-Packard Company, 1983. With permission.)

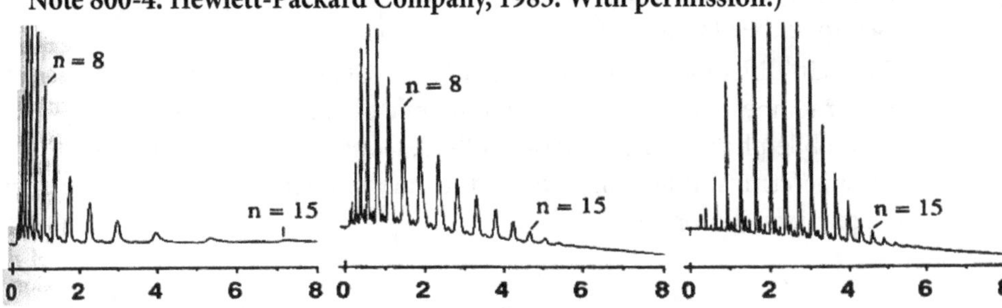

Fig. : Separation of Triton X-114 by SFC using programmed elution on a 10 cm x 2 mm I.D. Nucleosil Cig column, 3μm packing, at 170, with UV detection at 278 nm. Separation on the left: isobaric conditions at 210 bar with a mobile phase of carbon dioxide and methanol (2 + 0.125) mUmin. Center separation: composition gradient from 0.025 to 0.4 ml/min methanol over 8 min with other conditions as above. Separation on the right: pressure program from 130 to 375 bar over 8 min with the same mobile phase used for the isobaric separation. (From ref. [137]; ©Copyright Preston Publications, Inc.)

4. One major advantage of SFC is its ability to use detector available for either GC or LC, such as FID, UV-Vis, and Fluorescence detectors. This gives it a wide range of both universal and selective detections for use in either analytical or preparative-scale work.

GC detectors:	LC detectors:
Thermal conductivity detector	Refractive Index Detector
(TCD): 10-7 M (103-fold range)	(10-5 to 10-6 M)
Flame Ionization detector	Absorption Detector (UV/Vis)
(FID): 10-10 M (a 105-fold range)	(10-8 M)
Nitrogen-phosporus detector	Fluorescence Detector
(NPD): 10-10 M (a 106-fold range)	(10-10 M)
Electron capture detector	Conductivity Detector
(ECD): 10-14 ~ 10-16 M (a 103- 104 fold range)	(10-6 M)
Flame photometric detector	Electrochemical Detector
(FPD):10-14 M (P, S)	(10-11 M)
Electrochemical detector (S, halogen,nitrogen-)	Refractive Index Detector

a. Depending on which supercritical fluid is used, it is also possible to use SFC at lower T than GC. This makes it more useful in the separation of thermally unstable compounds.
b. The stationary phases used in SFC can be similar to those in LC as well as GC. Either packed or open-tubular columns may be used. Because of these advantages, SFC is commonly viewed as a technique which is complementary to both LC and GC.

14.2 INSTRUMENTATION

a. Instrumentation for SFC can be obtained commercially or adapting system used for either LC and GC.
b. The main difference of a SFC than a LC or GC system is the need to control bother temperature and pressure of mobile phase. This must be done to keep the mobile phase as a Supercritical fluid. Control of the pressure (density) of the supercritical fluid can also used to vary strength of mobile phase during the gradient elution in SFC .

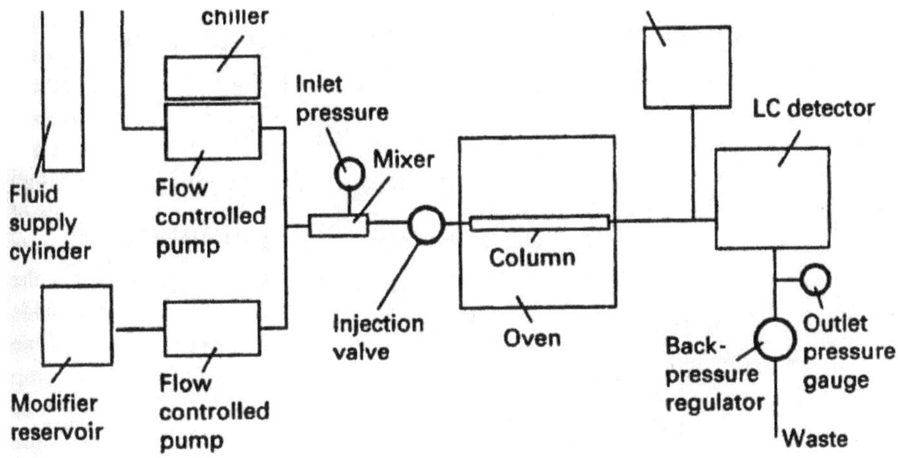

Fig. : Schematic diagram of a typical packed colum SFC indtrument capable of pressure (density and composition) mobile phase programming)

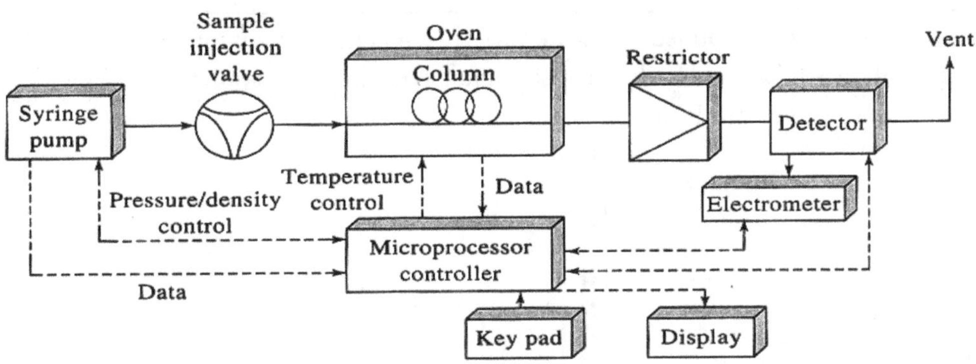

Fig. : Schematic of an instrument for supercritical fluid chromatography

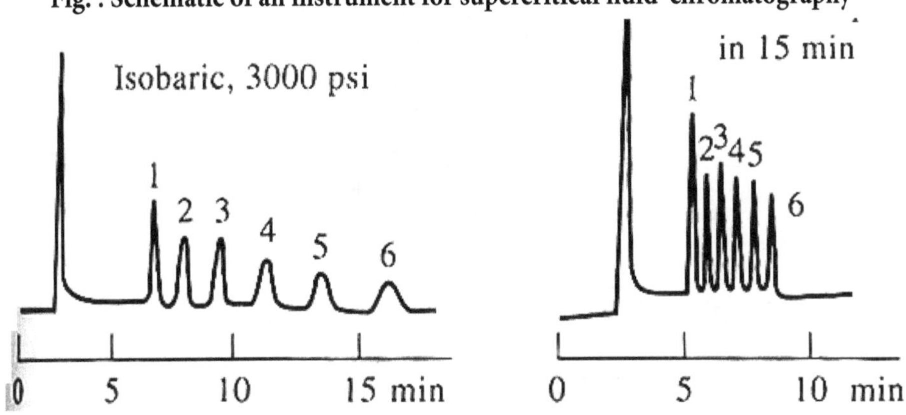

Fig. : Effect of pressure programming in supercritical fluid chromatography (Courtesy of brownlee labs , santa clara ,. CA)

Sample:	1. cholesteryl octanoate
	2. cholesteryl decylate
	3. cholesteryl laurate
	4. cholesteryl myristate
	5. cholesteryl palmitate
	6. cholesteryl stearate
Column:	DB−1
Mobile phase:	CO_2
Temperature:	90° C
Detector:	FID

14.3 APPLICATIONS

By now, SFC has been applied to a wide variety of materials, including Natural products, drug, foods, pesticides and herbicides, surfactants polymers, and polymer additives, fossil fuels, and explosives and Propellants.

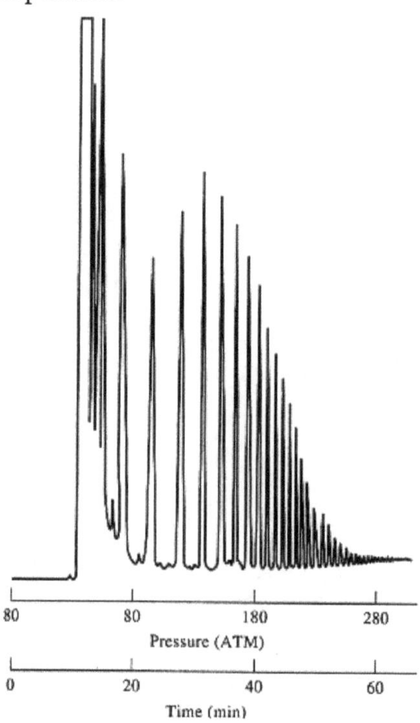

Fig. : Separation of oli gomers of dimethy, polysiloxane by supercritical fluid chromatography. (From C. M. White and R. K. Houck, HRC & CC, 1986, 9, 4. With , Permission

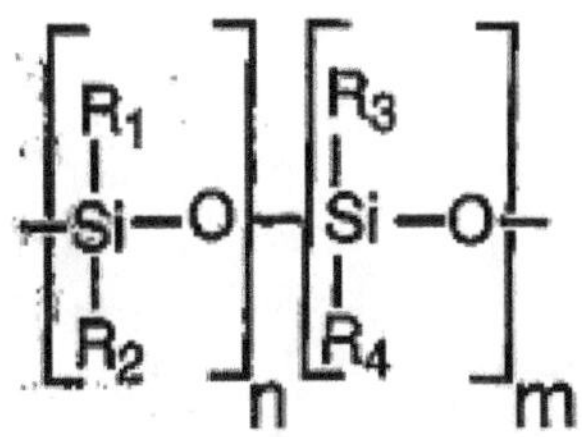

Dimethylpolysiloxane: non-volatile and special function groups

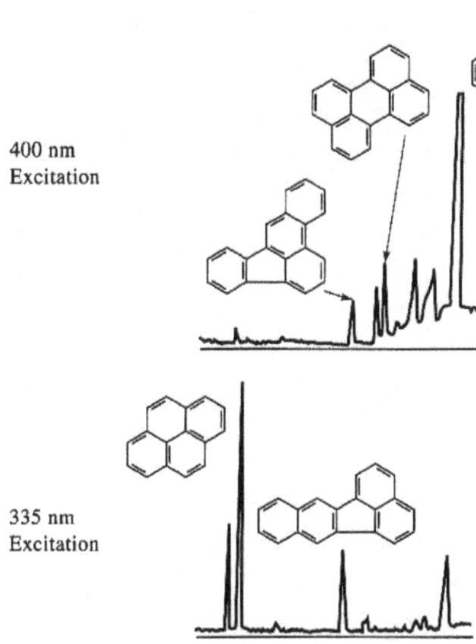

Fig. : Portions of the supercritical fluid chromatograms of polycyclic aromatics in a carbon black extract illustrating the selectivity achieved by fluorescence excitation at two wavelengths (From C. M. White and R. K. Houck, HRC & CC, 1986, 9, 4. With , Permission)

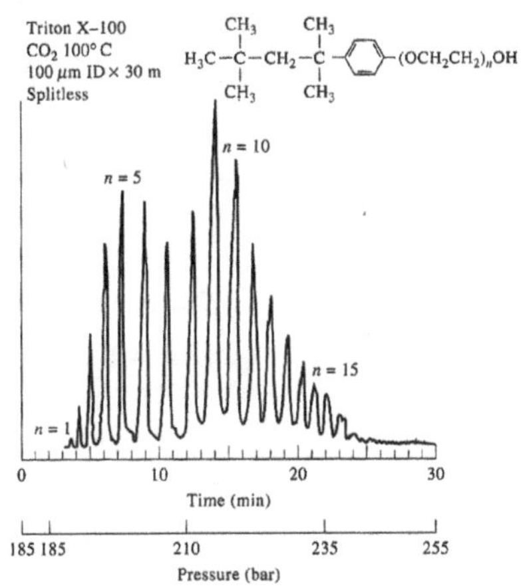

Fig. :Chromatograms for the nonionic surfactant Triton X-100 with total current mass spectrometric detection.

15

APPLICATION OF GAS AND LIQUID CHROMATOGRAPHY COUPLED TANDEM MASS SPECTROMETRY SYSTEMS FOR PESTICIDE QUANTIFICATION IN EXPORTED FRUIT AND VEGETABLES

15.1 INTRODUCTION: AN OVERVIEW

Monitoring pesticide residues in fruit and vegetables remains a key priority for international food safety. Crop growth and maintenance within emerging export economies such as China and India operate under substantially different regulations compared to their Western counterparts. Stringent pesticide monitoring is therefore essential to meet international regulatory requirements, ensure product safety and enable export to the widest possible market.

Over the past decade, the methods employed for pesticide monitoring have aligned with advances in the sophistication and availability of powerful analytical technology. Since 2004, the U.S. Food and Drug Administration (FDA) has recommended the QuEChERS (Quick, Easy, Cheap, Effective, Rugged and Safe) method for pesticide residue screening based on multi-reaction monitoring (MRM) with advanced gas or liquid chromatography coupled mass spectrometry. However, achieving accuracy and robustness with these routine quality control methodologies can be complex and time consuming, particularly when it comes to trace pesticide analysis within complex biological matrices. This chapter demonstrates how triple quadrupole mass spectrometry coupled gas and liquid chromatography systems [SCION gas chromatography/mass spectrometry (GC-MS) and EVOQ liquid chromatography/mass spectrometry (LC-MS)] provide the sensitive MRM analysis required for pesticide quantification in domestic and exported goods.

15.2 BREAKING DOWN THE BARRIERS TO INTERNATIONAL TRADE

Today, emerging economies around the world are benefiting substantially from the increased liberalization of global trade. Agricultural imports to the U.S. from China, for example, increased from $800 million in 2000 to approximately $3.4 billion in 2010. The imports continue to grow, exemplifying a strengthening commercial relationship that would have seemed incredible 30 years ago.] However, several barriers still exist that limit producers from accessing the full potential of the export market. Chief among these is the lack of global consensus or standardization on pesticide use. Products imported from regions where pesticide use is less restricted to those with stringent regulations are frequently subject to detention and often returned to their country of origin or disposed of, resulting in an immediate loss in investment.

In response, governments around the world are taking steps to bring their regulations guiding crop growth and maintenance in line with their Western counterparts. In Pakistan, the provincial government has announced its intention to harmonize rice production with FDA guidance, whereas China has recently moved to curtail the use of harmful pesticides widely banned on the international stage.[2, 3] Robust and accurate pesticide detection methods are essential to ensure consumable products comply with both international and domestic regulatory standards.

Extracting pesticide species from complex fruit and vegetable biomass samples is a complicated process. Moreover, achieving the sensitivity and specificity required to deliver and maintain accurate and reliable trace level analysis has been--until recently--beyond the scope of conventional analytical techniques. Finally achieving this high performance can be a laborious and time intensive task, which is undesirable for high throughput routine quality control. Advancing detection methodologies and analytical technology to accurately, robustly and rapidly qualify and quantitate trace pesticide residue is therefore a priority for regulatory bodies and instrument developers alike.

15.3 IMPROVING DETECTION PROCEDURES

In 2004, the FDA introduced QuEChERS sample preparation for pesticide screening. QuEChERS overcomes many of the limitations associated with routine quality control including lengthy analysis times, high sample volume and lack of assurance on complete pesticide recovery. Today, the method is standardized within AOAC 2007.01 EN15662 and has proven orders of magnitude more productive than conventional preparation techniques.

The QuEChERS protocol relies on homogenizing a sample through blending followed by centrifugation and extraction with a suitable reagent. The sample is then analyzed using a suitable analytical tool. The sophistication and availability of modern gas and liquid chromatography coupled mass spectrometry systems has led to them being adopted as the platform for QuEChERS screening.

In both GC-MS and LC-MS, analytes are eluted from the separation column into the MS component where ions are then detected and characterized in relation to their molecular mass. GC-MS is best suited for the analysis of nonpolar, volatile compounds while LC-MS is also able to separate nonvolatile species and is amenable for a greater range of pesticide products. Modern LC-MS instruments are generally considered more powerful, however where routine detection of known volatile analytes is required GC-MS systems are a suitable and lower priced solution.

Despite the increased sophistication, many commercially available GC-MS and LC-MS systems struggle to deliver the sensitivity and specificity for low level screening operations and many technical aspects of their functionality limit performance. Triple quadrupole Multi Reaction Monitoring (MRM) overcomes many of these limitations to deliver these improvements and is considered the gold standard for pesticide detection.

15.3.1 Innovations in LC-MS and GC-MS to Improve Pesticide Quantification Performance

Multiple reaction monitoring fragments analytes into multiple ions (see Figure 1). Incorporating this technique into GC-MS and LC-MS results in greater specificity, a dramatic increase in signal-to-noise ratio and better quantitative performance. In this way, MRM helps to maximize reliability in pesticide detection and ensure confidence in results.

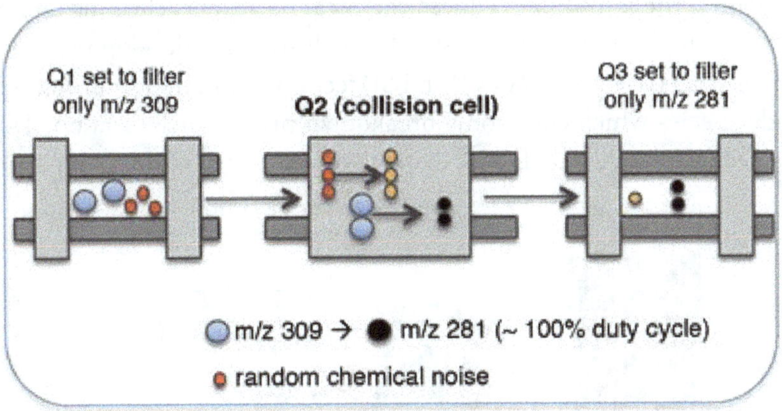

Fig. 1: Triple quadrupole mass spectrometry multi reaction monitoring fragments analytes into multiple ions to achieve greater specificity and confidence in analysis.

Innovations in hardware have refined the performance of triple quadrupole MS for pesticide quantification to ensure high performance is maintained throughout high throughput analysis. In GC-MS, implementing an axial ion source reduces build up in the ion source, as demonstrated by the Bruker SCION GC-MS Triple Quadrupole (see Figure 2). This maintains higher signal to noise across thousands of analyses, ensuring continued high performance and increased instrument up time with reduced cleaning.

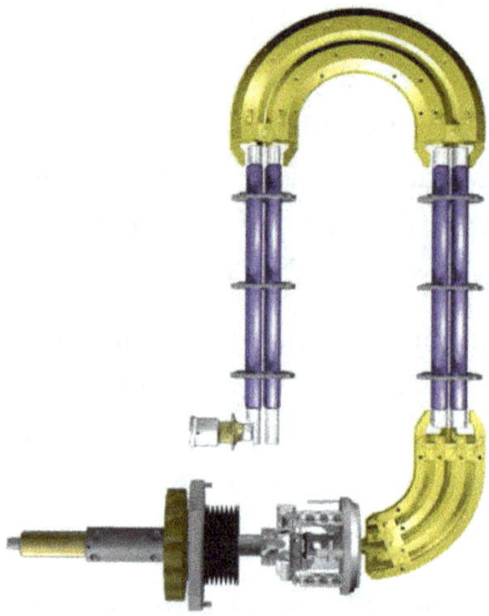

Fig. 2: An axial ion source provides higher transmission, allowing increased instrument up time with reduced cleaning.

Advances in LC-MS/MS design deliver similar improvements in sensitivity and robustness by optimizing ion transfer. For example, using an active exhaust to reduce gas recirculation within the ion source reduces chemical noise, cuts cleaning and maintains sensitivity during high throughput pesticide analysis. The development of vacuum insulated, probe-heated electrospray (VIP-HESI) technology moreover protects thermally labile pesticide species which commonly breakdown prior to nebulization. This further enhances the pesticide range that LC-MS can tackle, while again improving robustness.

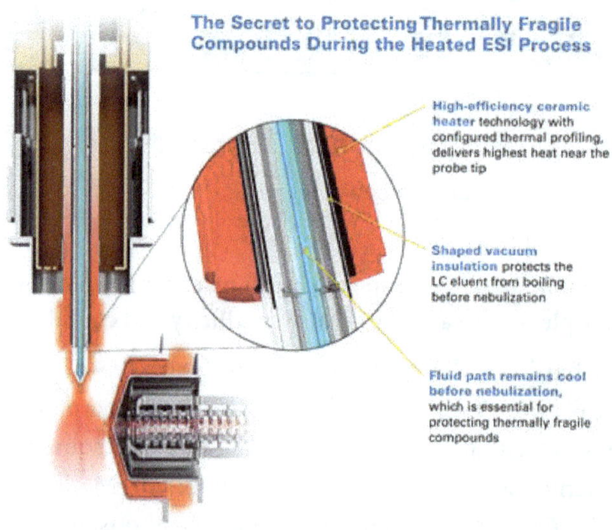

Fig. 3: Vacuum insulated, probe-heated electrospray technology ensures high signal-to-noise ratios and superior robustness of thermally labile species.

15.3.2 Easing Method Development with Compound-Based Scanning

As multi-residue pesticide analysis methods can have hundreds of targets to quantify in a single run, developing new methods can be a time-consuming task. However, compound-based scanning (CBS) streamlines method development for multi-residue analysis, using a compound library containing more than 2,500 MRM transitions covering more than 900 contaminants. CBS dramatically reduces the time taken to set the initial MRM methods to give accurate data while simultaneously improving workflow and productivity.

Software advances have also streamlined data review, to deliver the fastest sample-to-report time possible. Exception based data review significantly reduces the error rate for quantitative analysis by flagging any runs that fail to meet pre-set method criteria, as demonstrated by PACER from Bruker. This removes the need for nonexpert human interaction to judge chromatograms and enables analysts to focus on problem areas. In this way PACER instantly increases efficiency by eliminating unproductive time without compromising data quality. Ultimately, this works to ensure maximum productivity and reliability in pesticide screening.

The following case study explores how LC-MS/MS and GC-MS/MS systems incorporate these hardware and software developments to provide robust, fast and simple analysis of complex food matrices, while making use of the new QuEChERS protocol.

15.4 ANALYSIS OF FRUIT AND VEGETABLES WITH GC-MS/MS AND LC-MS/MS

Samples of rice, avocado and spinach--all representing low moisture, fatty and high moisture content vegetable groups--were extracted using a modified QuEChERS protocol (Figure 4).[4] Thirty pesticides amenable for both GC-MS and LC-MSwere spiked into the three vegetable matrices and prepared for analysis with the EVOQ LC-MS/MS and the SCION GC-MS/MS. The MRM method development workflow was set up using CBS. Target pesticides were selected from the software's MRM library before being exported to the CBS method editor. Optimal dwell times were then automatically calculated for each MRM based on its retention time window.

Excellent sensitivity was achieved for multi-residue pesticides within the vegetable matrices using both the GC-MS/MS and LC-MS/MS systems. This is shown in Table 1, which illustrates the R^2 values for the pesticides achieved, demonstrating excellent linearity was achieved.

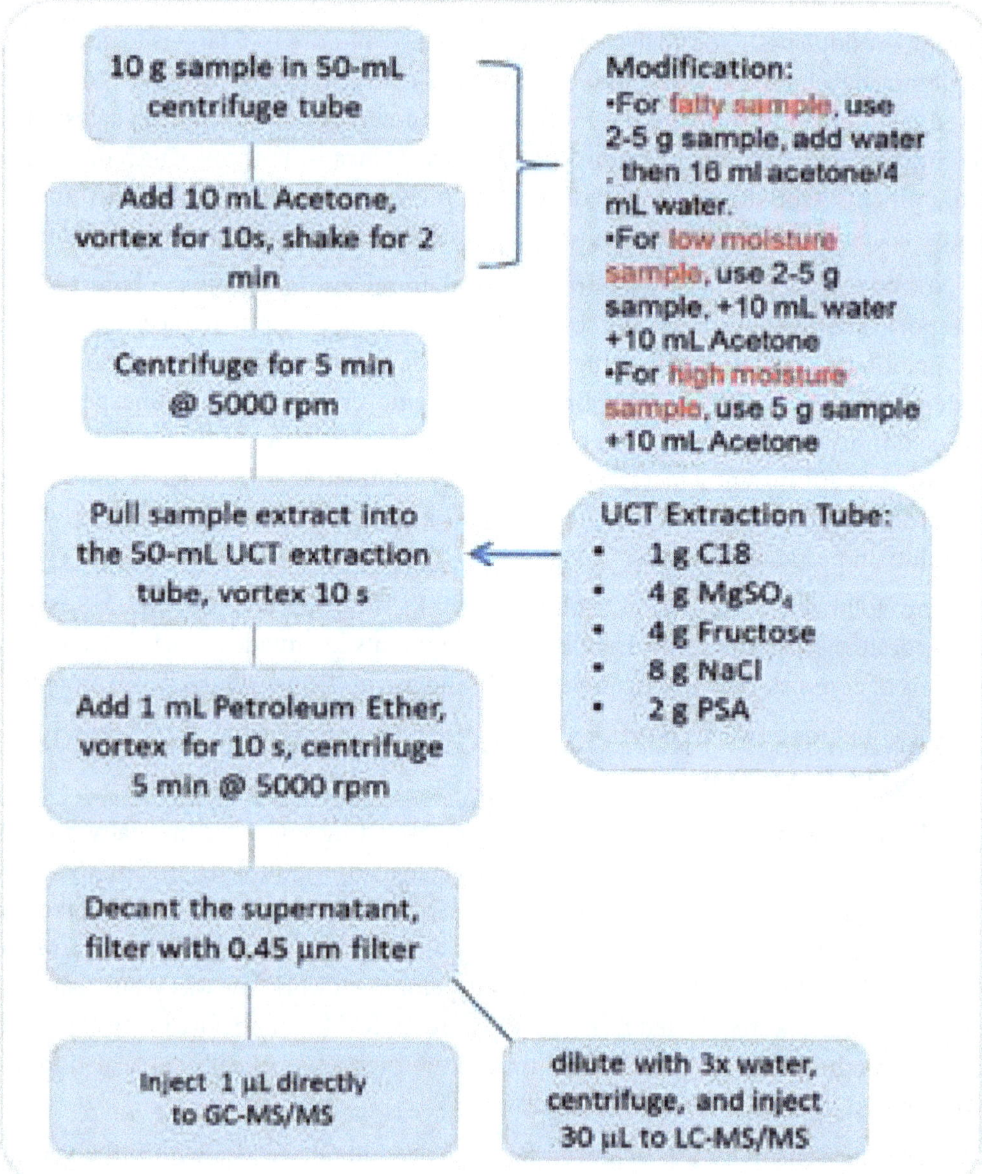

Fig. 4: The modified QuEChERS method further simplifies sample preparation of complex matrices.

Table 1: R2 values achieved for the calibration of pesticides in different matrices using GC-MS/MS and LC-MS/MS.

Compound name	Control		Rice		Avocado		Spinach	
	GC-MS/MS 1 - 100 ppb	LC-MS/MS 0.1 - 100 ppb	GC-MS/MS 1 - 100 ppb	LC-MS/MS 0.1 - 100 ppb	GC-MS/MS 1 - 100 ppb	LC-MS/MS 0.1 - 100 ppb	GC-MS/MS 1 - 100 ppb	LC-MS/MS 0.1 - 100 ppb
Ametryn	0.998	0.999	1.000	0.994	0.999	0.993	1.000	0.998
Azaconazole	0.997	0.999	0.999	0.995	0.999	0.994	1.000	0.999
Azoxystrobin	0.995	0.987	0.999	0.999	0.999	0.997	0.999	0.997
Benalaxyl	0.998	0.997	1.000	0.999	0.999	0.988	1.000	0.991
Bromacil	0.997	0.997	0.999	0.997	0.999	0.992	1.000	0.999
Butralin	0.995	0.999	0.999	0.998	0.997	0.998	1.000	0.993
Carboxine	0.997	0.992	1.000	0.998	0.996	0.999	0.998	0.995
Clomazone	0.998	0.991	1.000	1.000	0.999	0.998	1.000	0.990
Coumaphos	0.994	0.997	1.000	0.987	0.998	0.998	1.000	0.993
Diethofencarb	0.997	0.997	0.999	0.999	0.999	0.995	1.000	0.989
Diniconazole	0.996	0.998	0.999	0.999	0.999	0.986	1.000	0.998
Fenamidone	0.996	0.996	1.000	1.000	0.999	0.999	1.000	0.997
Fenamiphos	0.995	0.993	1.000	1.000	0.999	0.998	1.000	0.993
Fenbuconazol	0.993	0.997	0.999	0.999	0.999	0.995	0.999	1.000
Fenothiocarb	0.997	0.998	0.999	0.995	0.999	0.992	1.000	0.997
Fenpropimorph	0.997	0.995	0.999	0.997	0.999	1.000	1.000	0.992
Flusilazole	0.997	0.998	0.999	1.000	0.999	1.000	1.000	0.998
Hexaconazole	0.995	0.996	0.999	0.999	0.999	0.996	1.000	0.999
Hexazinone	0.997	1.000	1.000	0.989	0.999	0.992	1.000	0.996
Imazalil	0.993	0.989	0.999	0.992	0.999	0.997	0.987	0.992
Isoprocarb	0.998	0.994	1.000	0.998	0.999	0.997	1.000	0.991
Myclobutanil	0.997	0.996	0.999	0.998	0.999	0.998	1.000	0.990
Napropamide	0.999	0.995	1.000	0.995	0.999	0.994	0.999	0.998
Pendimethalin	0.993	0.998	0.998	0.999	0.997	0.998	0.999	0.994
Pyriproxifen	0.997	0.999	0.999	1.000	0.999	1.000	1.000	0.997
Tebuconazole	0.995	0.994	1.000	0.999	0.999	0.993	1.000	0.997
Thiabendazole	0.983	1.000	0.999	0.997	0.995	0.996	0.999	1.000
Thiamethoxam	0.998	0.997	0.999	1.000	0.989	1.000	0.998	0.997
Tricyclazole	0.994	0.991	0.999	0.995	N/A	0.999	0.992	0.996
Triflumizole	0.996	0.996	0.999	0.998	0.999	0.993	1.000	0.997

15.5 CONTROLLING CONCENTRATIONS

Meeting international pesticide regulations is one of the major challenges to exploiting the expanding consumables market. The need to sharpen the sensitivity, selectivity and ease of analysis for these routine pesticide analysis operations continues to drive innovation in instrument design. The advances in tandem mass spectrometry coupled chromatographic system hardware and software provide this high performance while simplifying method development, reducing instrument downtime for maintenance and maximizing ease of use, to meet the ongoing needs of pesticide monitoring.

16

LIQUID CHROMATOGRAPHY-TANDEM MASS SPECTROMETRY ANALYSIS OF ACETAMINOPHEN COVALENT BINDING TO GLUTATHIONE S-TRANSFERASES

16.1 INTRODUCTION: AN OVERVIEW

Hepatotoxicity induced by acetaminophen (APAP), a widely used analgesic and antipyretic, presents major health care challenges in Western societies. At therapeutic doses, APAP is considered harmless. However, at higher doses, it can act as a potent hepatotoxin, and cause cell necrosis and acute liver failure. Liver failure induced by APAP correlates often

Acetaminophen (APAP)-induced hepatotoxicity is the most common cause of acute liver failure in the Western world. APAP is bioactivated to N-acetyl p-benzoquinone imine (NAPQI), a reactive metabolite, which can subsequently covalently bind to glutathione and protein thiols. In this study, we have used liquid chromatography-tandem mass spectrometry (LC-MS/MS) to characterize NAPQI binding to human glutathione S-transferases (GSTs) in vitro. GSTs play a crucial role in the detoxification of reactive metabolites and therefore are interesting target proteins to study in the context of APAP covalent binding. Recombinantly-expressed and purified GSTs were used to assess NAPQI binding in vitro. APAP biotransformation to NAPQI was achieved using rat liver microsomes or human cytochrome P450 Supersomes in the presence of GSTA1, M1, M2, or P1. Resulting adducts were analyzed using bottom-up proteomics, with or without LC fractionation prior to LC-MS/MS analysis on a quadrupole-time-of-flight instrument with data-dependent acquisition (DDA). Targeted methods using multiple reaction monitoring (MRM) on a triple quadrupole platform were also developed by quantitatively labeling all available cysteine residues with a labeling reagent yielding isomerically-modified peptides following enzymatic digestion. Seven modified cysteine sites were confirmed, including Cys112 in GSTA1, Cys78 in GSTM1, Cys115 and 174 in GSTM2, as well as Cys15, 48, and 170 in GSTP1. Most modified peptides could be detected using both untargeted (DDA) and targeted (MRM) approaches, however the latter yielded better detection sensitivity with higher signal-to-noise and two sites were uniquely found by MRM.

with overdose or misusage, including chronic overuse. This often results in difficulties in diagnosis and treatment.

APAP is mostly detoxified via phase II metabolism; with direct glucuronidation (~55%) and sulfation (~40%). However, up to 10% undergoes bioactivation to N-acetyl p-benzoquinone imine (NAPQI), a highly reactive electrophile, which is known to covalently bind to nucleophilic centers, including cysteine thiols in liver proteins (Figure1). At higher doses, the fraction of oxidized APAP can increase up to 15%, due to saturation of the sulfation pathway. The formation of NAPQI is mediated by the cytochrome P450 (CYP) superfamily, with CYP1A2, 2D6, 2E1, and 3A4 isoforms as the major APAP oxidizing enzymes. NAPQI can be detoxified by subsequent phase II metabolism via conjugation with glutathione (GSH), partly mediated by different glutathione S-transferase (GST) isozymes.

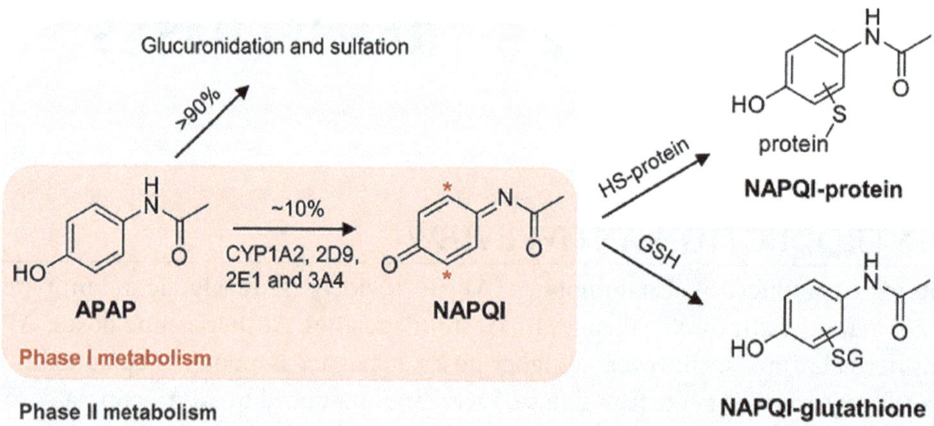

Fig. 1: Simplified scheme of APAP metabolism, leading to the formation of NAPQI with subsequent GSH conjugation. Hepatotoxicity of NAPQI is linked to protein binding to cysteine sites.

GSTs represent major detoxification enzymes with a protein size of ~25 kDa and multifunctional purposes, including induction of microsomal peroxidation and heavy metal resistance, as well as possible roles in diseases, such as pulmonary fibrosis. Members of cytosolic GSTs tend to dimerize with subunits of the same class, which results in a larger number of enzyme possibilities than genes. Different genetic variants of GST have shown varying effectiveness in the detoxification of NAPQI. The isozymes GST mu 1, pi 1, and theta 1 (GSTM1, P1, and T1) have been found to be most effective. However, due to staggered or inadvertent APAP overuse, cellular GSH can be depleted and the GSH conjugation pathway can be saturated, resulting in NAPQI accumulation, leading to increased protein binding. NAPQI modifications of certain proteins can alter critical cellular functions and NAPQI-protein adducts have been found to accumulate in hepatocytes. Thus, functional and conformational protein changes may ultimately lead to subsequent cell death in liver. In a previous study, we demonstrated a dose response to protein binding, using absolute quantitation of NAPQI-adducts to Cys34 in rat serum albumin in vivo . A subsequent related study quantifying modified human albumin in patient plasma samples was able to distinguish between moderate and severe APAP-induced acute liver failure.

Several NAPQI-protein adducts have been identified previously in mouse, rat, and human. These protein targets consist mostly of cytoplasmic liver proteins, most likely due to the increased solubility and detectability of these proteins. Included in these identified targets is mouse GSTP1. In vitro binding studies have also shown NAPQI-GST adducts, using 14C-APAP metabolism in mouse liver homogenates. In a previous study in rat liver microsomes (RLM), we reported that rat microsomal GST 1 (MGST1) was a target of NAPQI. A subsequent paper confirmed rat MGST1 as well as GSTM1 as targets. Binding to human MGST1 was observed by studying human liver microsomes incubations of APAP. found recombinant human GSTP1 modified after NAPQI formation by a CYPBM3 mutant in vitro. Certain of these previous reports were able to pinpoint the specific cysteine residues which were modified in the GSTs, namely Cys50 in rat and human MGST1, and Cys48 in human GSTP1. All of these cysteines are significant for enzyme function and alkylation would therefore result in inhibition. Since GSTs play a significant role in detoxification, further research into the susceptibility of different GST isozymes to being modified by NAPQI remains of high interest.

The analysis of proteins and their modifications benefits from continual advancements in liquid chromatography-tandem mass spectrometry (LC-MS/MS), a broadly used tool for proteomic investigations. Constant improvements in sensitivity and selectivity have made LC-MS/MS an invaluable method to characterize proteins and study their post-translational modifications. The development of high-resolution mass spectrometry (HRMS), data-dependent acquisition (DDA), targeted multiple reaction monitoring (MRM), and multidimensional LC have all contributed significantly to MS-based proteomics for highly sensitive detection of proteins and their modifications. As a result of its unique ability to distinguish changes in exact sites in the protein sequence, MS/MS has become especially vital in post-translational modification studies.

In this work, several analytical approaches (Figure 2) were investigated combining different techniques for sample preparation and LC-MS/MS analysis to identify APAP-related covalent binding to purified human GSTs (overview in Table 1, note: Met1 is considered here as the first amino acid in protein sequences throughout the whole manuscript). Strength of our study was not needing to use radioactivity or a fluorescence tag to selectively detect modified proteins or peptides (a brief overview of protein detection methods can be found in Table S1). In vitro NAPQI-protein binding to four GSTs was investigated (GST alpha 1 (A1), M1, M2, and P1). APAP oxidation was performed in vitro using RLM or CYP3A4 Supersomes. Then, NAPQI-modified proteins were digested via two proteases (trypsin and pepsin) in parallel and analyzed by two-dimensional LC-

MS/MS, using high-pH reversed-phase (RP) offline fractionation. Results of fractionated and non-fractionated samples were compared. MS/MS was employed with DDA on a high-resolution quadrupole-time-of-flight platform. High-sensitivity MRM measurements were then used for targeted analysis of modified peptides, using reference standards of all possible cysteine modified peptides.

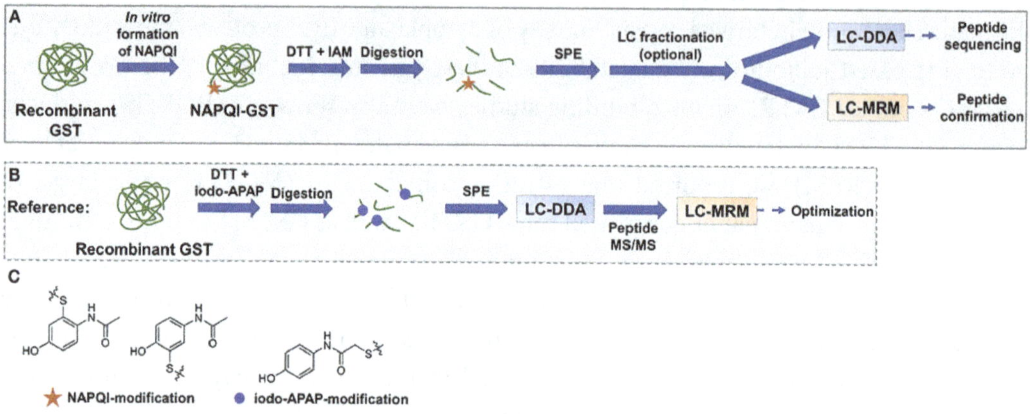

Fig. 2: Representative workflow of in vitro NAPQI binding to recombinant GSTs followed by reductive alkylation and digestion (trypsin or pepsin), SPE with optional LC fractionation, and LC-MS/MS analysis (A). NAPQI generation was achieved by activation of APAP with either RLM or CYP3A4 Supersomes. MRM methods were based on isomeric iodo-APAP-GST standard, digested and first analyzed by DDA to investigate ionization and fragmentation properties of iodo-APAP-peptides (B). Then, MRM transitions were built and optimized for each peptide individually. Cysteine alkylation reagent iodo-APAP yields positional isomer label to NAPQI-modified cysteine (C).

Table 1. Overview of studied human GST enzymesa.

GST	Accession number	Molecular weight [Da]	pI	Active site (function)	Cysteine site	References
A1	P08263	25,631	8.22	Tyr9 and Arg45 (GSH binding)	Cys112	Balogh et al. (2009)
M1	P09488	25,712	5.99	Lys50 (GSH binding) and Tyr116 (substrate binding)	Cys78, 87, 115, and 174	Johnson et al. (1993), Patskovsky et al. (2006)
M2	P28161	25,745	5.79	Lys50 (GSH binding), Tyr116 (substrate binding), and Thr210 (substrate specificity)	Cys87, 115, and 174	Johnson et al. (1993)
P1	P09211	23,356	5.29	Tyr8, Arg14, Trp39, and Lys45 (GSH binding)	Cys15, 48, 102, and 170	Ji et al. (1994), Oakley et al. (1997a,b), Prade et al. (1997), Ang et al. (2009), Federici et al. (2009)

[a]*Met1 is considered here as first entry in the amino acid sequence.*

16.2 MATERIALS AND METHODS

16.2.1 Chemicals and Materials

Aroclor 1254-induced male Sprague-Dawley rat liver microsomes (RLM) (part no.: M10001, lot no.: QEB and OTS) were purchased from BioreclamationIVT (Baltimore, MD). Human CYP3A4 Supersomes (also containing oxidoreductase and cytochrome b5, part no.: 456202, lot no.: 7304001) were from Corning (Corning, NY). Cysteine alkylating agent iodo-APAP (N-(4-hydroxyphenyl)-2-iodoacetamide) was synthesized in-house as previously described Escherichia coli cells were from laboratory stocks. Magnesium chloride and potassium phosphate (dibasic) were from Anachemia (Montréal, QC). Trypsin (TPCK-treated, from bovine pancreas), pepsin (from porcine gastric mucosa), dithiothreitol (DTT), acetaminophen (APAP), HPLC-grade acetonitrile (ACN) and methanol (MeOH), iodoacetamide (IAM), glucose-6-phosphate dehydrogenase (type XV, from baker›s yeast), and other chemicals were purchased from Sigma-Aldrich (St. Louis, MO). Ultrapure water was from a Millipore Synergy UV system (Billerica, MA).

GST Expression and Purification Protocols

Recombinant human GSTA1, M1, M2, and P1 proteins were expressed in E. coli and purified using a modification of published protocols. Overnight induction was proceeded when the culture reached $OD_{600} = 0.6$. Cells were lysed as previously described. Protein concentration was determined with the Bradford assay, using bovine serum albumin as reference standard. GST affinity purification was performed on a GSH-agarose column. Purified GSTs were dialyzed to remove GSH by 3 L changes against 500 μL into ammonium acetate for 48 h. Proteins were stored at −80°C.

Preparation of Alkylated GSTs Using iodo-APAP or IAM

Reductive alkylation of human GSTA1, M1, M2, or P1 (1.8–3.6 nmol) was performed in 100 mM ammonium bicarbonate (ABC, pH 8.5) using DTT (250 nmol; 20 min, 37°C), and either iodo-APAP (750 nmol in ACN; 45 min,) or IAM (750 nmol; 30 min) at 37°C (in dark). Alkylated proteins were incubated with either 10 μg trypsin (4 h at 37°C) in ABC buffer, or 10 μg pepsin (1 h at 37°C) with added 1% formic acid in 10% MeOH to obtain pH 2 during digestion. Trypsin digests were diluted with water; pepsin digests with ABC (100 mM), prior to solid-phase extraction (SPE) on 1 cc (30 mg) OASIS HLB cartridges (Waters, Milford, MA) eluted with 100% MeOH (1 mL). Eluates were dried and then resolubilized in 100 μL of 10% ACN prior to LC-MS/MS analysis.

NAPQI-GST Formation by RLM Incubation

RLM (1 mg/mL protein) and APAP (100 μM) were incubated (37°C and 500 rpm) in the presence of NADP+ (500 μM), and a NADPH-regenerating system: glucose 6-phosphate (10 mM), glucose-6-phosphate dehydrogenase (2 U/mL) and magnesium chloride (5 mM). Human GST (A1, M1, M2, or P1) in 100 mM phosphate buffer (pH 7.4, 1.8–3.6 nmol protein amount) was added to react with produced NAPQI for 3 h. The final incubation volume was 200 μL. Then, proteins were diluted in ABC (100 mM); 200 and 150 μL for tryptic and peptic digestion, respectively.

NAPQI-GST Formation by CYP3A4

CYP3A4 Supersomes (1 mg/mL protein, 125 pmol/mL CYP3A4 final concentration) and APAP (100 μM) were incubated (37°C and 500 rpm) in the presence of NADPH (1 mM in phosphate buffer). Then, GST (same as above) was added to react with produced NAPQI and diluted after 3 h (as above).

Digestion of NAPQI-GST

Reductive alkylation was performed using 10 μL DTT (25 mM; 20 min, 37°C) and 10 μL IAM (63 mM, 30 min, 37°C, in dark). All samples were then digested with either 10 μg trypsin (4 h at 37°C), or with 400 μL solution of 1% formic acid in 10% MeOH with 10 μg pepsin (1 h at 37°C). Trypsin digestion was diluted with 500 μL water, pepsin digestion with 200 μL ABC (100 mM). Digests were then cleaned-up by SPE and reconstituted as above.

High-pH RP Peptide Fractionation

Dried extracts were reconstituted in 10% ACN (120 μL) for injection (100 μL) onto a ZORBAX Extend-C18 column (250 × 4.6 mm; Agilent Technologies, Palo Alto, CA) with 5 μm (100 Å) particles on an Agilent 1200 series HPLC equipped with a temperature-controlled autosampler (at 8°C), binary pump, degasser, diode array detector, and cooled fraction collector (at 8°C). High-pH RP fractionation was performed at a flow rate of 0.6 mL/min with a gradient starting at 5% B for 2 min, increased to 50% B in 12 min, to 70% B within 0.5 min, and held for 6.5 min. Mobile phase A was 10 mM ammonium acetate (pH 10, adjusted with ammonium hydroxide) in water and mobile phase B was prepared with 10% A / 90% ACN. UV absorbance was monitored at 220 and 280 nm. Then, 16 (1 min) fractions were collected from 3 to 19 min. Fractions were concatenated (fraction 1+9, 2+10, …) into eight final fractions, dried and reconstituted in 100 μL 10% ACN.

LC-MS/MS Analysis With DDA

Samples were injected (30 μL) onto an Aeris PEPTIDE XB-C18 100 × 2.1 mm column, with solid core 1.7 μm particles (100 Å) fitted with a SecurityGuard ULTRA C18-peptide guard column (Phenomenex, Torrance, CA) using a Nexera UHPLC system (Shimadzu, Columbia, MD) with water (A) and ACN (B), both containing 0.1% formic acid, at a flow rate of 0.3 mL/min (40°C). The gradient started at 5% B (held for 2.5 min) and was linearly increased to 30% B within 37.5 min, to 50% B within 10 min, then to 85% B within 5 min (held for 3 min). MS and MS/MS spectra were collected on a high-resolution TripleTOF 5600 system (quadrupole-time-of-flight; Sciex, Concord, ON) equipped with a DuoSpray ion source in positive mode set at 5 kV source voltage, 500°C source temperature, and 50 psi GS1/GS2 gas flows, with a declustering potential of 80 V. The instrument performed a survey MS acquisition (TOFMS) from m/z 120–1,250 (250 ms accumulation time), followed by MS/MS (high sensitivity mode) from m/z 80–1,500 on the 15 most intense precursor ions (excluded for 20 s after two occurrences) using information dependent acquisition with dynamic background subtraction. Each MS/MS acquisition had an accumulation time of 50 ms and a collision energy (CE) of 30 ±10 V. The total cycle time was 1.05 s. MS and MS/MS calibration was performed after every four injections with a set of in-house standards, using an automatic calibrant delivery system. Sciex Analyst software version 1.7.1 (TripleTOF) was used for data acquisition. Raw data was visualized with PeakView 2.2.

Human files were searched against the UniProtKB/Swiss-Prot protein database (release date: 07/18/2018, including common protein contaminations) by ProteinPilot 5.0 software (Sciex) using the Paragon Algorithm To find potential APAP-related adducts, the feature probability was altered to 50% for NAPQI on cysteine (addition of $C_8H_7NO_2$). The search was performed for +2 to +4 charge states at a MS tolerance of 0.05 u on precursor ions and 0.1 u on fragment ions. Peptides and proteins were identified with a 1% false discovery rate (FDR) using a target-decoy database search algorithm.

LC-MS/MS Analysis in MRM Mode

LC-MRM experiments were performed on an identical LC platform and column, and a QTRAP 5500 (Sciex, Concord, ON) hybrid quadrupole-linear ion trap system with a TurboIonSpray ion source in positive mode. For fractionated samples, a shorter LC gradient (~28 min) was used. The gradient started at 5% B (held for 2.5 min) and was linearly increased to 30% B within 21.5 min, to 50% B within 2 min, then to 85% B within 0.5 min (held for 2 min). Source parameters were as follows: ion spray voltage, 5 kV; temperature, 550°C; GS1 and GS2, 50 psi; and curtain gas, 35 psi. Entrance and collision cell exit potentials were set at 10 and 13 V, respectively. Ion activation via collision-induced dissociation was performed at collision gas pressure of 5 (arbitrary units). Individual MRM transitions with corresponding CE settings can be found in Table S2. MRM transitions of modified peptides were integrated using Sciex MultiQuant 2.1. Sciex Analyst software version 1.6 was used for data acquisition. Raw data was visualized with PeakView 2.2.

16.3 RESULTS

16.3.1 Method Development

For unbiased analysis of binding to cysteines in recombinant GSTs, each GST was probed for covalent GSH-cysteine adducts, which could have occurred during GST purification over GSH affinity chromatography. GSTs were digested using trypsin and pepsin without reduction prior to alkylation, and then analyzed by DDA and subsequent database searching. Several cysteine GSH modifications were observed. However, these adducts were found with low absolute abundance compared to free cysteine (alkylated). An average relative signal of 0.03 was estimated, comparing glutathionylated peptides to alkylated peptides.

Optimal in vitro metabolism of APAP was examined by GSH trapping experiments (data not shown). Oxidizing RLM and CYP3A4 Supersomes were each tested with either NADPH, or NADP+ and a NADPH regenerating system. RLM incubation in combination with a NADPH regenerating system and 3A4 Supersomes in the presence of NADPH yielded optimal conditions for APAP in vitro metabolism. For both strategies, a combination of 1 h open tube incubation followed by 2 h closed tube was found best for oxidation. It must be noted that GST stock solutions contained reduced GSH, which also bound to NAPQI (data not shown). Denaturation and excessive buffer exchange of GSTs, to yield completely GSH-free stocks, did not permit the study of binding capabilities based on protein structure. Additionally, excessive treatment of proteins led to important sample loss, different for each protein (data not shown). Furthermore, inclusion of a His-tag during expression (for subsequent Ni-nitrilotriacetic acid purification) might alter protein folding and thus binding results. The use of non-mutant GSTs purified over GSH affinity combined with solvent-free elution seemed optimal in this study. These facts were considered and APAP concentrations in final incubations were adjusted accordingly to assure detectable amounts of NAPQI-GST being formed. Subsequently, protein digestion, SPE and offline fractionation were optimized for highest sequence coverage of target GST from database search results of DDA runs. Trypsin and pepsin digestion (4 and 1 h, respectively) in parallel was found optimal to yield a sequence coverage >83% (data not shown). Standard iodo-APAP-GST was used to test cysteine coverage and detection of all possible modified cysteine (see Table 2). Only the sites Cys15 and 102 were detected with a confidence <95%, based on small sequence size and low fragmentation of tryptic C15(iodo-APAP)AALR, and poor abundance of peptic GVEDLRC102(iodo-APAP) KYISL. For peptide separation, high-pH RP and fraction concatenation was used as an alternative strategy to conventional ion exchange offline fractionation. This enabled orthogonal separation to online (low-pH) RP chromatography without a desalting SPE step, thus reducing potential sample loss.

Table 2. Coverage of iodo-APAP-cysteine sites in DDA HRMS/MS.

GST	Cys	Peptide	z	Peptide confidence (%)	(DDA) absolute precursor area [cps]	RT [min]
Trypsin digestion						
M1	78	ITQSNALIC(iodo-APAP)Y	+2	95	6.48E+06	24.4
	87	HNLC(iodo-APAP)GETEEEK	+3	99	1.21E+06	7.0
	115	GMIC(iodo-APAP)YNPEFEK	+2	99	7.99E+05	21.6
	174	C(iodo-APAP)GESEK	+2	99	1.76E+06	23.8
M2	87	HNLC(iodo-APAP)YDPDFEK	+3	99	9.63E+05	3.4
	115	LC(iodo-APAP)YDPDFEK	+3	95	5.56E+07	20.8
	174	NQVFEPSC(iodo-APAP)LDAFPNK	+2	99	1.60E+07	32.0
P1	15	C(iodo-APAP)AALR	+2	36	1.92E+07	5.6
	48	ASC(iodo-APAP)LYGQLPK	+2	98	1.11E+07	19.4
Pepsin digestion						
A1	112	PVC(iodo-APAP)PPEEKDAKL	+2	99	3.34E+06	14.4
M1	115	GMIC(iodo-APAP)YNPEF	+2	99	6.89E+05	27.9
M2	87	IARKHNLC(iodo-APAP) FESEKEQIRe	+2	99	5.10E+05	7.4

GST	Cys	Peptide	z	Peptide confidence (%)	(DDA) absolute precursor area [cps]	RT [min]
P1	115	AKLC(iodo-APAP)YDPDF	+2	98	1.28E+07	21.9
	174	ERNQVFEPSC(iodo-APAP)L	+2	99	3.77E+07	23.9
	102	GVEDLRC(iodo-APAP)KYISL	+3	32	7.18E+04	23.6
	170	IHEVLAPGC(iodo-APAP)L	+2	99	3.30E+06	23.8

MRM experiments were based on initial DDA results of digested iodo-APAP-GST standards. After database searching, iodo-APAP-modified peptides were investigated for each cysteine site (data not shown). Based on signal abundance and MS/MS properties, up to three modified peptides per cysteine site were chosen (see Table S2); peptide precursor with more than seven residues and a charge state of +2 and +3 were preferred. Then, final transitions were based on the three most intense fragment ions (in DDA runs, with m/z <1250) with fragment m/z > precursor m/z. An additional transition was based on the most intense fragment (no m/z restriction). CE was further optimized for highest signal abundance of each transition.

16.3.2 Protein Analysis by DDA and Detection of NAPQI Binding

The number of found total proteins within 1% FDR ranged from 135 to 356 for RLM incubated, non-fractionated samples; from 40 to 80 for Supersomes incubations. Applying offline fractionation lead to an average increase in found proteins at 1% FDR of 23.7% and 28.5% for RLM and Supersomes incubated samples, respectively. NAPQI-alkylated peptides found by database searching were confirmed by using the following criteria: 1) precursor mass error <10 ppm, 2) MS/MS spectral confidence >95%, 3) the presence of one diagnostic production (see Figure 3), and 4) <0.5 min retention time deviation from iodo-APAP standard (for human GSTs). A total of six NAPQI-modified peptides of five distinct cysteine sites of three GSTs were found based on above criteria (see Table 3). Overlaid extracted ion chromatograms of all six peptides can be found in Figure 4A, Found diagnostic fragment ions can be found in Table 4. Additional MS and MS/MS spectra are available in Figures 4B,C and Figure S1. In addition, NQVFEPSC*LDAFPNLK from human GSTM2 and one peptide from rat MGST1 (from RLM background) were identified without detection of diagnostic fragment ions. MS/MS of modified rat MGST1 peptide, VFANPEDC*AGFGK, was confirmed by comparison to previously analyzed standard peptide. A comparison of NAPQI-modified peptides from human GSTs to their confidently identified carbamidomethylated variants can be found in Table 5. An overview of all carbamidomethylated GSTs can be found in Table S3.

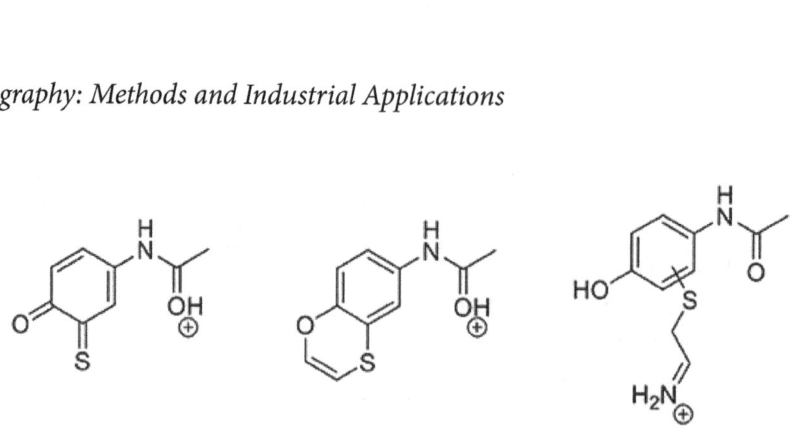

$C_8H_8NO_2S^+$ *m/z* 182.0270 — $C_{10}H_{10}NO_2S^+$ *m/z* 208.0427 — $C_{10}H_{13}N_2O_2S^+$ *m/z* 225.0692

Fig. 3: Proposed structures of diagnostic fragment ions.

Table 3. NAPQI-GST peptides confirmed by DDA and MRM methodsa.

GST	Cys	Peptide	DDA				MRM			
			RLM (1D)	RLM (2D)	3A4 (1D)	3A4 (2D)	RLM (1D)	RLM (2D)	3A4 (1D)	3A4 (2D)
A1	112	PVC*PPEEKDAL		✓			✓	✓		✓
		PVC*PPEEKDAKLAL	✓				✓	✓		
M1	78	ITQSNAILC*Y					✓		✓	✓
M2	115	AKLC*YDPDF						✓		✓
		LC*YDPDFEK		✓						
	174	ERNQVFEPSC*L				✓	✓	✓	✓	✓
		DVLERNQVFEPSC*								✓
		NQVFEPSC*LDAFPNLK	✓	✓			✓	✓	✓	✓
P1	15	C*AALR								✓
		ASC*LYGQLPK				✓				✓
		IHEVLAPGC*L	✓				✓	✓	✓	✓

aRLM (1D):RLM incubation without fractionation, RLM (2D): RLM incubation with fractionation, 3A4 (1D): CYP3A4 Supersome incubation without fractionation, 3A4 (2D): CYP3A4 Supersome incubatin with fractionation, DDA: confirmed by DDA, MRM: confirmed by MRM

Table 4. NAPQI-cysteine diagnostic fragment ions (MS2) detected in HRMS/MS experiments.

GST	Cys	Peptide	Diagnostic ion (ppm)					
			m/z 182.0270		m/z 208.0427		m/z 225.0692	
Al	112	PVC*PPEEKDAKL			✓	(-2.4)	✓	(-0.4)
		PVC•PPEEKDAKLAL			✓	(-1.9)	✓	(-1.8)
M2	115	LCIYOPOPEK					✓	(1.8)
	174	ERNQVFEPSC'L			✓	(-1.0)	✓	(-2.2)
PI	48	ArLYGGLPK	✓	(5.5)			✓	(5.8)
	170	IHEVLAPGC1			✓	(-9.6)		

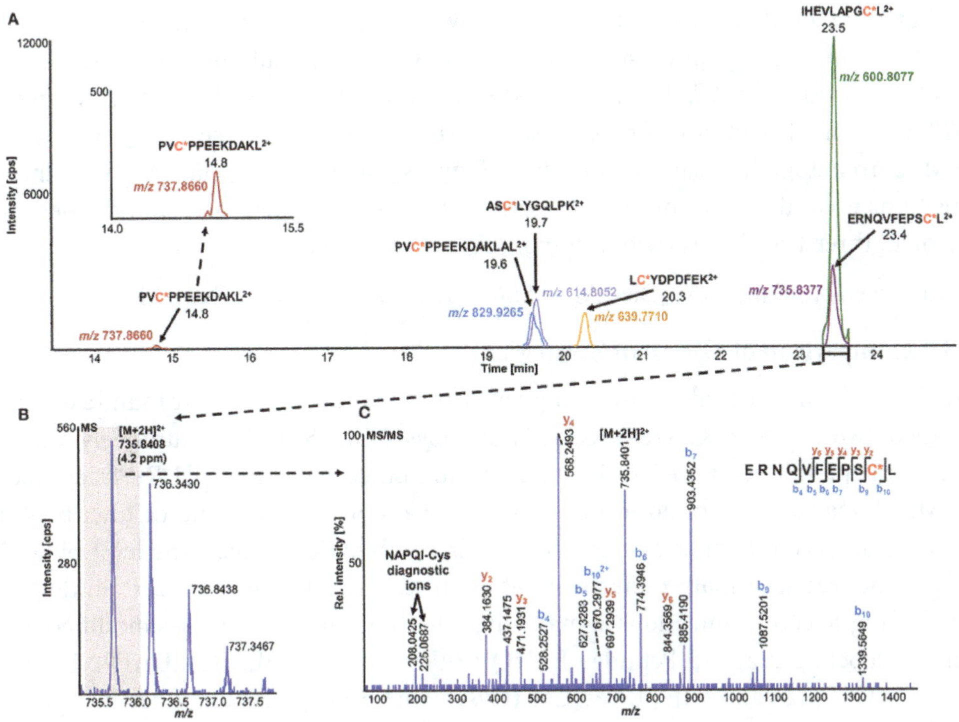

Fig. 4: Overlaid high-resolution extracted ion chromatograms (±0.01 u) of detected modified peptide precursors (A). Confirmation was based on high-resolution accurate mass TOFMS (B) and triggered (DDA) MS/MS (C) analyses, including diagnostic NAPQI-cysteine fragment ions.

Table 5. Comparison of NAPQI-peptide to the IAM-modified (CAM) version found with >95% confidence in the same DDA experiments.

GST	Cys	Peptide	NAPQI modification		CAM modification	
			R T [min]	Absolute peptide intensity [cps](z)	R T [min	Absolute peptide intensity [cps](z)
A1	112	PVC*PPEEKDAKL	14.8	9.96E+02(+3)	12.1	8.63E+03(+3)
		PVC*PPEEKDAKLAL	19.6	6.61E+03(+3)	17.1	1.77E+04(+2)
M2	115	LC*YDPDFEK	20.3	3.83E+02(+2)	17.2	5.86E+05(+2)
	174	ERNQVFEPSC*L	23.4	1.27E+03(+2)	19.3	3.19E+05(+2)
		NQVFEPSC*LDAFPNLK	31.6	1.84E+03(+2)	28.9	2.68E+05(+2)
P1	48	ASC*LYGQLPK	19.7	5.82E+02)(+2)	15.5	1.09E+4(+2)
		IHEVLAPGC*L	23.5	3.87E+03(+2)	19.5	1.00E+02(+1)

16.3.4 Confirmation of NAPQI Binding by MRM Strategies

NAPQI-modified peptides were identified in MRM experiments by following criteria: 1) signal-to-noise (S/N) of all transitions >10, 2) <17.2% deviation of relative abundance (see Equation 1) of two highest transitions (with fragment *m/z* > precursor *m/z*) from iodo-APAP-modified peptide standard, and 3) <0.5 min retention time deviation of all transitions from iodo-APAP standard. After data filtration, MRM strategies yielded 10 NAPQI-modified peptides of seven cysteine sites of four GSTs (see Table 3). Overlaid MRM chromatograms can be found in Figures S4–S7. The relative abundance of transitions was calculated through the ratio of its integrated peak area and the peak area sum of all three transitions (with fragment *m/z* > precursor *m/z*):

$$\text{Relative abundance} = \frac{A_x}{\sum_{i=1}^{3} A_i} \quad (1)$$ $$\text{Relative abundance} = \frac{A_x}{\sum_{i=1}^{3} A_i} \quad (1)$$

16.3.5 Comparison of Different Strategies

A total of 11 distinct peptides comprising seven unique cysteine sites were found as NAPQI-modified, two of which, Cys78 in GSTM1 and Cys15 in GSTP1, were uniquely found by MRM strategies. Thus, CYP3A4 incubations combined with high-pH RP fractionation and MRM analysis was found most suitable with a total of nine found distinct peptides, including all seven distinct cysteine sites. This combination reflects the least biological background, reduced matrix effects by offline fractionation and more certain detection by following selective transitions at matching retention times of peptides modified by the isomeric labeling reagent. Peptides ERNQVFEPSC*L, NQVFEPSC*LDAFPNLK (both Cys174 in GSTM2), and IHEVLAPGC*L (Cys170 in GSTP1) were identified using all four tested MRM strategies. Using MRM techniques demonstrated that dilution effects during offline fractionation, in the case where a specific peptide can elute over neighboring fractions (see Figure S8), could lead to target signal intensities below detection limits (S/N < 10) and thus missing identification in DDA and MRM experiments. This was observed in RLM incubations of GSTM1. Peptide ITQSNAILC*Y was only detected in MRM experiments without LC fractionation (see Table 3). Furthermore, DDA experiments showed higher susceptibility to interfering compounds, even when detection limits were not an issue. Co-eluting, high abundant interferences in TOFMS survey scans resulted in missed MS/MS acquisition of target peptides. A visual representation of missed fragmentation of a NAPQI-peptide target (IHEVLAPGC*L), compared to a positive MRM identification, can be found in Figure 5. Missing MS/MS acquisition of the IHEVLAPGC*L peptide precursor at m/z 600.8077 was caused by a higher abundant interference at m/z 601.3123, overlapping with the peptide [M+1+2H]2+ signal. In that case, only the highly abundant interference was analyzed in MS2 leading to a MS/MS spectrum of interference fragments and low abundant [M+1+2H]2+ fragments, not allowing confident identification. One peptide, LC*YDPDFEK (Cys115 in GSTM2), was found only through DDA efforts, which was based on susceptibility of MRM transitions to isobaric interferences. The peptide was ultimately identified through HRMS/MS sequencing only (see Figure S1C). These circumstances lead to differences between DDA or MRM analyses (e.g., RLM (1D) compared to 3A4 (1D) in Table 3), leading to more found modified sites with RLM incubations than with Supersomes, even though the latter were less complex incubations. Overall, NAPQI-peptide MRM signals reached higher signal abundance as extracted precursor ion (±0.01 u) signals in DDA survey scans (TOFMS) with comparable S/N. The use of both trypsin and pepsin as proteases was shown to be complementary. Several sites were uniquely found using either one of the two proteases. NAPQI-modified Cys15, 48, and 78 were uniquely found in trypsin digests, whereas Cys112 and 170 were only found after pepsin digestion. Cys115 and 174 in GSTM2 were the only sites with detectable NAPQI peptides from both protease treatments; AKLC*YDPDF and LC*YDPDFEK, and ERNQVFEPSC*L and NQVFEPSC*LDAFPNLK, respectively.

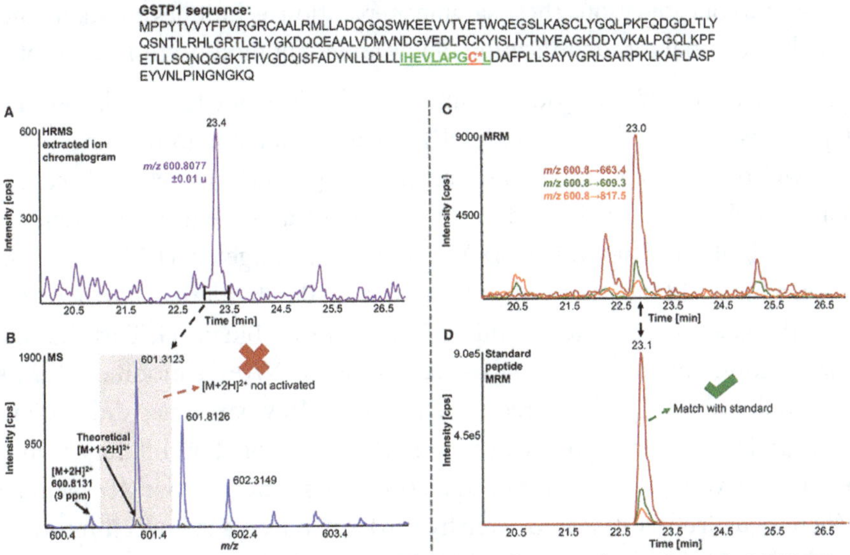

Fig. 5: Comparison of LC-DDA and LC-MRM results, used to identify NAPQI-modified GSTP1 peptide IHEVLAPGC170L (underlined) from in vitro incubation with CYP3A4 (Supersomes). DDA analysis of the target peptide (theoretical $[M+2H]^{2+}$ at m/z 600.8077) eluting at 23.4 min (A) did not yield MS/MS acquisition (B). Fragmentation was triggered by an interfering ion at m/z 601.3123, which resulted in a MS1 isolation window (red rectangle) not incorporating target peptide. MRM analysis (C) confirmed presence of IHEVLAPGC*L^{2+} by comparing to iodo-APAP-GST (D).

16.4 DISCUSSION

The GST family is an essential, multifunctional set of enzymes, crucial in many detoxification and binding reactions in liver and other tissues. GSTs are involved in the metabolism of many substances and xenobiotics, including reactive metabolites such as NAPQI. The understanding of NAPQI binding to GSTs is thus a logical step toward understanding its hepatotoxicity. This study used several approaches for targeted and untargeted identification of modified peptides from in vitro incubations. The use of different methodologies yielded complementary findings but also allowed a critical view on performance and inherent limitations of each method.

16.4.1 Performance of Different Strategies

This study highlighted the importance of using complementary proteases to better cover potential modification sites. Pre-fractionation of peptides was mostly beneficial for identifying adducted peptides. The double protease approach substantially increased protein sequence coverage. Multiple protease digestion can be easily applied in proteomics workflows and is highly flexible, due to wide range of specific and less specific enzymes with diverse cleavage behavior. Pre-fractionation increased peptide coverage and improved NAPQI-peptide discovery due to better resolution of peptides, thus reducing matrix effects in both MRM and DDA. Benefits from better peptide resolution in multiplexed LC outweighed the potential for sample loss and dilution effects introduced through fractionation and concatenation. These sample preparation techniques are viable approaches to broaden the potential for identification of low abundant modifications on proteins.

Targeted MRM-based methods are most easily developed with the availability of purified proteins, whereas untargeted DDA analyses offer a more flexible, universal application without time-consuming optimization steps. GSTs are relatively easy to purify via GSH affinity chromatography and thus represent ideal proteins for reference standard preparation to build and optimize MRM methods. Untargeted DDA with subsequent database searching did not require any custom-made standards. Furthermore, DDA allowed for the analysis of protein binding beyond recombinant GST as the only target during incubations. For example, database searching for NAPQI-modified RLM proteins detected modified rat MGST1. Moreover, DDA and MRM strategies did not only differ in terms of applicability but also performance. This was rooted in differences in both MS platforms, as well as the two different acquisition types. Low resolving triple quadrupole systems (here a quadrupole-linear ion trap hybrid) generally reach lower limits of detection based on a higher duty cycle in optimized MRM modes. In comparison, quadrupole-time-of-flight instruments are less sensitive but offer higher mass accuracy and the possibility for structural characterization by MS/MS. The difference in sensitivity is a crucial factor in characterizing low abundant protein modifications. In our findings, MRM strategies generally reached superior detection limits and thus led to two uniquely found modified peptides. In addition, even when detection limits were surpassed, low abundance of NAPQI-peptides resulted in missed peptide precursor fragmentation in DDA and thus were not identified. The final number of found modified sites, confirmed by DDA, was ultimately achieved here by using multiple attempts, which highlighted the limitations of unsupervised DDA in covalent adduct analysis. Furthermore, DDA relies on automatically triggered high quality MS/MS of modified peptides. This can be a challenge in complex biological samples, especially for low-abundant modifications, since so many peptides are eluting off the analytical column together. Even though high-speed high-resolution mass spectrometers, such as the one used in this work, have the ability to acquire spectra very fast and with high sensitivity, there are still many peptides not triggered for MS/MS acquisition under these conditions. MRM, on the other hand, does not suffer from this caveat. If a given peptide is present, above a certain threshold, it should be detected by MRM. This method still suffers from matrix effects (ion suppression) but is not competing with co-eluting species for MS/MS acquisition. In general, MRM methodologies offer higher robustness and reproducibility. However, it should be noted that the use of a semi-targeted DDA, using an inclusion list of putative modified peptides could potentially increase reproducibility. Furthermore, susceptibility of low mass resolving MRM to isobaric interferences might obstruct confident identification and quantitation in some cases, which is less likely in HRMS. Also constant improvements in instrumentation and software are improving DDA performances and the development of all-ion fragmentation in data-independent acquisition will certainly be viable avenues for future applications of HRMS in protein binding studies. In conclusion, MRM and DDA are powerful MS tools in proteomics, both of which have intrinsic advantages and disadvantages.

16.4.2 Putative NAPQI-cysteine Sites in GST

In this study we characterized seven cysteine sites from four cytosolic human GSTs by NAPQI *in vitro*. NAPQI binding to human GSTA1, M1, and M2 had not been reported previously. Only binding to Cys48 and 102 in human GSTP1 was previously confirmed. In one of their studies, also reported binding of clozapine metabolites to Cys48, as well as metabolized troglitazone to Cys15, 48, and 102 by *in vitro* incubation, tryptic digestion, affinity removal of background protein and LC-MS/MS analysis. In a follow-up study confirmed also diclofenac-related binding to Cys15 and 48. Using GSTP1 in a protein trapping study, binding of the reactive diquinone methide of raloxifene was also observed to Cys48. Here, we detected modified Cys48 and additional target sites of NAPQI in GSTP1 by using pepsin as an additional protease with different cleavage specificity than trypsin. We observed Cys15, 48, and 170 as NAPQI target sites in GSTP1. Stated that alkylation of Cys48 lead to complete inhibition of GSTP1, based on the close proximity to GSH binding sites, as well as subsequent disruption of the active GSTP1-1 dimer. Additionally, modification on Cys170 is believed to influence activity as well, which is also explained by its importance in dimerization. Furthermore, Cys15 is in direct proximity to Arg14 which is also involved in GSH binding. It is possible that alkylation of multiple sites combined leads to full enzymatic inhibition. Binding analysis of Cys102 presented several challenges in this study. The site was inaccessible via tryptic digestion and peptic peptides showed the lowest signal intensity of all screened sites. Sufficient recovery of Cys102 requires other proteases or potentially more limited proteolysis. used Asp-N to confirm binding of NAPQI to Cys102. However, it was reported that Cys102 binding occurred only with high molar ratios of NAPQI to protein. Alkylation reactions of Cys102, being more solvent exposed, was believed to play a minor role in enzymatic activity.

Previous alkylation studies of GSTA1 focused on in vivo binding of bromobenzene metabolites to Cys112 in rat. Detection of these low abundant peptide modifications was enabled by GSH affinity purification, tryptic digestion and capillary LC-MS. Human and rat GSTA1 share a high identity of 76%, including Cys112, which was found NAPQI-modified in our work. The relatively long distance between Cys112 and the active sites, Tyr9 and Arg45, was believed to ensure activity even after cysteine alkylation. However,

Cys112 is in range to Met94, Phe136, and Val139, which are associated to a hydrophobic pocket, important in dimerization. Alkylation of Cys112 could therefore compromise the formation of the GSTA1-1 dimer. However, further activity studies of NAPQI-modified GSTA1 are needed to investigate this.

In comparison to our findings of NAPQI-modified Cys78 in GSTM1, found rat GSTM1 (Cys87) modified in vivo by acrylonitrile using a combination of radiolabeling, affinity purification and LC-MS. Modification of Cys87 in rat was also observed by a recent study of , using in vitro metabolism of APAP and LC-MS/MS. Rat GSTM1 is missing Cys78, though Cys87 is identical in human and rat. Hence, an interspecies comparison of cysteine reactivity is not trivial, since both sites, Cys78 and 87, might compete for binding reactions. However, hypothesized that the reactivity of Cys87 is based on possible interactions with His85, which is only found in rat. Missing His85 could explain a lower reactivity of Cys87 in human GSTM1 (and M2). Nevertheless, the extent of enzyme inhibition after modification needs further study. Observed loss in rat GSTM1 activity only after IAM alkylation of Cys115 (identical in human GSTM1) and not Cys87. A possible explanation might be the close active site Tyr116, which is vital for substrate binding. Binding to Cys115 in human GSTM1 was not observed in our experiments. However, we could identify Cys115 and 174 as putative targets in human GSTM2. Studied in vivo binding of 4-bromophenol in rat GSTM2 and confirmed adducts by radioactive labeling, tryptic digestion and LC-MS/MS. Based on cysteine mutant studies of human GSTM2, it has been stated that Cys115 might be involved in enzyme activity, based on the mu class active site Tyr116. The effect of binding to GSTM2 needs further research, especially with regards to distinguishing both modified sites found in this study, Cys115 and 174. Furthermore, the difference of NAPQI binding properties of GSTM1 and GSTM2 are also crucial to investigate in future studies, since they share very high sequence homology.

16.5 CONCLUSION

In conclusion, we used in vitro incubations to identify adduct formation with NAPQI. Further research of in vivo and in vitro treated hepatocytes and liver fractions would complement our findings. The techniques shown here were designed to help study protein adduct formation and to guide subsequent research. Furthermore, a combination of our methods with targeted sample preparation and enrichment tools (e.g., immunoaffinity) could be applied to analyze the NAPQI binding affinity of GST and various proteins of interest in the future. A similar strategy can be used to find NAPQI-protein targets in more complex samples and unknown protein targets, while pinpointing binding sites. The novel use of an isomeric labeling reagent with identical chromatographic elution of NAPQI-modified cysteine containing peptides is invaluable for confirming the identity of modified peptides.

17

INSTRUMENTAL ANALYSIS OF POISON-BY CHROMATOGRAPHIC TECHNIQUES

17.1 INTRODUCTION

Chromatography may be defined as a technique of separating a mixture of components into individual components through equilibrium distribution between two phases. Chromatography is a non- destructive method for determining a multi- constituent mixture of trace, minor, or major constituents into its separate fractions. Chromatography was invented by a Russian Botanist Mikhail Semenovich Tswett in 1906. Chromatography has turn out to be one of the most significant analytical methods for the identification and quantification of compounds in the gaseous or liquid state. The simple principle is based on the concentration equilibrium of the components of concern, between two immiscible phases. One is known as the Stationary Phase, because it is immobilized within a column or fixed upon a support, while the second, called as Mobile Phase, is forced through the first. The phases are selected such that components of the sample have different solubilities in each phase. The differential migration of compounds leads to their separation.

Among all the instrumental analytical techniques this hydrodynamic system is the one with the comprehensive application. Chromatography conquers a dominant position that all laboratories involved in molecular analysis can confirm. Chromatography is a physico-chemical method of separation of components within mixtures, liquid or gaseous, in the same manner as distillation, crystallization, or the fractionated extraction. The applications of this technique are therefore several since many of heterogeneous mixtures, or those in solid form, can be dissolved by a suitable solvent which becomes, certainly, an additional component of the mixture.

17.2 SIGNIFICANCE OF CHROMATOGRAPHIC TECHNIQUE IN POISON DETECTION

Among the immense range of techniques available for biochemical study, chromatography subjugates an honoured position. Hardly any other analytical methods have had such a vital impression on the development of experimental chemistry and toxicology. Applications of this innovative technique to biochemical analysis were readily apparent. Chemical analysis of biological specimens is challenging because they comprise a complex mixture of constituents, many of which react correspondingly with reagents carefully chosen to detect a specific compound. Analytical chemistry is not exclusively intricate when dealing with pure solutions, but the presence of several possibly inquisitive substances, as in biological matrices, creates a challenge. Chromatography resolved that challenge by offering a method to separate components of a mixture, providing a pure specimen that could be measured by any of several analytical methods. At the same time as any technology adapts to the applications for which it is appropriate, chromatography has developed to come across the exceptional requirements of biochemical examination. Capillary gas chromatography modernized the analysis of drugs of abuse, with the capability to resolve congeneric compounds. In the same way, Liquid Chromatography has set way to the greater resolving ability of High Performance Liquid Chromatography (HPLC). High resolution derivatives of electrophoresis, a technique closely allied to chromatography, have made the evolution from research tools into commercial applications. Chromatography endures to advance, and the adaptability of this analytical method appears outstandingly suited to experimental chemistry and toxicology applications.

17.3 TYPES OF CHROMATOGRAPHIC TECHNIQUES

17.3.1 Liquid Chromatography

Chromatographic separations in Liquid Chromatography, except Size Exclusion Chromatograph, are the consequences of interactions developed between the functional groups of solute molecules, solvent molecules and the stationary phase. The interactions present in Liquid Chromatography are hydrogen bonding, Vander Waals forces and electrostatic forces, and the approaches of Liquid Chromatography are categorised as per the nature of these interactions.

The approach of chromatography is appropriate for a particular separation depends upon molecular mass, polarity and ionic characters of a solute. Different methods of Liquid Chromatography have developed for scrutinizing the range of mixtures in various kinds of matrices. The methods of Liquid Chromatography include Normal- Phase Liquid Chromatography (NPLC), Reversed Phase Liquid Chromatography (RPLC), Ion-Exchange Liquid Chromatography (IELC) and Size-Exclusion Chromatography (SELC). Selection of a liquid chromatographic approach for a particular analysis also requires selection of a column (stationary phase) and solvents for the mobile phase. Whichever mode is selected for development of a particular method, it is important that certain standards for method validation are fulfilled.

17.3.2 Adsorption Chromatography

The stationary phase is a solid medium to which the species observe through the dual effect of physisorption and chemisorption. The physico-chemical parameter involved here is the adsorption coefficient. It was established first and has a solid stationary phase and a liquid or gaseous mobile phase. Each solute has its own equilibrium between adsorption onto the surface of the solid and solubility in the solvent, the least soluble or finest adsorbed ones migrate more slowly. The outcome is a parting into bands comprehending different solutes. Liquid chromatography by a column comprising silica gel or alumina is an illustration of adsorption chromatography. The solvent that is placed into a column is called the eluent, and the liquid that flows out of the end of the column is called elute. Stationary phases have made considerable improvement since the time of Tswett, who used Calcium Carbonate or Inulin.

AT A GLANCE

- *Chromatography is a prominent separation tool that is used in all subdivisions of science, and is time and again the only means of separating components from complex mixtures.*
- *Russian botanist Mikhail Semenovich Tswett coined the word Chromatography in 1906 and the first analytical use of chromatography was described by James and Martin in 1952, for the usage of gas chromatography for the study of fatty acid mixtures.*
- *An extensive assortment of chromatographic procedures makes use of differences in size, charge, binding affinities and other properties. Many types of chromatography have been developed. These include Column chromatography, High performance Liquid Chromatography, Gas chromatography, Size Exclusion Chromatography, Ion exchange chromatography etc.*
- *Chromatography is probably the most important single analytical technique used today and probably continues to be so far the foreseeable future and it may be applied both quantitatively as well as qualitatively as it is primarily tool of separation.*
- *Applications of Chromatographic techniques to biochemical analysis were readily apparent. Chemical analysis of biological specimens is hard since they comprise a complex combination of constituents, several of which respond in the same way with reagents chosen to detect a specific compound.*
- *Analytical chemistry is not exclusively complicated when dealing with pure solutions, but the manifestation of several possibly interfering substances, as in biological matrices, produces a challenge.*

17.3.3 Size Exclusion Chromatography

Size Exclusion Chromatography (SEC) is a method by which molecules can be separated according to their size in solution, as a consequence linking indirectly to their molecular masses. To attain this, stationary phases hold pores through which compounds are able to diffuse to a certain amount. Size Exclusion Chromatography is based upon the capability of the sample molecules to penetrate into the highly porous bead- like structure of the stationary phase. Separation occurs only as a result of the different degrees of penetration. Molecules of relatively smaller weight are decelerated in their progression in the column because they can pass into the stationary mobile phase within the pores of the packing. This method is referred to as Gel Filtration Chromatography (GFC) when the stationary phase is hydrophilic and as Gel Permeation Chromatography (GPC) when the stationary phase is hydrophobic.

17.3.4 Partition Chromatography

In this, the immobile phase is a non-volatile liquid that is apprehended as a thin coating on the surface of an inert solid. The mixture to be separated is transported by a gas or a liquid as the mobile phase. The solutes dispense itself among the moving and the motionless parts, with the more solvable constituent in the mobile phase accomplished the end of the chromatography column first. The stationary phase is an immobilized liquid upon an inert and porous material, which has only a mechanical role of support. Impregnation, the oldest procedure for immobilizing a liquid on a porous material, is a technique now out-dated because of the eminent possibility of washing out the column, which is called bleeding.

17.3.5 Supercritical Fluid Chromatography (SFC)

SFC works, as its tag proposes, a liquid in the supercritical state as its movable phase. This indicates to developments in the separations of ther molabile compounds and largely for compounds of high molecular weight. The instrument is theoretically a hybrid of gas chromatograph and liquid chromatography. The instrumentation for Supercritical Fluid Chromatography represents hybrid assemblies of Gas Chromatography and High Performance Liquid Chromatography instrument. To assure the flow rate of the supercritical fluid a syringe pump or a reciprocal pump is used, which is maintained below the critical temperature by means of a cryostat regulated at around 0°C. Gases above their critical pressure and temperature are in a supercritical state, intermediate between that of a gas and liquid. Supercritical fluids have strong extraction properties because the solubility of compounds in fluid is close to that of a true solvent and much lower viscosity permits it to percolate through packed bed of sample. So, not only there is an efficient contact between the extracting fluid and the sample but the fluid is easily removed when it is released from its supercritical state.

17.3 .6 Thin Layer Chromatography (TLC)

Thin Layer Chromatography (TLC) is an invaluable method used in chemistry and biochemistry. Being economical and sensitive, this technique that is simple to use, can be automated. It has become important essentially since it is possible to carry out several separations in parallel. The principle of the separation between phases of the sample components is though the migration of the constituents through the stationary phase is different. Separation is conducted on a thin layer of stationary phase, usually based upon silica gel and deposited on a rectangular plate made out of glass, plastic or aluminium of a few centimetres in dimensions. To maintain the stationary phase on the support and to assure the cohesion of the particles, an inert binder like gypsum is mixed into the stationary phase during the manufacture of the plate. The constituents can be identified by simultaneously running standards with the unknown.

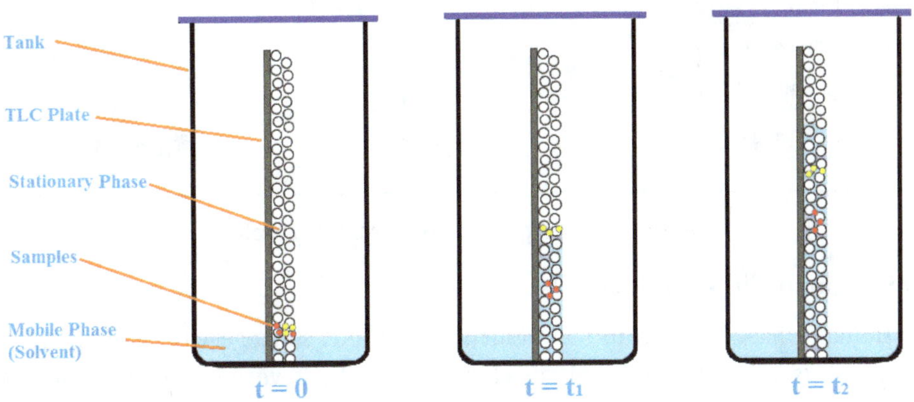

17.3.7 Gas Chromatography (GC)

Gas Chromatography comprises separation of a mixture of volatiles by partitioning between a solid or liquid stationary phase and a gaseous mobile phase and finds diverse application in toxicological analysis of alcohols, volatile poisons, drugs, plant poisons. Gas chromatography, especially Gas Liquid Chromatography involves a sample being vaporised and injected onto the head of the chromatographic column.

The sample is transported through the column by the stream of inert, gaseous mobile phase. The column contains a liquid stationary phase, which is adsorbed onto the surface of an inert solid. On passing through the column volatile components impart different affinities to stationary and mobile phases and separated in fraction.

17.3.8 Gas Liquid Chromatography (GLC)

As indicated above the mobile phase here is a gas and the stationary phase is an immobilized liquid, either by impregnation or by bonding to an inert support which could be, rather basically, the inner surface of the column. This is the technique commonly called Gas Phase Chromatography (GC). The gaseous sample must be brought to its vapour state. It was Martin and Synge who, in 1941, suggested the replacement of the liquid mobile phase by a gas in order to improve the separations. From this time arises the exact start of the expansion of analytical chromatography

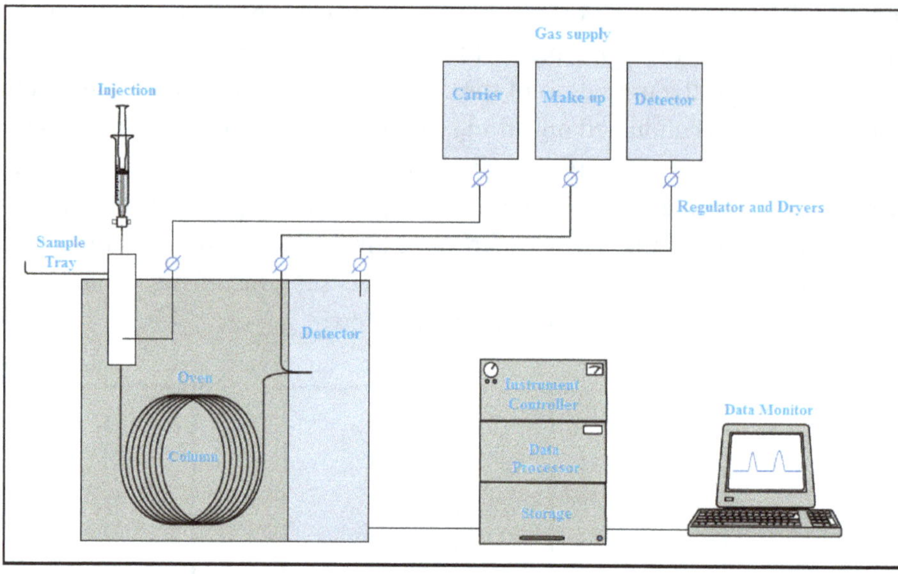

There are two types of Gas Chromatography:

17.3.9 Gas Solid Chromatography (GSC)

The stationary phase is a porous solid (such as graphite, silica gel or alumina) while the mobile phase is a gas. This type of gas chromatography is very effective for analysis of gaseous mixtures or of compounds that have a low boiling point. The parameter concerned is the adsorption coefficient.

17.3.10 High Performance Liquid Chromatography (HPLC)

High Performance Liquid Chromatography constitutes a wide-ranging purpose analytical technique derived from the most ancient form of preparative Liquid Chromatography. The contemporary technique is significantly improved in terms of selectivity, resolution, through miniaturization and the use of very elaborate stationary phases. These phases comprise spherical micro-particles with diameters of 2 - 5μm, or porous monolithic material that leads to a significant pressure drop on the column. A large pressure needs to be exerted upon the mobile phase to obtain a continuous flow. This is a type of column chromatography of which a solvent with the chemicals of interest are forced through a thin column at high pressures. The compounds are then separated out based on molecular size and polarity. There are two types of High Performance Liquid Chromatography; one is Normal-Phase High Performance Liquid Chromatography and Reverse-Phase High Performance Liquid Chromatography. The most common type is Reverse-Phase High Performance Liquid Chromatography in which the silica inside the column is actually non-polar. Normal-phase HPLC uses a polar column. High Performance Liquid Chromatography can be used for both quantitative and qualitative information.

17.3.11 High Performance Thin Layer Chromatography (HP-TLC)

High Performance Thin Layer Chromatography is an improvement of the technique where the sorbent material (e.g. Silica Gel 60) has a finer particle size and a narrower particle size distribution than conventional Thin Layer Chromatography. HPTLC plates have an improved surface homogeneity and are thinner. The resolution is enhanced, analysis times are shorter and it is sufficient to apply nanolitres or nanograms of sample (Nano-TLC). After development of the chromatogram to a certain distance, the thin layer chromatography plate is removed and mobile phase is evaporated. The track is scanned in a densitometer with a light beam in visible or ultraviolet range of the spectrum. Depending upon the mode of scanning absorbance or fluorescence is measured by diffuse reflectance.

17.3.12 Ion Chromatography

This chromatographic technique is concerned with the separation of ions and polar compounds. Stationary phases contain ionic sites that create dipolar interactions with the analytes present in the sample. If a compound has a high charge density, it will be retained a longer time by the stationary phase. This exchange process is much slower when compared with those found in other types of chromatography. In this technique the mobile phase is a buffered solution while the solid stationary phase has a surface composed of ionic sites. These phases allow the exchange of their mobile counter ion with ions of the same charge present in the sample. This type of separation depends on ionic distribution coefficients.

18

ION CHROMATOGRAPHY AN ADVANTAGEOUS TECHNIQUE IN SOIL ANALYSIS

18.1 INTRODUCTION: AN OVERVIEW

The content of organic and inorganic substances in soils plays a very important role in their quality and use. The importance of inorganic and organic anions and cations in soil quality is huge, thus we need a suitable analytical methods for their determination in such complex matrices. In this range ion chromatography is an advantageous instrumental analytical technique. It has been used in most areas of analytical chemistry and has become a versatile and powerful technique for the analysis of a vast number of ions present in the environment, food and many other matrices, including soils. The key to valuable data when carrying out ions analyses in soils is the use of correct and reproducible sample preparation methods. The paper reviews the principles of ion chromatography and its application in soil research. Ion chromatography is used to provide information on the fate of nutrient anions resulting from agricultural practices and to determine soil nutrient retention and leaching rates. It is widely accepted as the standard reference methodology for many analytes in different samples matrices including soil analysis.

In soils there are organic and inorganic substances in ionic forms. Soil nutrients exist as anions and cations when dissolved. The nutrients which exist as cations are mostly Ca_2^+, Mg_2^+, NH_4^+,K^+, Na^+, Al_3^+, Fe_2^+, Mn_2^+, Zn_2^+ and Cu_2^+. Some of the nutrients which exist as anions are: Cl-, NO_3^-, SO_4^{2-},H_2PO_4+, $PO4_3^-$, BO_3^-, and MoO_4^-. Ion exchange capacity is the soils ability to hoild anions and cations.

When soil particles are negatively charged they attract and hold on to cations stopping them from being leached down the soil profile, while negatively charged soil particles repel anions. The implication of this is that negatively charged nutrients such as nitrates, sulfates and chlorides are vulnerable to leaching down the soil profile.

Chromatographic methods are used widely in soil analysis. They help to separate and determine polar and non-polar components; acidic, neutral and alkaline compounds, as well as organic and inorganic substances. It is necessary to use an appropriate chromatography type, which depends on the physicochemical properties of the examined sample and its components. Gas chromatography (GC) and liquid chromatography (LC) can be used to separate and determine about 20% and 80% of the known compounds, respectively. Ion chromatography (IC) is a part of high-performance liquid chromatography used to separate and determine anions and cations, and also other substances after converting them into the ionic forms.

Ion chromatography is very useful method used for the determination of common inorganic anions and cations, as well as ionic forms of some metals/metalloids and organic substances. Since its introduction in 1975 ion chromatography has been used as a versatile and powerful technique for the analysis of a vast number of inorganic and organic ions present in samples with different matrices.

18.2. THE PRINCIPLES OF ION CHROMATOGRAPHY

There are two main types of ion chromatography: suppressed ion chromatography and non-suppressed ion chromatography. In ion chromatography with suppressed conductivity detection an eluent containing a suitable electrolyte is passed through a specialty high performance ionexchange resin to a suppressor and then to a conductivity detector. Analyte ions are separated on the ion-exchange column and the separated sample ions (together with the eluent) move to the suppressor. In the suppressor, the conductance of the eluent is lowered and the conductance of the eluted analytes is increased, leading to a large increase in the signal to noise ratio of the detection signal. In nonsuppressed ion chromatography separation column is connected directly to the conductivity detector.

The background signal is reduced to a manageable level by using an ion-exchange separation column of low exchange capacity that lowers the eluent concentration needed for separation. Both suppressed and non-suppressed modes have been applied to the analysis of various sample matrices. However, suppressed ion chromatography application to anions analysis is much more popular. Ion exchange remains the primary separation mode used in ion chromatography today, although the apparatus used for the separation of the ionic species includes moreover ion-exclusion, ion-pairing chromatography and reversed phase liquid chromatography.

The key concept in the evolution of ion chromatography was suppressor. In the suppressor, the conductance of the eluent is lowered, or suppressed, and the conductance of the sample ions is increased, leading to a large increase in the signal-to-noise ratio of the detection signal. The stationary phase used in ion chromatography consists of polystyrene (PS), ethylvinylbenzene (EVB), or methacrylate resins copolymerized with divinylbenzene (DVB) and modified with ionexchange ionexchange groups. Furthermore polyvinyl, silicabased and many other materials were used [8]. At the beginning of ion chromatography development stationary phases utilized particles greater than 40 plates efficiency in relation to column dimension. In the first commercial column used for ion chromatography (Dionex AS-1), the particle size ncies increased to 700 theoretical plates. Modern high-capacity columns have efficiencies over 8000 plates and Next, the combination of positively and negatively charged sites in a single particle, or within the functional groups of a single molecule attached to the surface of an absorbent, provides unique opportunities to vary the selectivity of separation. Carbonate and bicarbonate eluents have been used as the mainstay of anion-exchange ion chromatography for many years.

However, the ideal eluent is hydroxide since, after suppression, it forms water that has virtually zero conductance and, therefore, provides the perfect conductivity baseline. Unfortunately, hydroxide eluent is difficult to use because it readily absorbs carbon dioxide and forms carbonate. The type of eluent for cation-exchange chromatography depends on the detection method being used. The most popular eluents used in cation analysis are low concentration mineral acids such as HCl, HNO_3, and H_2SO_4, sometimes containing organic modifiers (e.g., ethylenediamine, 2,3-diaminopropionic acid), these have significant influence on the retention time and resolution. Although the conductivity detector is still the most popular, other types of detection modes such as amperometric, potentiometric, UV/Vis, chemiluminescence, fluorescence, as well as atomic absorption spectrometry, atomic emission spectrometry, inductively coupled plasma optical emission spectrometry, inductively coupled plasma mass spectrometry and mass spectrometry can be applied for different analytes.

18.3 SAMPLE PREPARATION METHOD

As in other analytical methods, the key stage of analysis (especially of such complex samples as the soil) is the proper preparation of samples for analysis. The analysis of soils by ion chromatography requires either the transfer of the whole sample, or at least the ions of interest, into an aqueous phase. Sample preparation methods for ion chromatography are very numerous and the apparatus and time required for sample preparation vary considerably.

Dissolution of the sample or extraction of the ions to be determined is normally carried out at room temperature, but can also be accelerated by gentle warming or by heating the solvent up to the boiling point. The choice of extracting solution is dependent on both the soils matrix and upon the nature of the determined ions.

Water is preferred, in order to avoid introducing extraneous peaks into the final chromatogram. Sometimes, water combined with a miscible solvent such as methanol, solutions of dilute acid or base, dilute salt solutions, or even the eluent, can be used . If an aqueous extraction is not sufficient, it may be necessary to consider a wet chemical digestion. Acid digestion with concentrated mineral acids is basically inappropriate for ion chromatography because the excess of the acid coanion can lead to the appearance of a large interfering peak in the final chromatogram and can also cause column overloading.

If mineral acids such as HCl, H_2SO_4, or HNO_3 have to be used, the difference in retention times between the sample anions and the solvent anion should be as large as possible. For acid extraction followed by the determination of the standard anions chromatography with direct conductivity measurement, tartaric acid, perchloric acid and/or formic acid can be used. For chemical suppression ion chromatography, acetic acid, or formic acid can be used for determining common anions. In general, acid digestion is better suited to preparing samples which are to be analyzed for cations (e.g., transition metals and rare earth elements) using ion chromatography with post column reaction detection.

An alternative to the acid digestion of solid samples is fusion under alkaline conditions. The procedure is usually very labor intensive, takes a long time, and involves a high risk of errors from contamination and loss of analyte. This method can be used for the determination of fluoride and chloride in geological materials after fusion with sodium carbonate and subsequent injection into an ion chromatography column.

18.4 APPLICATIONS OF ION CHROMATOGRAPHY IN SOIL ANALYSIS

Ion chromatography offers several advantages over conventional methods for determination of ions: high selectivity in samples with complex matrices; small sample volume; possibility to use different detection modes; simultaneous determination of anions and cations, or inorganic and organic ions; species analysis; use of cheap, safe, and environment friendly chemicals. In the range of applications of ion chromatography, inorganic anions are, by far, the most important. The primary reason for this is the lack of alternative methods for anion analysis, which is not the case for cation analysis where many other instrumental techniques are readily available.

However, ion chromatography has an advantage over spectroscopic techniques for cation analysis in the area of metal/metalloids ions, and ammonia determination. An example chromatograms of anions and cations in soil are given on Figure 1 and 2, respectively.

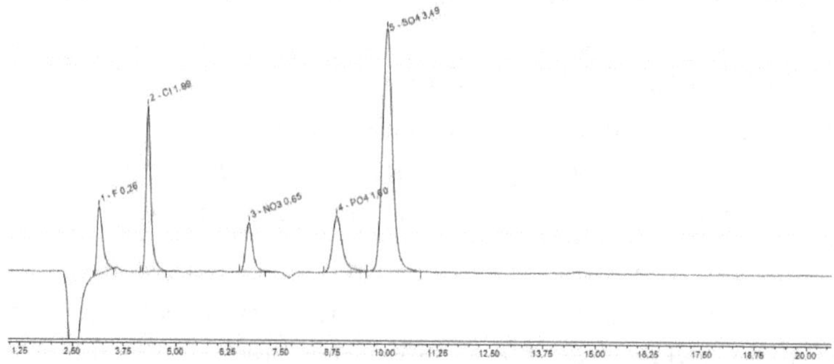

Fig. 1: Chromatogram of anions in soil from Pyskowice (Poland) Sample preparation: Air dried soil samples were ground and sieved.

To prepare a water extract 10g of soil were flooded 100ml of deionized water and shaken for 24 hours. Obtained extract were filtered by syringe filters.

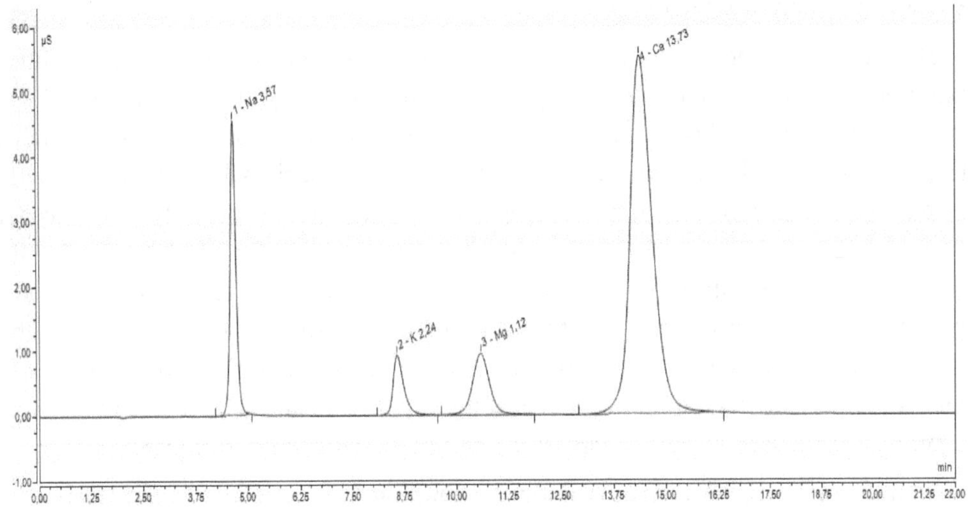

Fig. 2: Chromatogram of cations in soil from Pyskowice (Poland)

Analytical column	:	Dionex IonPac CS16 5 x 250 mm
Eluent	:	30mM Methanesulfonic acid (MSA)
Eluent flow rate	:	1,5mL/min
Detection	:	Suppressed conductivity
Injection volume	:	25µL

Sample preparation: Air dried soil samples were ground and sieved. To prepare a water extract 10 g of soil were flooded 100 mL of deionized water and shaken for 24 hours. Obtained extract were filtered by syringe filters.

Both ion chromatography and ion exclusion chromatography have been used extensively for the determination of low-molecular-weight organic acids in soil extracts. To separate complex mixtures (e.g. inorganic and organic acids) two dimensional chromatographic techniques can be use.

The organic acids are separated in the first dimension, whereas the fraction of the nearly unresolved, highly dissociated inorganic acids were passed to an anion concentrator column by means of a column switching valve. These techniques require the compatibility of the eluent with both columns. Recently, ion chromatography has also found increased application in the field of species analysis. Toxicological test results show that in many cases specific forms of a given element, rather than its total content, have a decisive impact on living organisms. For that reason, the information on the total element content in a given sample is less important than understanding the occurrence of various element forms. One of the most popular examples of species analysis is simultaneously separation and determination of nitrogen ions (NO^{-}_{2} NO3 $^{-}$, NH4 $^{+}$). Studying low analyte concentrations, particularly in complex matrix samples, requires meticulous and sophisticated analytical techniques.

The latest trends embrace the so-called hyphenated methods, which combine different separation and detection methods. Hyphenated methods (e.g. IC-MS and IC-ICP-MS) has incredibly developed over the last years, it is still a relatively new field of the analytical chemistry. Its further development depends on many factors, such as developing new methods of separation, detection and sample preparation as well as the availability of new certified reference materials. Hyphenated methods offer great possibilities and their main advantages include: extremely low limits of detection and quantification, and very good precision and repeatability of determinations. On the other hand, they have their specific limitations, such as high price and complexity of the apparatus. The application of hyphenated methods requires perfect understanding of analytical methodologies and apparatus. Those systems are expensive and used in scientific research, rather than for routine analyses.

Manufacturers of IC instrumentation are taking great efforts to offer users a wide range of instrumentation which, with respect to their flexibility, robustness and cost, generally satisfy the considerably different requirements in the various fields of application, including soil analysis. Real world applications of IC vary greatly but there is virtually no scientific field in which this method cannot be used for the routine determination of ions or for solving special problems. Besides the methods recommended by ISO, EN and U.S. EPA, there exist many methods developed by manufacturers of chromatography equipment. Examples of IC standards and application notes recommended dedicated to soils analysis are given in Table 1.

Table 1. Examples of ISO, US EPA, Dionex and Metrohm application notes for soil analysis by using IC Method

Method number	Method name
ISO/EN	
EN13368-1: 2001	Fertilizers — Determination of chelating agents In fertilizers by ion chromatography. Part 1: EDTA, HEDTA and DTPA
EN13368-2: 2001	Fertilizers — Determination of chelating agents In fertilizers by ion chromatography. Part 2: EDDHA and EDDHMA
EN 15192 (2006)	Characterization of waste and soil. Determination of chromium (VI) in solid material by alkaline digestion and ion chromatography with spectrophotometric detection
EN 16318 (2013)	Fertilizers. Determination of trace elements. Determination of chromium(VI) by photometry (method A) and by ion chromatography with spectrophotometric detection (method B)

US EPA	
6860	Perchlorate in water, soils and solid wastes using ion chromatography/electrospray ionization/mass spectrometry (IC/ESI/MS OR IC/ESI/MS/MS) ionization/mass spectrometry (IC/ESI/MS OR IC/ESI/MS/MS)
300.0	Determination of inorganic anions by ion chromatography
9056	Determination of inorganic anions by ion chromatography
9056A	Determination of inorganic anions by ion chromatography
3060A	Alkaline digestion for hexavalent chromium
Dionex (Thermo Scientific)	
AN69	Determination of aluminum in complex matrices using chelation ion chromatography
AN1113	Determination of chloride and sulfate in water and soil
AU165	Separation of Cr(III) and Cr(VI) by ion chromatography
AN1079	Slimultaneous determination of trivalent chromium and hexavalent chromium using accelerated solvent extraction and ion chromatography
AN123	Determination of perchlorate levels in food and soil samples using accelerated solvent extraction and ion chromatography
Metrohm	
AN U-77	Chromate in strongly alkaline digestion solution according to EN ISO 15192 (digested soil) and EN 16318 (fertilizer)
AN S-360	Soluble tungsten and molybdenum in soil
AN M-11	Determination of soluble Cr(III) and Cr(VI) in alkaline soil extract applying IC-ICP/MS
AN M-12	Iron speciation analysis in soil using IC-ICP/MS in accordance with EPA SW846 Method 6800
AN S-142	Silicate, sulfate, and phosphate in a clay extract
AN S-58	Iodide in the presence of the standard anions in a leach solution
AN S-9	Six standard anions in soil eluate
AN CIC-12	Halogens and sulfur in clay using Metrohm combustion IC
AN C-99	Lanthanides in a rock sample by ion chromatography using non-suppressed conductivity detection
AN C-40	Calcium and magnesium in rock extracts

18.5 CONCLUSIONS AT A GLANCE

Until 1975, only a small range of analytical parameters could be measured automatically; therefore, it was necessary to develop and validate new methods to extend the scope of such parameters. An alternative that has almost replaced most of the wet chemical methods used in anions and cations analysis is ion chromatography. It is an attractive technique, especially for laboratories that need to determine numerous anions and cations in several thousand samples but do not have the throughput to justify the purchase of large automatic analyzers that are usually based on colorimetric procedures. Ion chromatography eliminates the need to use the hazardous reagents that are often integral to wet chemical methods.

The most important recent challenges related to the development and application of ion chromatography in soils analysis are: new sample preparation methods; improving the speed and selectivity of the separation of analytes; lowering of limits of detection and limits of quantification; development of new standard methods and extending the scope of the analysis for new group of substances.

19

CHROMATOGRAPHIC APPLICATIONS IN MEDICINE

19.1 INTRODUCTION

Chromatography has been preferred in the medical field as one of the most important analytical methods used for the identification and quantification of a drug and its metabolites in recent years. Various chromatographic techniques have been developed to differentiate drugs according to their characteristics and interaction types. Among these techniques, particularly High Performance Affinity Chromatography (HPAC), Cell Membrane Chromatography (CMC), Mixed Mode Chromatography (MMC), High Performance Liquid Chromatography (HPLC) are used for pharmaceutical purposes for bio analysis of drugs in preclinical and clinical studies. The success of Chromatography in the development of rapid and precise analytical methods provides superior specificity and sensitivity in the research of pharmacokinetic properties of drugs in R&D. In personalized medicine, it is important how the dosage and effect of drugs varies from person to person. In recent years, chromotography has been preferred as a promising technique in the analysis of drug protein binding and in the examination of clinical or pharmaceutical samples.

Chromatographic methods have become widespread practices for clinical analysis both in routine and research laboratories. Clinical analysis refers to testing of biological materials intended to determine the reliability and efficacy, diagnostic, and therapeutic processes. Chromatographic techniques play an irreplaceable role, particularly in drug, toxicology and biomarker analysis. In these analysis, such biological materials as whole blood, serum, plasma, urine, feces and tissues and macromolecules, such as lipids and proteins are used.

Generally, clinical analysis can be categorized as therapeutic drug monitoring, biomarkers' analysis for laboratory-based diagnostic purposes, and clinical toxicology. Therapeutic drug monitoring (TDM) is commonly used to determine the difference between the minimum active concentrations and minimum toxic concentrations of a drug. Biomarkers are biological compounds that can be operationalized as an indicator of normal biological processes, pathogenic processes, and pharmacological responses. They include low molecular weight metabolites, vitamins, hormones, lipids, peptides, and proteins. They are important specifically for early disease diagnosis, assessment of the disease process, monitoring of drug response, and therapeutic intervention. Clinical toxicology analysis focuses on various toxic agents associated with diseases or clinical symptoms after short-term or long-term exposure.

Clinical analysis requires fast, highly efficient and reliable methods. Low efficiency and sensitivity, and nonspecific interactions in the analyzes used in routine clinical applications cause the reliability of these techniques to decrease. As a result, more specific analyzes should be preferred today than traditional methods. Chromatographic techniques have become more popular in both routine and research laboratories in recent years. The purpose of this review is to provide an overview of the state of the existing chromatographic methods used in clinical analysis. This review summarizes the chromatographic trends in clinical analysis, discusses their advantages and disadvantages, and lastly offers a novel perspective to future research. For this purpose, the chromatography techniques described in the review have been preferred because of their excellent properties in terms of sensitivity, selectivity, time and efficiency compared to the routine analysis and other chromatography techniques.

19.2 SOME CHOROMATOGRAPHIC METHODS IN CLINICAL ANALYSIS

A variety of chromatographic techniques have been developed to differentiate the properties and interactions of drugs in clinical analysis. In this section, sensitivity, selectivity, analysis time and yield characteristics of chromatographic methods used in clinical analysis are discussed in detail.

19.2.1. High Performance Liquid Chromatography (HPLC) Applications

HPLC-MS has been employed as a selective alternative technique use in clinical laboratories, especially in endocrinology and toxicology, and to overcome problems related to the analysis of TDM, steroid hormones, and vitamins, new-born screening, and immunoassay. HPLC-MS's basic features are selectiveness, precision, efficiency, and analysis time.

HPLC-MS is a reliable and effective quantitative method in drug and biomarker analysis for the purpose of disease diagnosis and assessment. For the development of the method, the employment of a universal LC method is preferred. Reversed-phase mode is the most commonly used for the analysis of a mobile phase consisting of acetonitrile/methanol and formic/acetic acid, drug, and biomarker especially in the C18 column and gradient mode In drug and biomarker analysis, the C18 column and the mobile phase consisting of acetonitrile/methanol and formic/acetic acid in the gradient mode are most commonly used ones (Fig 1).

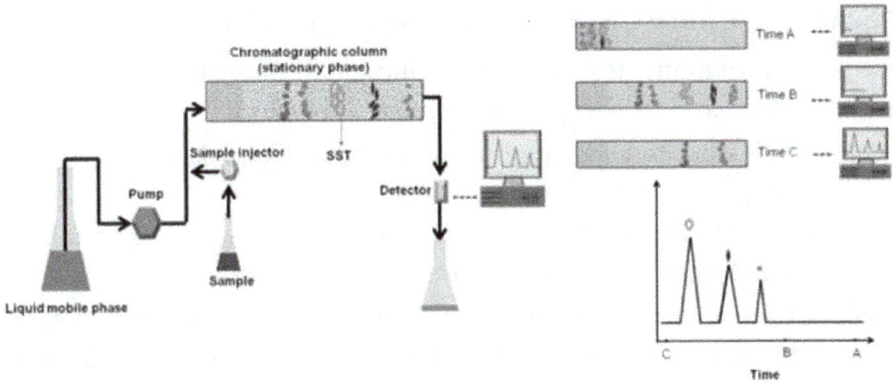

Fig. 1: LC-MS's schematized presentation

19.2.2 Gas Chromatography (Gc) Applications

Due to its high precision and widespread applicability in the analysis of various physicochemical agents, GC is commonly used as an alternative to GC-MS/MS, HPLC-MS/MS techniques in forensic toxicology, doping analysis, and routine bioanalytical laboratories. Most of the GC practices use a liquid film coated on a column wall and open tubular capillary columns (Fig 2). GC is also used for metabolomics and lipidomics in targeted analyses of a wide variety of volatile compounds. Such endogenous compounds as fatty acids, steroids, and other hormones, anesthetics, analgesics, antidepressants, antipsychotics and antiepileptics, and such narcotics as cocaine, cannabinoids, amphetamines, MeOH and breath volatile compounds are the compounds that are typically identified in clinical GC routines. Clinical analysis determines biomarkers of tuberculosis, cancer, neurodegenerative diseases, and diseases associated with metabolic disorders. This analysis is performed in a period of ten minutes.

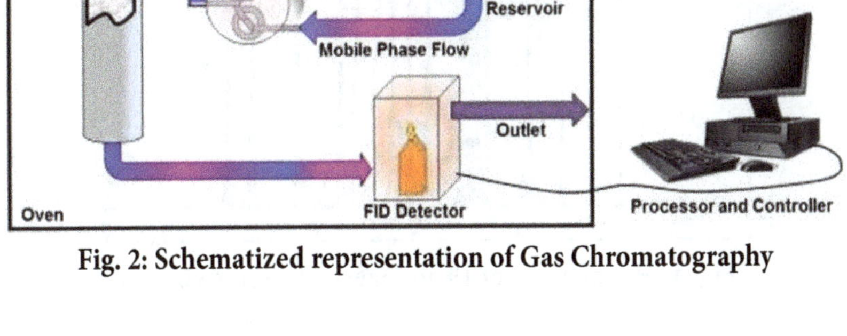

Fig. 2: Schematized representation of Gas Chromatography

20.2.3 Supercritical Fluid Chromatography (Sfc) Applications

Open tubular columns are used in supercritical fluid chromatography as in GC. The advantages of this technique over GC and HPLC are as follows:

1. Lower viscosity and higher distribution allowing for faster and more efficient separation,
2. Environment-friendly CO_2- based mobile phase mixed with organic solvents,
3. Broad-spectrum stationary phase.

Thanks to these features, supercritical fluid chromatography allows for a faster analysis with advanced resolution. Moreover, this technique is complementary to GC and HPLC commonly used in clinical analyses. Therefore, SFC has the potential to become a method drawing substantial attention in clinical analyses (Fig 3). Furthermore, it is widely used in drug discovery and development.

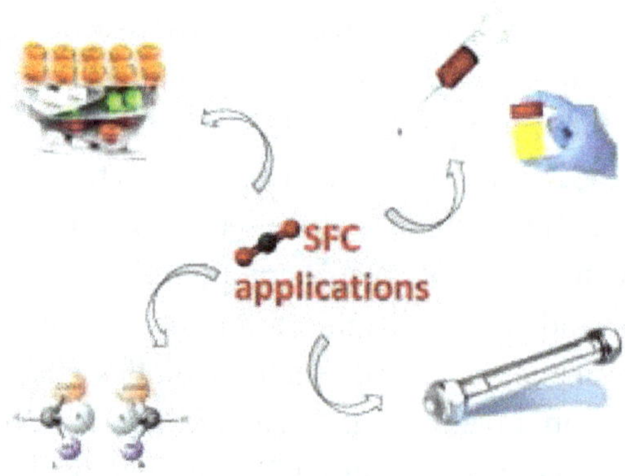

Fig. 3: Supercritical chromatography in pharmaceutical analysis

SFC is carried out in clinical analyses to monitor the drugs with serious side effects including citalopram in plasma and ketamine metabolites in urine and antiepileptics and chemotherapeutics. While lipophilic compounds, carotenoids and fat-soluble vitamins E and D can be generally identified by HPLC in a period of ten minutes, it can be performed by SFC in a few minutes (Fig 4).

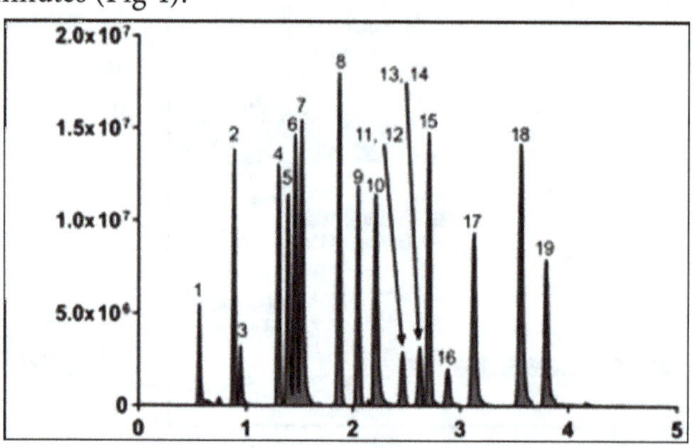

Fig. 4: Min analysis of 19 endogenous steroids by SFC

SFC is effectively used to screen endogenous hormones, and especially steroid, and to quantify neurotransmitters. The structural separation of similar steroids is presented in Fig 4. These results can be used for the diagnosis of many diseases.

19.2.4. High Performance Affinity Chromatography Affinity Chromotography (HPAC)

Biochemical and chemical agents in the body, antibody's binding with antigen, hormones' interaction with their receptors and drugs' binding with their biological targets or their carrier agents are important in many clinical processes. High Performance Affinity Chromatography has become a highly preferred technique to research and characterize interactions between proteins and drugs.

Affinity chromatography is a method relying on specific and reversible absorption of a targeted biomolecule by ligands immobilized on an insoluble support material (matrix) and incorporating binding ends that are complementary to the target molecule (Fig 5). Elusion of the target molecule from the medium can be achieved either by the use of a racing ligand or with a change in the pH, ionic power or polarity of the medium (mobile phase). The mobile phase used during the elusion of a compound from the column should not denature the compound and should not cause any change in its specific activity and function.

Affinity chromatography relies on the reversible interaction between a protein and a specific ligand immobilized in a chromatographic matrix. Affinity chromatography is a type of liquid chromatography that uses a biologically related agent as a stationary phase to purify or analyze specific sample components. High performance affinity chromatography allows for the use of HPLC supports and equipment for an affinity-based separation or analysis.

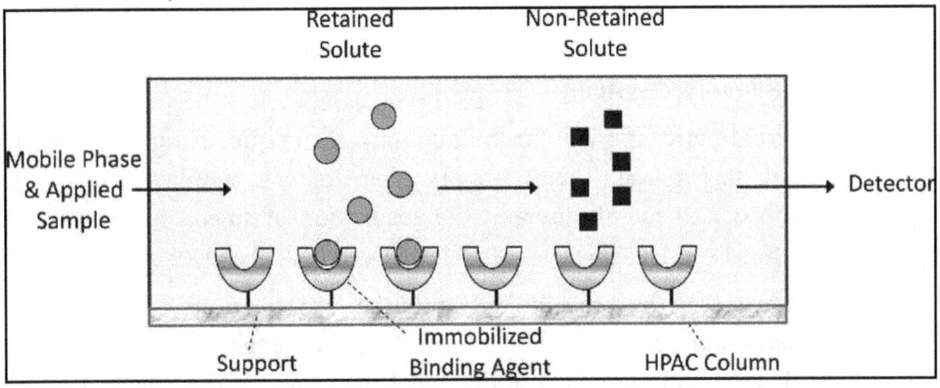

Fig. 5: Basic components of High Performance Affinity Chromatography.

Among the fields of application are high-performance drug screening, studies on modified protein, studies on personalized medicine and drug-drug interaction. HPAC has some characteristics that make it interesting for drug-protein binding studies. For example, This method has a good sensitivity and is relatively high and easy to apply to the routine. Besides, it can be used together with various types of detectors and it was shown by many studies that it achieves a good correlation with reference methods when used to examine the interaction between serum proteins and drugs. Another useful feature of HPAC is its capability to reuse the same protein or binding agent for a multitude of experiments.

There are various ways a disease can affect the drug-protein interactions in blood. The most important of these ways is the change in the concentration of a protein or a binding agent. For example, the changes in the concentration of other binding agents, e.g. HSA, in blood and in the compounds of these agents, e.g. lipoproteins,. It is known that the level of AGP (α1-acid glycoprotein) may increase from 2 folds to 5 folds during the acute phase of some diseases.

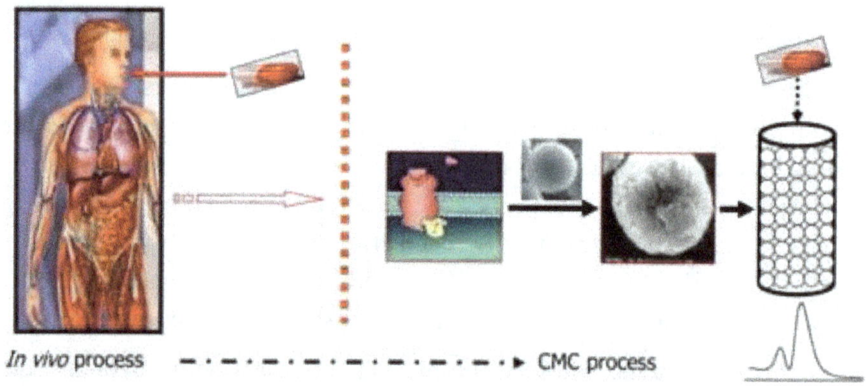

Fig. 6: Recent advances in Cell Membrane Chromatography

19.2.5. Cell Membrane Chromatography (CMC) Applications

Cell Membrane Chromatography has recently become a practical method to screen active components and to investigate the interaction between drugs and receptors (Fig 6). The interaction between drug and cell membrane receptors is analyzed in vitro. CMC offers several advantages, such as fast and high precision and is suitable for the screening of active components in mixed systems.

CMC is a biomimetic affinity chromatography technique using cell membrane receptor as a stationary phase. The stationary phase of cell membrane is developed through the fusion of cell membrane and the adsorption of silicon hydroxyl (Si-OH) groups on the silica gel surface. Silica gel can be used as the ideal carrier of cell membrane.

Paraformaldehyde can be cross-linked with the amino group on the surface of protein and its membrane protein facilitates sustaining the spatial configuration. Besides, paraformaldehyde as a protein cross-linking agent may lead to damages to the protein structure and decrease the activity incapable of ensuring the reliability of screening results. Thus, there is need for a novel method to improve cell membrane's ability to attach to the silica gel without damaging the cell membrane.

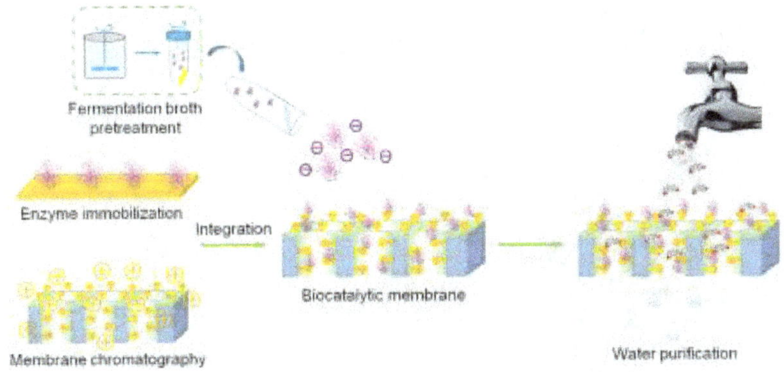

Fig. 7: Cell Membrane Chromatography analysis

One of CMC's disadvantages is that colons have a short life. The failure of membrane proteins adsorbing on silica gels to maintain their fast movements or activities affects the stability and replicability of the screening results. When using cell membrane columns, receptors on the surface of a cell membrane may gradually lose their original biological activity. Yet a membrane adsorbed on the surface of a mobile phase can be easily separated when the mobile phase is rinsed. These two reasons cause the column of cell membrane to be short-lived (Fig 7). Generally they fail after a constant use of 48-72 hours. CMC's high-performance drug receptor is promising for interaction characterization. CMC as a drug analysis method with separation and activity screening properties is expected to improve rapidly in terms of integration and automation.

19.2.6. Mixed-Mode Chromatography (MMC) Applications

It incorporates superior features for the separation of compounds not binding well nor solving well by typical reversed phase LC methods particularly for polar and charged molecules. Because a single MMC column has multiple-binding feature, it contributes an additional dimension to the separation method with the adjustment of MMC's mobile phase conditions. Additionally, MMC is an effective purification method to separate the complex compound matrices or to examine trace amount of analytes precisely. Mixed-mode columns are advantageous because they have multiple retention processes. The most common of the columns exhibiting both reversed-phase and ion exchange (anionic, cationic or zwitterionic) behavior is the mixed-mode column.

MMC allows adjusting the mobile phase/eluant conditions to regulate the interactions of certain analytes. As a new-generation mixed-mode stationary phase, it offers a better separation (Fig 8).

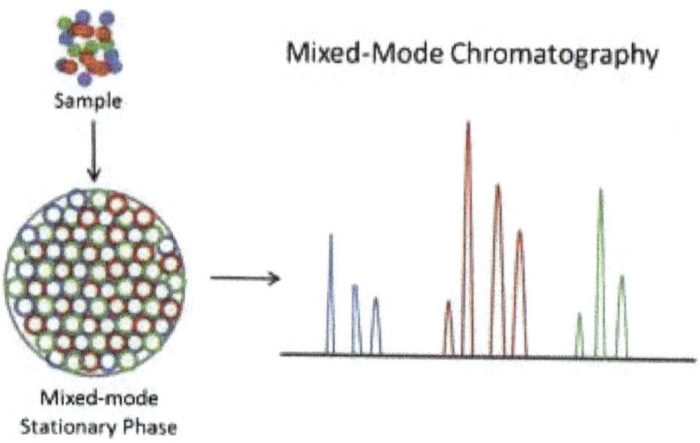

Fig. 8: Mixed-mode Chromatography

MMC is a chromatographic method incorporating multiple interaction mechanisms of solutes in the stationary phase. MMC is used as a method alternate and complementary to traditional reversed-phase (RP) chromatography, ion exchange (IEX) chromatography, hydrophilic interaction liquid chromatography (HILIC), and normal phase (NP) chromatography (Fig 9). In the past years, MMC has been attracting considerable scholarly attention in both biopharmaceutics and industry (Lämmerhofer et al., 2008). Peptide and protein analysis and purification play a remarkable role in the characterization of buildup of antibodies and heterogeneity.

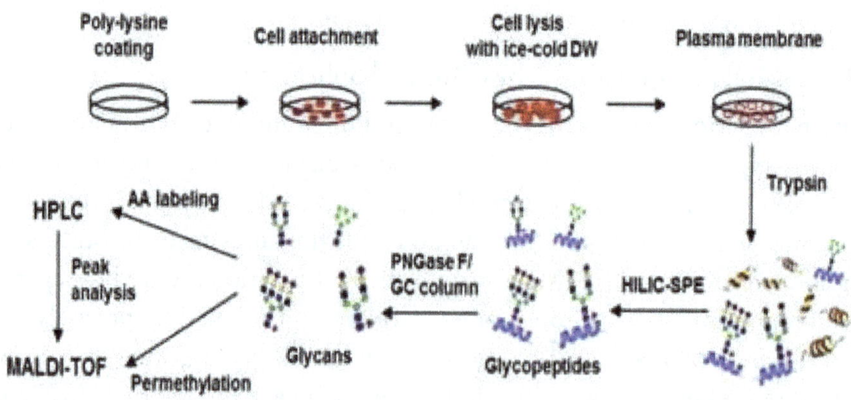

Fig. 9: Isolation of plasma membrane protein by MMC method

19.3 RESULTS AND ADVICES

This review highlights the significant characteristics of some chromatographic techniques employed in clinical practices in consideration of their precision, selectivity, analysis time, and performance.

Clinical analysis must be offered at a high speed with a high level of efficiency and requires routine application. Spectrophotometric, electrochemical and immunological tests are commonly used in the analysis of various compounds for their high efficiency and universality. In addition, chromatographic techniques also play an indispensable role in the analysis of clinical analysis, particularly in drug screening, toxicology and biomarker analysis. The combination of the novel techniques with the available ones must result in more advanced practices related to affinity chromatography and HPAC in the characterization of biological interactions for the purpose of clinical and pharmaceutical analyses.

There are many advantages that HPAC has contributed to drug-protein interactions and personalized medicine. These advantages include this technique's speed and precision, ability to work with various formats and sensing modes, easiness of automation, and ability to work with a small amount of sample and binding agent. GC is used in more than 30% of clinical chromatographic practices in toxicology. CMC is a simple, specific and time-saving technique to examine drug-receptor interactions, screening active compounds from complex mixtures, and quality of a product. In recent years, MMC has become more popular in the separation of high-polarity molecules by integrating with conventional reversed-phase and other chromatographic methods. In conclusion, there is a considerable need in the clinical field for well-devised universal chromatographic methods to simplify the processes and methods to analyze a great many biological compounds.

20

PAPER CHROMATOGRAPHY OF ANTIFUNGAL ANTIBIOTICS

20.1 INTRODUCTION: AN OVERVIEW

Paper chromatography is very helpful in the identification and comparison of antibiotics, but very few results from the application of this technique have been published so far. The awakening of interest in the search for antibiotics which are active against fungi pathogenic to plants and animals makes the investigator aware of the scant information that is available to help characterize such agents. Knowledge of the Rf values of the known antifungal agents would help advance the search for such therapeutic materials. Often these values, in conjunction with the biological inhibitory activities, are the only criteria by which one can determine whether the unknown agent appears to be a new material. When the Rf data are obtained for shake cultures, one can eliminate many previously described antibiotic-producing organisms from further study, thus allowing more research to be applied to antibiotics which are new and different. The studies reported in this paper should serve as a guide to the identification of some of the known antifungal antibiotics.

20.2 MATERIALS AND METHODS

The one-dimensional, ascending paper chromatogram technique was used. The paper strips (18 x 35 cm) were cut from large sheets of Whatman filter paper No. 1 (47 x 57 cm) by dividing the long dimension into widths of 18 cm and then cutting these sections to a length of 35 cm. Solutions of the antibiotic were placed 3 cm from the lower edge of the strip with a micropipette in 5-lambda (0.005-ml) portions until the desired quantity of substance had been added. The amounts placed on the paper varied with the sensitivity the solvent system used for developing the chromatogram.

TABLE 1. Details of Pyrex dish technique

Test Organism	Antibiotics	Agar	Incubation Time	Incubation Temp	Spore Suspension
Penicillium oxalicum 99*	Filipin Nystatin Fradicin Ascosin Thiolutin Rimocidin sulfate Trichomycin	Trypticase soy agart	hr 18-24	C 26	5 ml with a light transmission of 50% in 1 20 ml agar
Glomerella cingulata 100$	Antimycin A	Trypticase soy agart	40	26	5 ml with a light transmission of 50% in 120 ml agar
Bacillus subtilis Cohn 23	Actinomycin Gliotoxin Nigericin Streptothricin Sulfate	Emerson's agar§	18	30	400,000 spores per ml agar
Candida albicans 40	Candicidin A Endomycin	Emerson's agar	18	30	1 ml 1-day-old Emerson's broth culture in 120 ml agar
Saccharomyces sp. 451	Cycloheximide	Emerson's agar	18-24	30	1 ml 1-day-old Emerson's broth culture in 120 ml agar

* Numbers for test organisms correspond to numbers used in the collection of D. Gottlieb, Horticulture Field Laboratory, University of Illinois, Urbana, Illinois.
t Baltimore Biological Laboratories, Baltimore, Md.
$ To prevent bacterial contamination 100 Ag/m1 Chloramphenicol were included in the agar.
§ Yeast extract, 1 g; bactopeptone, 4 g; beef extract, 4 g; NaCl, 2.5 g; cerelose, 10 g; agar, 15 to 20 g; distilled water, 1000 ml ¶ Probably S. pastorianus.

Four materials were tested on each strip. Eight hundred ml of solvent mixture were placed in the bot- tom of cylindrical glass jars (25 x 45 cm). After the spots of solution on the paper strip had dried, the sheet was suspended in the jar so that approximately 1 cm dipped into the solvent system. The papers were then developed at 26 C until the solvent front had moved up about 27 cm. The strips were then air-dried and placed on seeded assay plates for 10 to 15 minutes.

TABLE 2. Solvent systems and their development time

Number	Solvent Systems	Development Time
		hr
I	Water saturated *n*-Butanol	15-16
II	*n*-Butanol:acetic acid: H_2O (2:1:1)	15-16
III	*n*-Butanol:acetic pyridine: H_2O (1:0:6:1)	15-16
IV	3% Ammonium chloride	2-3
V	50% hydrous acetone	6-7
VI	Benzene: acetic acid: H_2O (2:2:1) (organic layer)	7-10
VII	*n*-Butanol saturated water	8-10

The position of the antibiotic on the paper was determined in the following manner: The Pyrex dish technique was used and 120 ml of a seeded agar particularly suited to each organism were placed in the sterile dish. All plates were prepared immediately before putting the paper strip on the seeded agar. After incubation, clear zones occurred wherever the antibiotic had diffused from the paper into the agar. The distance that the solvent system had moved in the same time was measured on the original paper strip. An Rf value was calculated as the ratio of movement of the antibiotic to that of the solvent front.

Antibiotic	Rf Value in Solvent System No.:t						
	I	II	III	IV	V	VI	VII
With antifungal and antibacterial activity Actinomycin	0.96	0.94	0.96	0.00-0.21	0.96	0.99	0.53t or 0.70§
Endomycin	0.69	0.93	0.86	0.00	0.88	0.00	
Gliotoxin	0.90	0.95	0.96	0.54	0.88	0.98	
Nigericin.	0.95	0.98	0.98	0.00	0.86	0.00 & 1.00	0.28
Streptothricin sulfate	0.00	0.13	0.09	0.94	0.18	0.00	
Thiolutin	0.83	0.90	0.93	0.28	0.77	1.00	0.48
With antifungal activity only							
Antimycin A	0.93	0.98	0.97	0.00	0.86	1.00	0.20
Cycloheximide	0.90	0.96	0.96	0.87	0.91	0.96	
Filipin	0.91	0.95	0.96	0.00	0.72	0.18	0.00
Fradicin	0.84	0.85	0.94	0.00	0.36-0.67	0.89	0.07
Ascosin	0.33	0.88	0.79	0.00	0.43-0.73	0.00	
Candicidin A	0.44	0.86	0.80	0.00	0.38-0.69	0.00	
Trichomycin	0.35	0.85	0.76	0.00	0.68	0.00	
Nystatin	0.25	0.76	0.73	?	0.82	0.00	
Rimocidin sulfate	0.39	0.84	0.77	?	0.91	0.00	

* Average of at least two determinations.

†Composition of these systems is given in table 2.

‡With amounts of 2 to 10 Aig.

§ With amounts of 0.2 to 0.5 jig.

Seven solvent systems. were used for 15 different antifungal agents, 6 of which also show antibacterial activity. These antibiotics were. actinomycin; antimycin A ascosin candicidin A; cycloheximide, endomycin, filipin, fradicin, gliotoxin, nigericin, nystatin, rimocidin sulfate obtained from Chas. All further information concerning the test organisms, solvent systems, concentrations of different antibiotics, and so forth, is given in tables 1, 2, and 3.

20.3 RESULTS AND DISCUSSION

The Rf values of the 15 antibiotics are given in table 4. In a few cases certain solvent systems were unsatisfactory for specific antibiotics. For example, ascosin, candicidin A, and fradicin in 50 per cent acetone gave Rf values which varied over a rather wide range.

Nystatin and rimocidin sulfate were inactive when placed in 3 per cent ammonium chloride. Nigericin, even though crystalline, always gave two spots in the benzene-acetic acid-water system.

The results in table 4 provided the information for preparing two flow sheets (figures 1 and 2), on which most of the 15 antibiotics can be separated on the basis of their Rf values. In some cases the differences are not very pronounced and other properties of the material must also be considered in order to determine the identity of the antibiotic. Separation of the ascosin-candicidin- trichomycin and nystatin-rimocidin groups was very difficult with the solvent systems used in this study and ascosin could not be separated from trichomycin.

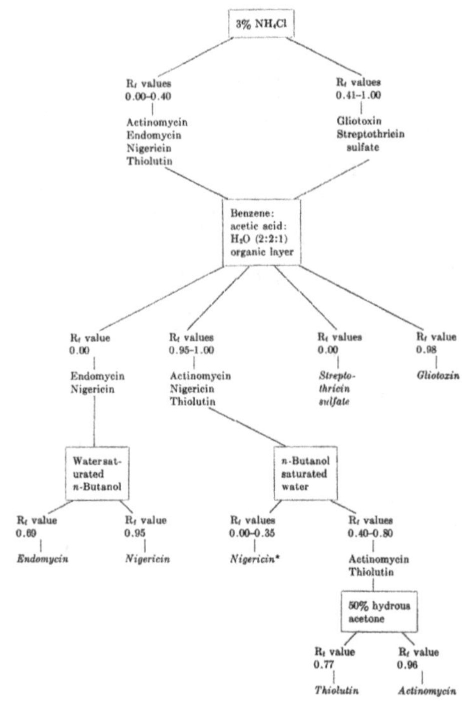

* In 50% hydrous acetone R_f value is 0.86.

FIG. 1. Flow sheet for the separation of six antibiotics with antifungal and antibacterial activity on the basis of their R_f valu n five solvent systems.

In general, the paper chromatography technique has been helpful in establishing the identity of these antibiotics.

But it is even more important that the information can be used to establish differences. This is especially necessary in the search for new antibiotics, for all other known materials must be "ruled out." Consistent differences among known materials and the unknown, when run under the same conditions, would indicate the presence of a new material. While paper chromatography is a useful technique, it is not a specific criterion, for it sometimes fails to differentiate some closely related antibiotics. It can best be used in combination with the other methods of differentiation available to the microbiologist and chemist.

21

METABOLOMICS OF GASTRIC CANCER METASTASIS DETECTED BY GAS CHROMATOGRAPHY AND MASS SPECTROMETRY

21.1 INTRODUCTION

Gastric cancer is one of the most common malignancies and the second cause of cancer-related death worldwide and in most Asian countries, such as China. So far, surgical resection remains the only curative treatment option. However, because of its asymptomatic properties, most patients are frequently misdiagnosed until local and distant metastases occur, leading to a poor prognosis of gastric cancer patients with a 5-year survival rate of less than 30%. Most gastric cancer-related deaths occur as a result of metastasis. Metastatic recurrence is the main obstacle to the improvement of therapy for gastric cancer. No effective treatment modality is available for this deadly disease at present. Similarly, almost no prophylactic therapies can block dissemination of gastric cancer cells and prevent its metastasis. Currently, histological staging of gastric cancer is mainly based on the depth of its invasion and metastasis, both of which are considered the most important indicators of recurrence and prognosis of gastric cancer after curative resection. It has been shown that some gene candidates, such as cell adhesion molecules, are involved in the process of gastric cancer metastasis. However, no routine molecule markers for predicting gastric cancer metastasis and prognosis are available because of their high variability in expression levels. The prognosis of patients with advanced gastric cancer remains very poor because the molecular mechanism underlying its metastasis is not fully understood. Gastric cancer metastasis, which is a complex and multistep process, involves release, migration and penetration of its cells through the vessel walls, arrest of its cells in microcirculation of distant organs and their subsequent migration and growth at the metastatic sites. To get a better insight into the mechanism of such a process, metastasis and non-metastasis animal models of gastric cancer were established using the human gastric cancer cell line SGC-7901. Because of the same genetic backgrounds, the animal is a suitable comparative system for studying the molecular changes in gastric cancer metastasis. At present, most investigations are focused on the identification of altered genes and proteins that play an important role in gastric cancer progression. However, only a few reports are available on the identification of key metabolites characterizing gastric cancer metastasis. Metabolomics, an omic science in systems biology, is the comprehensive and simultaneous profiling of metabolic changes occurring in living systems in response to genetic, environmental, or lifestyle factors. The missing link between genotype and phenotype can be established, which may provide information about gastric cancer that is complementary to genomics and proteomics analysis, thus improving our understanding of the pathogenic mechanisms and metabolic phenotype of gastric cancer. It has been recently shown that metabolomic method has great potentials in identifying the new diagnostic markers and therapeutic targets for different cancers, such as breast, prostate, pancreatic, liver, colon and gastric cancers, suggesting that metabolic alterations play a role in the biology of cancer. A more recent metabolomic analysis showed that increased sarcosine synthesis is an important metabolic change during prostate cancer progression. However, to our knowledge, there is no metabolomic study on metastasis of gastric cancer. In this study, metabolomic difference in metastasis and non-metastasis models of gastric cancer was detected by gas chromatography (GC) and mass spectrometry (MS), respectively, in accordance with our hypothesis that there were metabolite clusters associated with metastasis of gastric cancer.

21.2 DESCRIPTION OF MATERIALS AND METHODS

Chemicals Tetrahydrofuran (THF), N-methyl-N-t-butyldimethylsilyltrifluoro-acetamide (MBDSTFA) and chromatographic pure were purchased from Sigma Chemical Co. (St Louis, MO, USA). Pyridine, sodium hydroxide, chloroform, anhydrous ethanol, and anhydrous sodium sulfate, purchased from China National Pharmaceutical Group Corporation (Shanghai, China), were of analytical grade. Vacuum dryer was the product of Shanghai NOTED Technologies (China). Distilled water was obtained from the Milli-Q System (Millipore, MA, USA). Animals Male mice with severe combined immune deficiency (SCID) at the age of 6 wk, weighing 20-25 g, were obtained from Shanghai Experimental Animal Center, Chinese Academy of Sciences (China). The mice were housed under specific pathogen-free conditions with free access to food and drinking water. Experiment was performed in accordance with the Chinese National Guidelines for the Care and Use of Laboratory Animals and the relative ethical regulations of our university. Animal treatment Twenty-two SCID mice were randomly divided into metastasis group (n = 8), non-metastasis group (n = 8), and control group (n = 6). An animal model of metastasis was induced by orthotopic implantation of histologically intact tissue of human gastric carcinoma as previously described with some minor modifications. Human gastric cancer SGC-7901 (Shanghai Cancer Institute), a poorly-differentiated adenocarcinoma line, was originally derived from a primary tumor and maintained by passage in subcutis of nude mice. Tumors were removed aseptically with necrotic tissues cleaved away. The adjacent healthy tissues were cut into pieces (about 3 mm × 4 mm) in Hank's balanced salt solution, weighed and adjusted to 100 mg. After the SCID mice were anesthetized with 4.3% trichloraldehyde hydrate, an incision was made through the left upper abdominal pararectal line of mice in metastasis group with the peritoneal cavity carefully exposed. A part of the serosal membrane in middle of the greater curvature of stomach was mechanically injured using scissors. One hundred mg of tumor pieces was fixed at each injured site on the serosal membrane surface. After the stomach was returned to the peritoneal cavity, the abdominal wall and skin were closed. After the SCID mice were anesthetized with 4.3% trichloraldehyde hydrate, an incision was made in the left oxter of mice in non-metastasis group. Then, 100 mg of tumor tissue pieces was implanted into the subcutis (ectopic implantation) of mice. Meanwhile, the mice in non-metastasis group underwent the same orthotopic operation as those in metastasis group but with no tumor implantation into the gastric wall. The mice that underwent the same procedure with no tumor implantation served as a control group. All animals tolerated the surgical procedure well with no anesthesia-related death. Sample collection and pathological examination After anesthesia, all mice were sacrificed and subjected to autopsy. Tumors growing at the orthotopic or subcutaneous sites were harvested. Half of each tumor was snap-frozen at -80°C, while the other half was fixed in 4% formalin and embedded in paraffin. Tissues were collected from lymph nodes and all organs, fixed in 4% formalin, and processed for routine paraffin embedding after careful macroscopic examination. The tissue was cut into 4 μm-thick sections which were stained with hematoxylin and eosin, and evaluated histologically for metastasis of lymph nodes, liver or other organs by microscopy. Sample pretreatment and derivation One mL stilled water and one mL anhydrous ethanol were added to each 100 mg tissue sample. Gastric carcinoma tissue samples were ground with stilled water and anhydrous ethanol (1:1, v/v). The mixture was ultrasonicated at 4°C for 30 min and vortexed for 2 min. The tissue sample (0.1 g/mL) was centrifuged at 18 000 ×g (10 000 r/min) for 10 min. The aqueous layer was adjusted to pH 7.8 with 0.1 mol/L potassium phosphate buffer solution, and 1.5 mL supernatant was obtained from each sample. The collected supernatant (500 μL) was evaporated to dryness at 50°C for 24 h in a vacuum dryer. Then 150 μL tetrahydrofuran was added to each of the dried tissue extracts, vortex-mixed for 10 min and evaporated to complete dryness in a nitrogen evaporator. MBDSTFA (100 μL) was added to each sample and derived at 60°C for 30 min. The samples were vortexed for 30 s after derivation, for GC and MS analysis. GC and MS analysis Each of the samples was injected into an Agilent 6980 GC system equipped with a HP5MS capillary column (30 mm × 0.25 mm, i.d., 0.25 μm), and a quadrupole mass spectrometric detector (Agilent Technologies, Palo Alto, CA, USA). Helium was used as the carrier gas with a constant flow rate of 1.0 mL/min. One μL sample was injected into the Agilent 6980 GC system at 280°C. The column temperature was initially kept at 100°C for 3 min, and then elevated to 220°C at an increasing rate of 10°C per min, followed by 10°C per min to 280°C for 5 min. Both of the interface and ion source temperature were 200°C. MS was conducted in an electron impact ionization mode at 70eV. Mass data were obtained in a full scan mode from m/z 100 to 600. Total ion chromatograms (TIC) and fragmentation patterns of GC were acquired using the GC/MSD Chem Station Software (Agilent Technologies, Palo Alto, CA, USA). Compounds were identified by comparing the mass spectrum with a standard mass spectrum in the National Institute of Standards and Technology (NIST) Mass Spectra Library. Peaks with a similarity index more than 70% were the assigned compound names, while those with a similarity index less than 70% were considered unknown metabolites. Chromatograms were subjected to noise reduction prior to peak area integration. Any known artificial peaks due to derivatization of column bleed and BSTFA should be excluded from the data set. The resulting three-dimensional matrix included sample information, peak retention time, and peak area. Statistical analysis Data, normalized by dividing the sum of all peak areas in the sample (1 mg) before multivariate analysis, were expressed as mean ± SD and introduced into SPSS16.0 for Windows. Metabolite levels were compared by independent t test for the detection of significant differences in metastasis and non-metastasis groups. $P < 0.05$ was considered statistically significant. Principal components analysis (PCA) was performed to differentiate the metabolic patterns in metastasis and non metastasis groups. The differentiation performance was validated by the area under the curve (AUC) of receiver operating characteristic (ROC) curves. Similarly, metabolomic differences in metastasis group, non-metastasis group and control group were analyzed by t test and PCA.

21.3 RESULTS

The mean weight of mice in three groups was 23.81 ± 0.16, 23.87 ± 0.19 and 23.98 ± 0.19 g, respectively. Microscopy showed localized poorly-differentiated adenocarcinoma in all mice of non-metastasis and metastasis groups at the implanted sites. The average tumor weight was 4.28 ± 0.20 g in non-metastasis group and 4.30 ± 0.3 g in metastasis group, indicating that the tumor grows at a similar rate in the subcutis or stomach. However, the metastatic rate was significantly different when the tumor was implanted into the ectopic or orthotopic sites. Two thirds of the mice bearing an orthotopic tumor developed metastatic tumors in the region of lymph nodes, half in liver and one fourth in other organs. In contrast, no tumor metastasis was observed in the non-metastasis group after ectopic implantation. Macroscopic and histological examination showed no gastric cancer in the mice without tumor implantation. These results indicate that GC and MS are of a high reproducibility in the retention time of metabolites. Of the 152 signals detected in the tissue samples, some were not consistently found in other samples or could not be assigned to the unique metabolites because the abundance was too low. Fifty-eight signals could be identified by comparing a standard mass chromatogram with that in the NIST Library (Table 1). According to the NIST Mass Spectra Database, most of the chromatograms were identified as endogenous metabolites, such as amino acids, organic acids, inorganic acids, fatty acids, and pyrimidines, which were involved in several metabolic pathways, including glycolysis (lactic acid, alaline), serine metabolism (serine, phosphoserine), proline metabolism (proline), glutamic acid metabolism, glutamine metabolism, tricarboxylic acid (TCA) cycle (succinate), nucleotide metabolism (pyrimidine), fatty acid metabolism (docosanoic acid, and octadecanoic acid), and methylation (glycine).

Table 1 Tissue metabolites in mice of control, non-metastasis and metastasis groups.

Peak No.	Percentage	RT (min) metabolites	Match
1	6.1128	Propanoic acid	87
2	7.9894	Acetic acid	97
3	9.8780	Silanol	93
4	10.9850	Lactic acid	88
5	11.4610	Propanedioic acid	96
6	12.1110	2-methylglutaconic acid	80
7	14.9138	Butanoic acid	90
8	15.4200	L-Alanine	89
9	15.9640	Glycine	92
10	16.4966	Urea	93
11	16.7655	Acetamide	81
12	17.5162	1, 2-Butanediol	93
13	17.9103	L-methioninamide	82
14	18.1660	L-Valine	80
15	18.3168	Ethanol	73
16	18.6734	5-Nonanol	74
17	19.1113	D, L-leucine	98
18	19.3427	Pyrimidine	93
19	19.4178	Dimethylglycine	96
20	19.7869	L-isoleucine	86
21	20.4186	Succinate	85
22	21.0066	Propanamide	87
23	21.5133	Butanedioic acid	81
24	21.8323	2-butenoic acid	95
25	23.0521	Hexanethioic acid	89
26	23.8090	Malic acid	92
27	24.3032	L-Proline	91
28	24.4971	Pyrrolidine	72

Peak No.	Percentage	RT (min) metabolites	Match
29	24.5909	L-Methionine	83
30	25.0350	L-Serine	97
31	25.5230	L-Threonine	84
32	25.9421	Silanol	95
33	26.2298	1, 3-propanediamine	91
34	26.4800	L-phenylalanine	96
35	26.5676	Tetrame-thylenediamine	83
36	27.1681	Hexanoic acid	97
37	27.3933	L-Aspartic acid	90
38	27.9563	Glutamine	94
39	28.1940	Phosphoserine	98
40	28.3942	Mannofuranose	83
41	28.7319	L-Glutamic acid	84
42	28.8383	Hypoxanthine	91
43	29.5264	Glucofuranose	75
44	29.7953	L-Lysine	95
45	30.0330	2-furancarboxylic acid	99
46	30.7462	Glucose	97
47	31.6532	L-histidine	87
48	31.7845	L-arginine	89
49	32.0472	L-Tyrosine	92
50	32.1598	D-ribofuranose	83
51	32.4726	Decanoic acid	81
52	32.6540	Hexanedioic acid	95
53	33.5861	5-pyrimidinecarboxylic acid	89
54	34.6932	Myo-inositol	94
55	34.9560	Hexadecanoic acid	90
56	35.0435	Docosanoic acid	92
57	35.3250	Octadecanoic acid	79
58	35.4439	Maltose	92

The GC and MS data about tissue metabolites in metastasis and non-metastasis groups were analyzed by Student's t test. Marker metabolites selected by Student's t test are presented in Table 2. Among these metabolites, proline was the most up-regulated tissue metabolite in metastasis group, which was 2.45-fold higher than that in non-metastasis group. Glutamine was the most down-regulated tissue metabolite in the metastasis group, which was 1.71-fold lower than that in the non-metastasis group. The lactic acid, L-alanine, L-valine, leucine, malic acid, L-aspartic acid, serine, phosphoserine, dimethylglycine,

glycine, L-glutamic acid, L-lysine, myo-inositol, propanedioic acid, docosanoic acid, octadecanoic acid, arginine, pyrrodine, and pyrimidine were significantly up-regulated, while the glucose, succinate, L-isoleucine, L-methionine, propanamide, L-threonic acid, and butanedioic acid were remarkably down-regulated in the metastasis group compared to the non-metastasis group. The main metabolic pathways associated with metastasis of gastric cancer included glycolysis (lactic acid, alaline), serine metabolism (serine, phosphoserine), proline metabolism (proline), tricarboxylic acid (TCA) cycle (succinate, malic acid), fatty acid metabolism (docosanoic acid, and octadecanoic acid), and methylation(glycine).

Table 2 Different metabolites identified in metastasis and non-metastasis groups (mean ± SD).

No.	Metabolites	Metastasis group (n = 8)	Non-metastasis group (n = 8)	P	Fold
1	Lactate	4.194 ± 0.595	2.662 ± 0.218	0.000	1.58
2	Propanedioic	0.058 ± 0.004	0.0324 ± 0.002	0.000	1.80
3	Alanine	2.643 ± 0.145	1.903 ± 0.195	0.000	1.39
4	Glycine	0.762 ± 0.054	0.456 ± 0.172	0.001	1.67
5	Valine	1.036 ± 0.113	0.650 ± 0.047	0.000	1.59
6	Leucine	2.447 ± 0.306	1.756 ± 0.204	0.000	1.39
7	Pyrimidine	0.214 ± 0.030	0.166 ± 0.031	0.008	1.29
8	Dimethylglycine	0.372 ± 0.047	0.263 ± 0.020	0.000	1.42
9	Isoleucine	0.772 ± 0.087	0.992 ± 0.104	0.000	-1.29
10	Succinate	0.0794 ± 0.006	0.123 ± 0.018	0.000	-1.56
11	Propanamide	0.101 ± 0.007	0.147 ± 0.006	0.000	-1.45
12	Butanedioic	0.050 ± 0.008	0.073 ± 0.015	0.004	-1.46
13	Malicacid	3.053 ± 0.348	1.788 ± 0.116	0.000	1.71
14	Proline	1.173 ± 0.093	0.479 ± 0.072	0.000	2.45
15	Pyrrolidine	0.105 ± 0.016	0.067 ± 0.0083	0.000	1.57
16	Methionine	1.169 ± 0.099	1.518 ± 0.101	0.000	-1.30
17	Serine	1.742 ± 0.108	1.175 ± 0.074	0.000	1.48
18	Threonine	0.538 ± 0.040	0.632 ± 0.063	0.003	-1.17
19	Aspartic	5.870 ± 0.485	3.839 ± 0.453	0.000	1.53
20	Glutamine	0.090 ± 0.010	0.153 ± 0.016	0.000	-1.71
21	Phosphoserine	0.169 ± 0.0096	0.109 ± 0.012	0.000	1.56
22	Glutamate	3.533 ± 0.310	2.727 ± 0.200	0.000	1.30
23	Hypoxanthine	1.673 ± 0.150	1.841 ± 0.138	0.035	-1.10
24	Lysine	2.304 ± 0.137	1.808 ± 0.114	0.000	1.27
25	Glucose	1.338 ± 0.118	1.956 ± 0.142	0.000	-1.46

No.	Metabolites	Metastasis group (n = 8)	Non-metastasis group (n = 8)	P	Fold
26	Arginine	1.342 ± 0.178	0.987 ± 0.088	0.000	1.36
27	Inositol	0.409 ± 0.030	0.314 ± 0.022	0.000	1.30
28	Docosanoic	0.123 ± 0.012	0.076 ± 0.016	0.000	1.61
29	Octadecanoic	0.756 ± 0.103	0.555 ± 0.134	0.005	1.36

P-value was calculated by student t test. $P < 0.05$ was considered statistically significant. Fold change with a positive value indicates a relatively higher level of metabolites while a negative value indicates a relatively lower level in metastasis group than in non-metastasis group. The levels of lactic acid, propanedioic acid, L-alanine, L-valine, glycine, leucine, malic acid, pyrrodine, serine, L-proline, L-methionine, L-threonine, L-aspartic acid, phosphoserine, L-glutamic acid, L-lysine, arginine, myo-inositol, docosanoic acid, and octadecanoic acid were higher in non-metastasis group and metastasis group than in control group, while the levels of glucose, L-isoleucine, succinate, and glutamine were lower in control group than in non-metastasis group and metastasis group. PCA was used to identify the metabolic patterns associated with gastric cancer metastasis. The PCA scores of non-metastasis group and metastasis group were scattered into two different regions. ROC analysis, which was performed using the values determined by the first two components of the PCA model, confirmed the robustness of the PCA model. The sensitivity and specificity trade-offs were summarized for each variable with the AUC. The AUC value for the PCA model was 1.00, demonstrating a good differential value for metastasis of gastric cancer. The PCA model was also validated by ROC analysis (AUC = 1.00), showing that the growth of human gastric cancer in mice was of distinct metabolic properties compared to that in the host.

To elucidate the underlying mechanisms of metastasis and to identify the metabolomic markers of gastric cancer metastasis. Gastric tumors from metastatic and non-metastatic groups were used in this study. Metabolites and different metabolic patterns were analyzed by gas chromatography, mass spectrometry and principal components analysis (PCA), respectively. Differentiation performance was validated by the area under the curve (AUC) of receiver operating characteristic curves. Twenty-nine metabolites were differentially expressed in animal models of human gastric cancer. Of the 29 metabolites, 20 were up-regulated and 9 were down-regulated in metastasis group compared to non-metastasis group. PCA models from the metabolite profiles could differentiate the metastatic from the non-metastatic specimens with an AUC value of 1.0. These metabolites were mainly involved in several metabolic pathways, including glycolysis (lactic acid, alaline), serine metabolism (serine, phosphoserine), proline metabolism (proline), glutamic acid metabolism, tricarboxylic acid cycle (succinate, malic acid), nucleotide metabolism (pyrimidine), fatty acid metabolism (docosanoic acid, and octadecanoic acid), and methylation(glycine). The serine and proline metabolisms were highlighted during the progression of metastasis.

22.4 DISCUSSION

Metastasis is one of the leading causes for poor prognosis of patients with gastric cancer. The approaches used in tumor staging and prognosis assessment have a large number of limitations. Recent advances in the field of metabolomics have provided the new opportunities to identify novel diagnostic markers and therapeutic targets for gastric cancer. It is very important to establish a relevant model of cancer metastasis. Cell lines are often used as experimental models to investigate the genetic and biochemical alterations in tumor cells. However, these models have limitations due to the lack of interactions between tumor cells and host. Our previous study showed that human gastric cancer growing at orthotopic sites produces metastasis, and does not result in metastasis in SCID mice when it is implanted into the ectopic sites. It has been shown that a mouse gastric cancer model develops metastasis similar to human tumor, which is in favor of research on gastric cancer metastasis. Consistent with our results, an orthotopically implanted pancreatic cancer model is a prerequisite for identifying a differential gene expression pattern unique to metastatic human pancreatic cancer. In the present study, a highly metastasis model was successfully induced by orthotopic implantation of histologically intact tissue of SGC-7901 human gastric carcinoma while a non-metastasis model was established by subcutaneous implantation of the same cancer tissue in SCID mice. Tissue samples from gastric and subcutaneous tumors were derived from the same human gastric cancer with the same genetic background. The main difference in the two tumor models was their metastatic abilities. Therefore, metabolic analysis of these samples is valuable for the elucidation of metastatic mechanisms. To our knowledge, this is the first metabolomic investigation of gastric cancer metastasis in animal models. The comparative metabolomic strategy was used to identify the marker metabolites which were differentially expressed in animal models of human gastric cancer with distinct metastatic phenotypes. Of the 29 differential metabolites detected in the tumor models, 20 were up-regulated and 9 were down-regulated in the metastasis group compared to those in the non-metastasis group, indicating that these metabolites are involved in several metabolic pathways. The lactic acid level was higher while the glucose level was lower in metastasis group than in non-metastasis group. Lactic acid is the end product of glycolysis. Increased glucose uptake and consumption are frequently observed in many cancer cells even under normoxic conditions, known as the Warburg effect. It has been demonstrated that the lactic acid level is increased in various metastatic cancers, including renal, uterine cervix, head and neck, colorectal cancers. Moreover, the high lactate level in tumor tissue is associated with its metastasis and poor prognosis. In fact, increased lactic acid produced by tumor cells can result in acid-mediated matrix degradation, T cell inactivation, up-regulation of VEGF and HIF-1 alpha, and enhancement of cell motility, thus providing favorable conditions for metastatic spread. Therefore, high lactic acid levels reflect an increased energy demand for tumor progression. In the present study, GC and MS showed that proline was the most up-regulated tissue metabolite, indicating that increased proline in metastatic gastric cancer tissue may be correlated with the increased turnover of extracellular matrix in metastatic cancer cells. Tumor cells need increased degradation of collagen during the process of invasion and metastasis, thus producing a large amount of proline. Pyrroline-5-carboxylic (P5C) is the precursor of proline and also its degradation product. Proline oxidase, also known as proline dehydrogenase, catalyzes the first step of proline to P5C in mitochondria and the latter is converted to proline by the cytosolic P5C reductase. It has been demonstrated that proline oxidase can be induced by p53 due to genotoxic stress and initiates apoptosis by the mitochondrial and death receptor pathways. Proline is a stress substrate and matrix metalloproteinases can degrade collagen in the extracellular matrix. So far little attention has been paid to the correlation between proline metabolism and tumor progression. It has been shown that proline consumption is increased in patients with metastatic renal cancer. Increased proline biosynthesis has been recently observed in metastatic breast cancer cell lines. Furthermore, proline metabolism is linked with arginine and glutamate metabolism, TCA cycle and pentose phosphate pathway (PPP) due to P5C, suggesting that the significantly up-regulated metabolism of proline is highly correlated with cancer metastasis. In this study, serine metabolism was involved in the metastatic process of gastric cancer, showing alterations in the pathway. The serine and phosphoserine increased 1.48-fold and 1.56-fold, respectively in metastatic tumors. It was reported that all the three genes involving the serine biosynthesis pathway, phosphoglycerate dehydrogenase (PHGDH), phosphoserine aminotransferase 1 (PSAT1), and phosphoserine phosphatase (PSPH) are up-regulated in highly metastatic breast cancer cells, which is agreement with our results. Although most amino acids are increased in gastric cancer and colorectal cancer tissue compared with their adjacent normal tissue, little is known about the alterations related to metastatic behaviors. In the present study, the levels of leucine, valine glutamate and lysine were higher while the levels of methionine and threonine were lower in metastatic specimens, indicating that the demand for energy is increased in metastatic progression. Glycine can derive from serine synthesized by the glycolytic intermediate phosphoglycerate and its elevated levels may be associated with glycolysis. Sarcosine, a methylation derivative of the amino acid, glycine, is related to prostate cancer progression. Dimethylglycine is another methylation product of glycine and how it is up-regulated remains unclear. The disturbed TCA cycle, observed in a large number of tumors including gastric cancer, is considered to be related to carcinogenesis. In this study, the TCA cycle intermediates such as succinate and malic acid were remarkably perturbed in metastatic specimens, suggesting that enhanced glycolysis contributes to metastatic progression. In this study, the expression patterns of metastatic and non-metastatic human carcinoma were compared by metabolomic analysis, and several important metabolic pathways associated with metastasis of gastric cancer were identified. Of note, proline and serine metabolism were highlighted in this study. Further functional and clinical sample analysis of the metabolic pathways is needed to demonstrate their role in gastric cancer metastasis. The metabolic pathways may be exploited as biomarkers for gastric cancer progression.

22.4.1 Background

Gastric cancer is the second cause of cancer-related death worldwide, and its metastasis is one of the leading causes of cancer-related death. The molecular mechanisms underlying gastric cancer metastasis are still not fully understood. Recent metabolomic studies have shown that metabolic alterations play a role in the biology of gastric cancer. The metabolic profiling of tumor tissue is used to elucidate the underlying mechanisms and identify the metabolomic markers of gastric cancer metastasis for improving its diagnostic and therapeutic strategies.

22.4.2 Research Frontiers

Metabolomics, an OMIC science in systems biology, is the comprehensive and simultaneous profiling of metabolic changes occurring in living systems in response to genetic, environmental, or lifestyle factors. Recently, metabolomic method has shown great potentials in identifying the new diagnostic markers and therapeutic targets for cancers. However, metabolomic study on gastric cancer metastasis remains scarce.

22.4.3 Innovations and Breakthrough

Most investigations have been focused on identifying the altered genes and proteins that play a role in cancer progression. In this study, the expression patterns of metastatic and non-metastatic human carcinoma were compared by metabolomic analysis, and several important metabolic pathways associated with metastasis of gastric cancer were identified. This is the first report on metabolomic investigation of gastric cancer metastasis.

22.4.4 Applications

The results of this study indicate that the metabolic pathways can be exploited as biomarkers for gastric cancer progression, which can be used in diagnosis and treatment of gastric cancer.

22.4.5 Terminology

Metabolomics, an OMIC science in systems biology, is the comprehensive and simultaneous profiling of metabolic changes occurring in living systems in response to genetic, environmental and lifestyle factors. Gas chromatography (GC) and mass spectrometry (MS) have been widely applied in metabolomic investigation because of their high sensitivity, peak resolution and reproducibility.

22

ROLE OF CHROMATOGRAPHY FOR MONITORING OF BREAST CANCER BIOMARKERS

22.1 INTRODUCTION: AN OVERVIEW

An early detection of cancer may provide the best eradicable treatment, hence, detection methods play a crucial role in cancer treatment. This article describes the role of chromatography for the separation and identification of breast cancer biomarkers. Biomarkers provide an assessment of prognosis, monitoring the effectiveness of therapy and predicting cancer reoccurrence. The important biomarkers of breast cancer detected by chromatography are genetic (single nucleotide polymorphism), proteomics (prostate-specific antigen, estrogen receptor, Human Kallikrein 14, progesterone receptor, Galectin-3-binding protein, alpha-1- Antichymotrypsin, Elafin, CA15.3, CEA, CA27.29, Her-2/neu, AAG, GCDFP-15 and GCDFP-24 etc.), metabolites (glucose, cholesterol, carbohydrate antigen 15-3, etc.) and breathing agents (2-propanol, 2, 3-dihydro-1-phenyl-4(1H)-quinazolinone,1-phenyl-ethanone, heptanal, isopropyl myristate etc.). In view of these points, attempts have been made to discuss the state-of-art of chromatography for biomarkers detection in breast cancer patients. Some modern chromatographic techniques such as nano-HPLC, LC-MS and DPHPLC have been discussed. Besides, efforts have been made to discuss sensitivity, specificity and protein profiles of these biomarkers.

Cancer is the most devastating disease causing about 6 million deaths globally per year. It is estimated that the number of cancer patients death may increase by 9 and 11.4 millions by the year of 2015 and 2030, respectively which is a serious issue for all of us. Globally, cancer is the second largest lethal disease accounting for millions of deaths; after cardio-vascular diseases. Among to various types of cancers, breast cancer is the second leading cause of deaths among women. According to a new report the number of breast cancer patients may increase up to 1.45 million in the year 2010. Early detection of breast cancer is the urgent need of today to decline the rate of mortality. It is estimated that the chances of 5 years survival rate of breast cancer patients were 96% in localized condition and 86 % in metastases stage in 2006, which increased to 98.3% in localized condition. Non-invasive techniques such as mammography, ultrasound, MRI, CT scan, PET scan and invasive technique (biopsies) are common methods for the detection of breast cancer. In spite of effective and frontline screening tool, the sensitivity of mammography significantly decreases in heterogeneously dense breast cancers. It does not diagnose almost half of the palpable tumors in extremely dense breasts. Magnetic resonance and ultrasound imaging are quite good approaches to detect malignant and invasive cancers as compared to mammography.

These imaging strategies are supposed to overcome the drawbacks of mammography. But some research groups have shown that these screening methods result in over diagnoses, leading to increased recall and negative open surgical biopsy rates. The molecular screening methods in combination with these imaging techniques are helpful in alleviating this situation. In addition to these, biomarkers technique is an innovative approach emerging in the last decade, which can detect cancers in early stage too. This technique is also useful for monitoring the progression and regression of cancer after treatment. The biomarker techniques of cancer detection involve the search of tissues and bio-fluids. The biomarkers may be genes, proteins, metabolites and breathing ingredients which provide information of abnormal growth of cells in the cancerous cells. The issue arises to detect these biomarkers and various techniques have been used for their detection, which include chromatography, immunohistochemistry (IHC), enzymelinked immunosorbent assay; (ELISA), two dimensional gel electrophoresis (2DE), fluorescence in situ hybridization (FISH), polymerase chain reaction (PCR), real-time polymerase chain reaction (RT-PCR), matrix-associated laser desorption/ionization time-of-flight MS (MALDI-TOF-MS), surface-enhanced laser desorption/ionization time-of-flight MS (SELDI-TOF-MS), liquid chromatography coupled with various detectors. Among these techniques, chromatography is the best approach due to its ease of operation, inexpensiveness, selectiveness, reproducibility and low limits of detections. Moreover, the development of nano liquid chromatography on micro-chips is an innovation in chromatography as it requires samples at nano gram level and provides detection at nano or low levels. Keeping all these facts into consideration, it is considered worthwhile to describe the state-of-art of the chromatographic detection of breast cancer biomarkers, which is a useful and effective tool for controlling breast cancer in early as well as later stages.

22.2 BIOMARKERS

Defined the biomarkers as cellular, biochemical or molecular alterations in biological media such as human tissues, cells, or fluids. But, Naylor specified the definition of biomarkers as "a factor that can be objectively measured and evaluated as an indicator of normal biological processes, pathogenic processes, or pharmacological responses to a therapeutic intervention". Biomarkers are used in screening, diagnosis and monitoring of cancer progression and control. For the reduction of the rate of mortality, the discovery of biomarkers is very demandable and they are highly sensitive, specific and therapeutics for breast cancer.

The different types of biomarkers are described in the following sub-sections. Recently, low molecular weight peptides have been patented as a biomarker for breast cancer.

22.2.1 Genetic Biomarkers

It is well known that all biological activities are controlled by genetic materials i.e. DNA and RNAs. The carcinogenesis is initiated by the change in DNA or, sometimes, in RNAs structures. Therefore, the abnormalities in the degree of functionalities of normal cells can be monitored by the changes in these genetic materials; as genetic indicators. According to the genomic technology, the determination and monitoring of the genetic materials are indicators of normal biological process, pathogenic process and/or response to therapeutic or other intervention. It is interesting to observe that the genomic indicators are helpful for the diagnosis and treatment of breast cancer at early stages. Tomita et al. discovered the degree of the breast cancer by measuring the genetic and epigenetic modification in cfp-DNA and reported mutation in genes, which was responsible for breast cancer. Studied the epigenetic methylation of DNA for the early detection of breast cancer. Also identified 29 μ-RNAs that were differentially expressed in breast cancer tissue as compared to normal tissue. HER2/neu is a very essential proto-oncogene for the detection of early breast cancer and used for the development of trastuzumab as a clinically relevant biomarker. Novel miRNA sequences biomarkers, which may be used as a base for molecular targeted treatment of cancer.

Observed that HER2/neu is over expressed in the breast cancer, and, hence, a potential biomarker for the detection of breast cancer. It also carried out an exhaustive study of the whole genome changes, including mutations, methylation, translocation, presence of viral genes, or allelic imbalance (AI). Based on these observations, the authors concluded that all changes were present in the majority of the primary breast tumors and they were also encountered in the plasma of breast cancer patient also identified two tumor suppressor genes i.e. BRCA1 and BRCA2, for an early detection of breast cancer. Discovered kallikrein gene 14(hK14) as the potential genetic biomarker for breast cancer.

22.2.2 Proteomic Biomarkers

Almost all enzymes and hormones are proteins and control various biological functions of a cell. A slight variation in these proteins may modulate the biological activities of a cell, resulting into an abnormal growth of tissues. Besides, these proteins alterations may affect cell metabolism and physiology, cell growth and death, and secretion of molecules that signal other cells and tissues. Therefore, the abnormal behavior of any cell or tissue in body may be identified by the change in the proteins and may be called as protein biomarkers. This is also true in case of the changes in proteins that are associated with breast cancer cells and termed as breast cancer protein biomarkers. As in case of the genomic technology, the proteomic biomarkers are studied under proteomic research. Most important identified protein biomarkers are CA15.3, CA27.29, CA125, carcinoembryonic antigen CEA, CA19.9, BR27.29, Her2/Neu, tissue polypeptide antigen (TPA) and tissues polypeptide specific antigen (TPS). It has been reported that these biomarkers are helpful for the early detection of breast cancer. Twenty two protein biomarkers can be used as a diagnostic tool for determining the stage of breast cancer. Sauter et al. [44, 45] studied the proteomic breast cancer biomarkers in nipple aspirated fluid (NAF) and observed that high levels of insulin-like growth factor binding protein-3 (IGFBP-3) were associated with breast cancers. Scientist identified BC1, BC2, and BC3 as breast cancer biomarkers in serum. It was also observed an elevated level of HER2/neu in the saliva of breast cancer patients. In another study, a European group working on tumor markers identified MUC-1 mucin glycoproteins, CA 15.3 and BR 27.29 as remarkable serum markers for breast cancers, but did not recommend these for diagnosis due to their low sensitivities. However, the newly developed nano-liquid chromatography may overcome this limitation.

22.2.3 Metabolic Biomarkers

Millions of metabolites are produced by the body cells and they are of definite structures and compositions in normal cells. A slight variation in their structure is an indication of abnormal behavior/growth of cell. In this way, the abnormal growth of body cell can be monitored by analyzing cell metabolites. Some workers used this approach to identify breast cancer and called these metabolites as breast cancer biomarkers. It has been reported that the metabolic rate and metabolites are different in cancerous cells in comparison to normal ones. The metabolomic technology when deals in these sorts of biomarkers (as metabolites) also exposes the changes due to the alteration of proteins. It was observed the presence of low serum high-density lipoprotein cholesterol (HDL-C), which is an important indicator for breast cancer in overweight and obese women. As identified four breast cancer biomarkers i.e. homovanillate, 4-hydroxyphenylacetate, 5-hydroxyindoleacetate and urea in the urine of breast cancer patients. It was observed high concentrations of 8- hydroxy-2-deoxyguanosine and 5-hydroxymethyl-2-deoxy uridine in the urine of breast cancer women. Contrarily, the same group observed low concentrations of androsterone and 16α-hydroxydehydroepiandrosterone in these patients. Due to the advancement in paraphernalia, the traces of these can be detected and, hence, breast cancer can be identified in early stage too. It was compared the ratio of 2-hydroxyestrone and 16-α-hydroxyestrone and concluded that this ratio was moderately correlated in premenopausal women and poorly in postmenopausal women, who gave an indication of breast cancer.

22.2.4 Breath Biomarkers

Recently, an interesting observation has been reported citing certain volatile organic compounds (VOCs) in the breath of breast cancer women, which have been considered as breath biomarkers. Scientists identified different types of VOCs as breath breast cancer biomarkers in the patients of breast cancers. The most common and reported biomarkers are 2-propanol, 2, 3-dihydro-1- phenyl-4(1H)- quinazolinone,1-phenyl-ethanone, heptanal and isopropyl myristate. The sensitivity and specificity of these biomarkers reported were 93.8% and 84.6%, respectively. Furthermore, these workers concluded that the breath testing is a valuable tool for the breast cancer patients due to high sensitivity and specificity of these biomarkers.

22.3 DETERMINATION OF BIOMARKERS

The separation and identification of a biomarker is a very important issue but is a tedious job due to their presence in trace amount. Therefore, highly sensitive, selective and efficient methods are required for this purpose. The monitoring of the biomarkers includes several steps comprising selection of patients, sample collection, sample preparation, analyses and their identification. A systematic protocol for the detection of biomarkers is shown in Fig. (1). The choice of sample collection from patients depends on several factors such as age, pre- and post-menopause, stage of cancer etc. The samples are collected by specific techniques and we are not discussing them here in detail as it is out of the domain of this article. The sample preparation is a crucial issue and some methods have been developed and used, which are discussed in the following section.

22.3.1 Sample Preparation

It is obvious that sample preparation is an integral part in bio-analyses, which is true in monitoring of breast cancer too. It is important to mention here that sample preparation for bio-markers is essential due to the presence of other several biological identities. Moreover, sample handling is essential due to the presence of biomarkers at trace levels. It is important to mention here that it is a tedious job as it requires the removal of undesirable biological identities. However, this job may be completed with ease by spiking, standardizing and validating of sample for reproducible results. The sensitivity, selectivity and reproducibility of analytical techniques depend on the sample collection, handling and storage. The sample preparation of biomarkers is carried out from serum/plasma as all the biological components are present in plasma. Sometimes, serum is preferred due to the interference of anti-coagulant added into plasma. It has been observed by some workers that protein profiles obtained from plasma and serum are very different. The commonly used techniques for sample preparation of bio-markers are centrifugation, ultrafiltration, disk plates, solvent extraction, solid phase extraction etc. Besides, some workers used gel and capillary electrophoresis and chromatographic modalities for sample preparations. For example, separated serum proteins biomarkers from plasma by using 2D-HPLC with anion displacement column. Evaluated 14 different solvents including methanol, ethanol, acetonitrile and acetone as well as different combinations of methanol and acetonitrile or ethanol to select the best solvent system for the comprehensive extraction of metabolites from serum. They concluded that methanol was the best solvent for the extraction of metabolites and useful for precipitation of serum proteins in breast cancer patients. One of the most frequently used techniques is solid phase extraction for sample preparation. Scientists compared different types of SPEs for sample preparation of biomarkers and concluded that multiple affinity removal cartridges (MARC) were the best for this purpose due to low loss of non-targeted proteins.

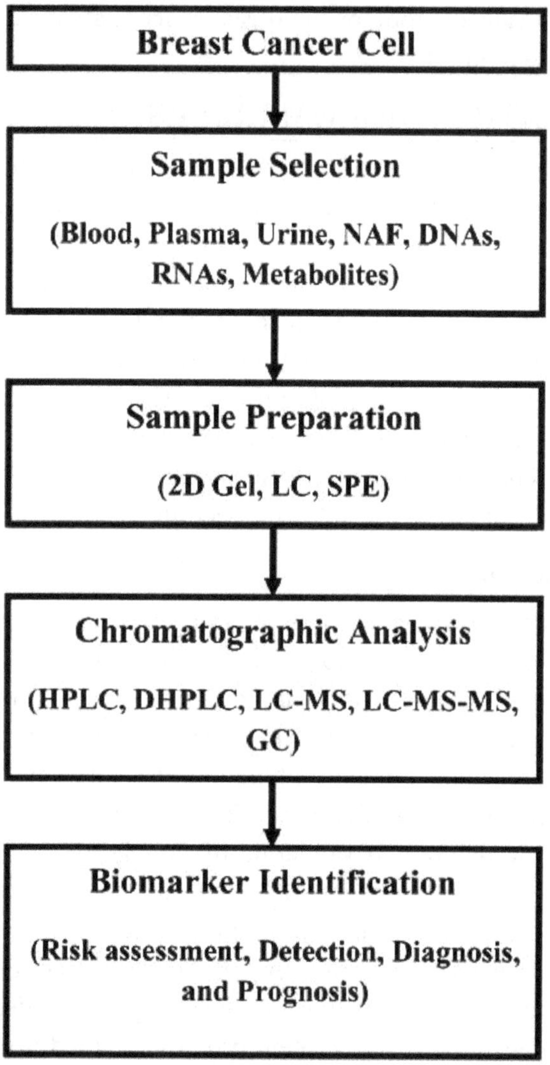

Fig. 1: Protocol of biomarkers diagnosis by chromatography.

22.3.2 Separation and Identification of Biomarkers

In biomarkers monitoring, sample preparation is followed by separation and identification of biomarkers, which are carried out by different techniques such as immunohistochemistry (IHC), enzyme-linked immunosorbent assay; (ELISA), two dimensional gel electrophoresis (2DE), fluorescence in situ hybridization (FISH), polymerase chain reaction (PCR), real-time polymerase chain reaction (RTPCR), matrix-associated laser desorption/ionization time-offlight MS (MALDI-TOF-MS), surface-enhanced laser desorption/ionization time-of-flight MS (SELDI-TOF-MS), liquid chromatography and stable isotope labeling. During the preparation of this article it has been observed that among various analytical methods, chromatography may be an ideal technique for analyzing biomarkers due to several advantages that are already discussed in the introduction section. Therefore, the analyses of breast cancer biomarkers by chromatography are described.

22.3.3 Separation and Identification of Biomarkers by Chromatography

The analysis is the second part for the monitoring of biomarkers after sample preparation. Among various methods used for analyses of biomarkers; as discussed above; the recent evolution of nano-liquid chromatography has become the choice of biomarkers monitoring. The most important chromatographic modalities used for analyses of biomarkers are gas chromatography (GC) and liquid chromatography (LC). Again LC has achieved good reputation in bio-markers analyses due to its wide range of application, ease of operation and reproducible results. Liquid chromatographies used for this purpose are High Performance Liquid Chromatography, generally, coupled with MS detector, micellar electrokinetic chromatography (MEKC) and capillary electro chromatography (CEC). The analysis of biomarkers by chromatography is discussed in the following subsections.

22.3.4 Separation and Identification of Biomarkers by Gas Chromatography

Mostly metabolic biomarkers are volatile at the working temperature of Gas Chromatography (GC). Therefore, GC has been used for monitoring these types of breast cancer biomarkers.Some workers also used Gas Chromatography- Mass Spectrometry (GC-MS) for the analysis and validation of 2-hydroxyestrone and 16-alpha -hydroxyestrone biomarkers in the urine of breast cancer patients. Furthermore, the presence of these biomarkers was verified by enzyme immunoassays (EIAs) method. The authors compared the results of both methods and concluded that the correlation between EIA and GC-MS was excellent; with better results of GC-MS; as EIAs is a slightly time consuming technique. Similarly also used GCMS for the analysis of volatile organic compounds (VOCs) as breathe biomarker for breast cancer patients with 93.8% and 84.6% sensitivity and specificity, respectively. Recently, scientists analyzed more than 30 metabolites and confirmed 8-hydroxy-2-deoxyguanosine and 5-hydroxymethyl-2-deoxyuridine in the urine of breast cancer women. As it is reported low concentrations of androsterone and 16μ-hydroxydehydroepiandrosterone in breast cancer patients by GC-MS hyphenated with partial least squares discriminate analysis. Used GC-MS for the identification of biomarker in the biological sample.

22.3.5 Separation and Identification of Biomarkers by Liquid Chromatography

Of course, GC is a quite good technique but suffers from certain limitations i.e. it is only useful for volatile biomarkers. Therefore, liquid chromatography (LC) has achieved a great reputation in analytical science due to its wide range of applications with least limitations. LC has been used for analyzing some biomarkers and the work reported till today is described in the following sub-sections.

Liquid Chromatography-Mass Spectrometry (LC-MS)

Among various detectors mass spectrometer (MS) is the best one in LC due to its selective, sensitive and low limit of detection nature. Therefore, LC-MS is an ideal combination for analyzing any type of biomarkers. In view of these facts, this hyphenation of chromatography has been used for determining biomarkers in breast cancer patients. Analyzed the nipple aspirate fluid (NAF) proteome by using LC-MS for identifying breast cancer biomarkers. Used metal affinity chromatography for the identification of Kallikrein 14(hK14) breast cancer biomarker and also used multi lactin affinity chromatography (M-LAC) for the characterization of multiple glycoprotein biomarkers in breast cancer patients.

As discussed chromatographic separation combined with mass spectrometry for the proteomic analysis of serum proteins and reported several potential breast cancer biomarkers including heat shock protein HSP27, transcriptional regulator 14-3-3 sigma and the derivatives of the complement component C3a. Described micro fluidic liquid chromatography (μ-LC) system for proteomic investigations of MCF7 breast cancer cells. For readers, the microchip nano liquid chromatogram is shown in Fig. (2) indicating sharp peaks of proteins within 45 mints. Also screened Galectin-3-binding protein and alpha -1-antichymotrypsin as breast cancer biomarkers by using LC-MS-MS. Scientist also used liquid chromatography mass spectrometry (LC-MS-MS) for the separation and identification of breast cancer biomarkers in nipple aspirated fluid (NAF). Few workers used LC-MS method for the determination of sucrose in urine. As per the authors, this technique can be used at large-scale for epidemiological studies; this suggests a link between the intake of refined sugars and an increased risk for colorectal, breast, pancreatic and endometrial cancer. Scientist also developed a quantitative proteomic LC-MS approach for the identification of protein changes in plasma samples obtained from a mouse model of breast cancer. Recently, also used affinity liquid chromatography-mass spectrometry (ALC-MS) for screening urinary ribonucleosides to differentiate between breast cancer patients and healthy volunteers. The authors collected urine samples for nucleosides from 113 breast cancer patients and 99 volunteers. The authors reported 87.67% and 89.90% as sensitivity and specificity, respectively. As performed the proteomic analysis of breast cancer proteins (biomarkers) by fluorogenic derivatization-liquid chromatography/ tandem mass spectrometry (FD-LC-MS/MS). In this study, authors performed the proteomic analysis of seven breast cancer cells and one normal cell as shown in Fig. (3).

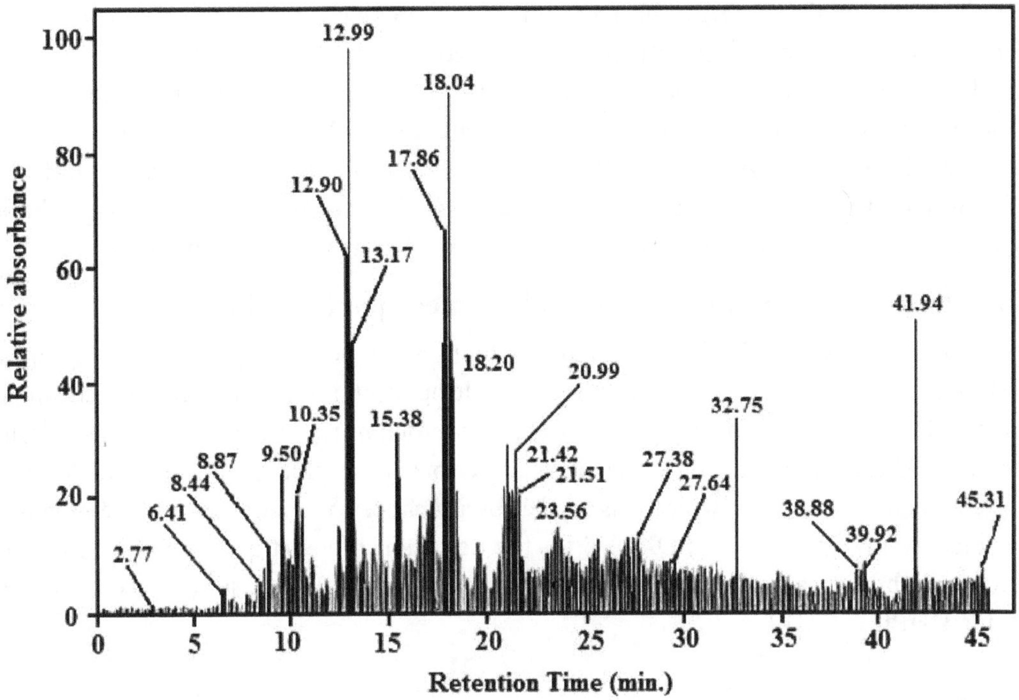

Fig. 2: Nano LC-MS/MS analysis of a protein digest prepared from the MCF7 breast cancer cell line on microchip .

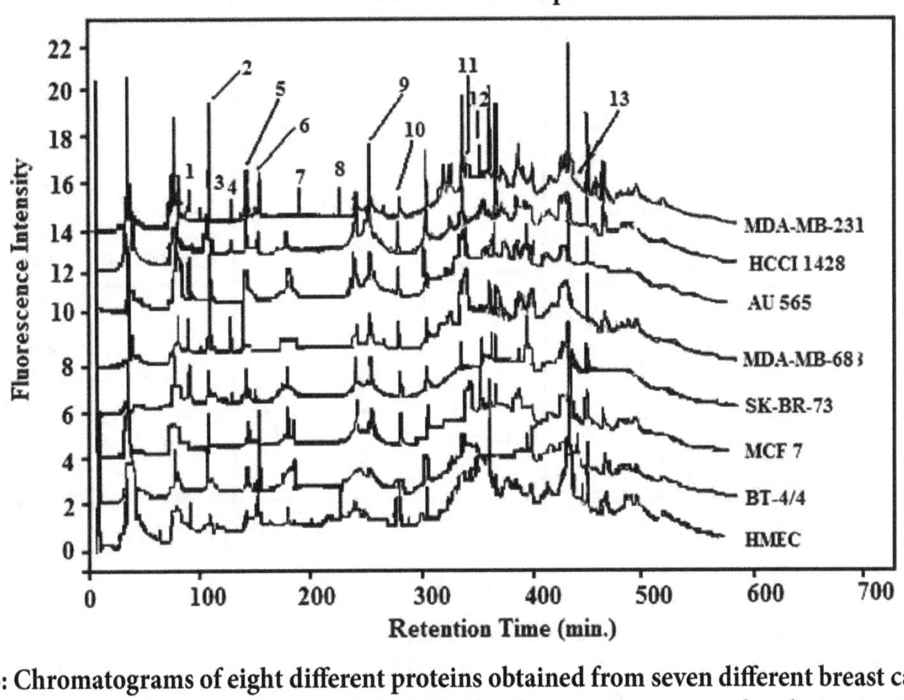

Fig. 3: Chromatograms of eight different proteins obtained from seven different breast cancer patients and one normal woman (HMEC). The chromatograms are after derivatization of proteins with DAABD-Cl .

A perusal of this Figure indicates that there are some changes in the chromatograms of proteins in breast cancer patients; in comparison to normal volunteers. All thirteen peaks in the chromatograms showed the different variations in breast cancer biomarkers and authors concluded that RanGAP was solely detected in all the cancer cells. The cancer cells express high levels of RKIP and low levels of annexin-2, galectin-1 and EF-Tu. Besides, the presence of co-operatively expressed annexin-2 and galectin-1 without tropomyosin-1 in a breast cancer cell was suggestive for its metastatic property.

Denaturing High Performance Liquid Chromatography (DHPLC)

A new modality of HPLC i.e. denaturing high performance liquid chromatography (DHPLC) provides a useful platform for high-throughput mutation detection and single nucleotide polymorphisms (SNP) discovery due to denaturing of DNA, which is helpful in determining mutation sites. Scientist described a specific HPLC method for the determination of HbA1c, which was used as breast cancer biomarker in postmenopausal women. Many workers used DHPLC for the detection of genetic mutation utilized for analyzing breast cancer. Also observed mutation of ATM in breast cancer patients by using DHPLC.

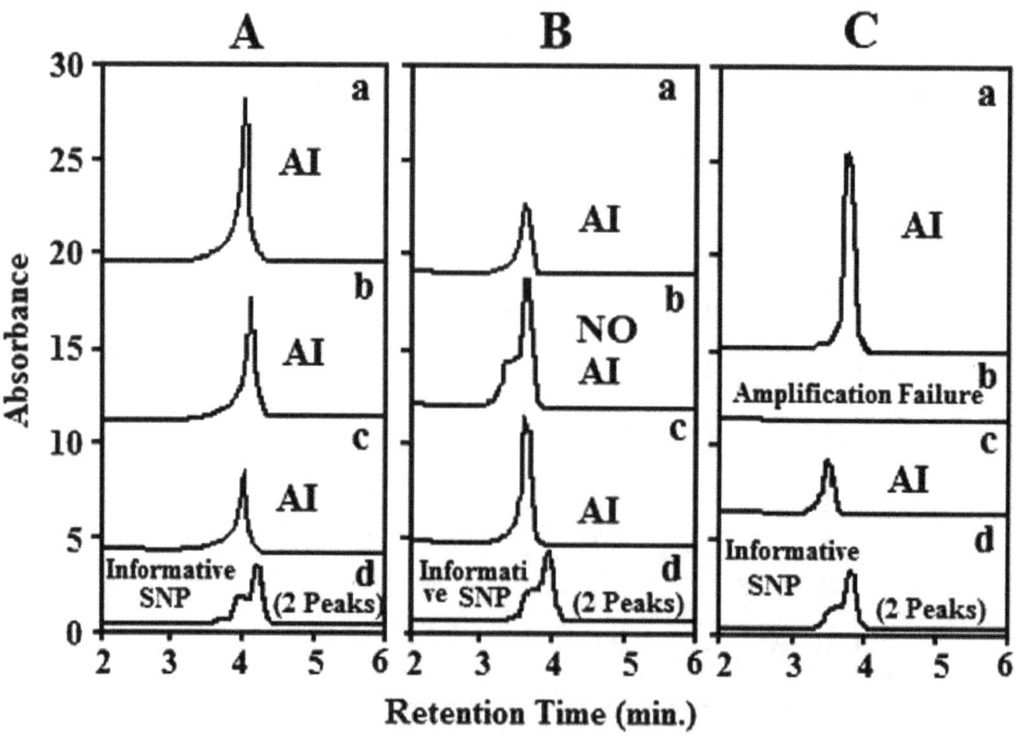

Fig. 4: Detection of allelic imbalance (AI) by DHPLC pre- and post amplification of plasma circulating DNA through BL-WGA and MDA techniques. A: Patient 1 (SNP-S14), B: Patient 2 (SNP-S14) and C: Patient 3 (SNP-S14). Chromatograms: a: Amplification of plasma circulating DNA via- BL-WGA, b: Amplification of plasma circulating DNA via-MDA, c: Plasma DNA before amplification and d: Normal genomic DNA.

It is clear from this Figure that BL-WGA technique of amplification is more efficient for apoptotic and necrotic DNA fragments from plasma circulating DNA and has a reliable downstream screening for all other genetic alterations as compared to MDA technique.

22.4 FUTURE PERSPECTIVES OF CHROMATOGRAPHY IN BIOMARKER ANALYSES

Presently, among various cancers, breast cancer is havoc for females and millions of patients die every year globally. Therefore, the treatment of this sort of cancer; even in late stage should be our priority. Of course, the first step in cancer treatment is its diagnosis and from the ongoing discussion we can confirm that chromatography may be an ideal technique for the separation and identification of breast cancer. Of course, the present developed chromatographic modalities are being used for the separation and identification of biomarkers but are not fully capable due to trace concentrations of biomarkers. However, the recently developed micro-chip based chromatography may be an ideal tool for this purpose. Besides, sample preparation methods are not fully developed and the recoveries of biomarkers are poor again due to low concentration of these species.

Therefore more advanced paraphernalia is required including on-line hyphenation of sample preparation. Briefly, the wide range of applications and low limits of detection of nanoliquid chromatography dictate us a bright future of breast cancer patients due to an early diagnosis and treatment of this disease by chromatographic methods. We believe that after few years chromatography will be a routine analytical tool of cancer pathological laboratories.

22.5 CURRENT & FUTURE DEVELOPMENTS

Obviously, scientists, oncologists, pathologists and regulatory authorities are fighting with cancer and developing new methods and technologies to treat it. The role of chromatography in the identification of breast cancer biomarkers is a land mark in the area of oncology. During the completion of this article our experience dictates us that chromatographic modalities may be a remarkable tool for the separation and identification of breast cancer biomarkers but are not fully developed, which need further more advancements. We appeal to the scientists, oncologists, pathologists and regulatory authorities to encourage the development of chromatographic methods for the diagnosis of biomarkers. Furthermore, chromatographic technique must be used for the separation and identification of other sorts of cancers as well. Liquid-liquid partition chromatography is used for the purification of Taxane, which seems useful in the treatment of breast cancer too [88]. We hope that in the coming year's chromatography will play a crucial role for the eradication of cancers. Let us hope for the best future of chromatography in cancer patient's treatment.

RELATED TERMINOLOGY

Accuracy: Measurement result **accuracy** Indicates its closeness to the true value.

Adsorption chromatography: Adsorption chromatography mode of separation in which a solute or sample components are attracted to a solid surface, the stationary phase, by adsorption retention forces, the mobile phase may be a gas or liquid.

Adsorption retention forces: Adsorption retention forces attraction of a solute onto a solid stationary phase due to microporosity (pores 5~ 50 nm) and polar character (formation of van der Waal's forces and hydrogen bonding) of the surface, described by Langmuir isotherms (see isotherms).

Affinity chromatography separation: Affinity chromatography separation effected by affinity of solute molecules for a bio-specific stationary phase consisting of complex organic molecules bonded to an inert support material, e.g. separation of proteins on a bonded antibody stationary phase. The technique is really selective filtration rather than chromatography.

Asymmetry: Factor describing the shape of a chromatographic peak. The chromatographic theory assumes a Gaussian shape and that peaks are symmetrical. A quantitative measure is the peak asymmetry factor, which is the ratio of the distance from the peak apex to the backside of the chromatography curve over the distance from the peak apex to the front side of the chromatography curve at 10 % of the peak height.

Back-flushing: Back-flushing used to remove components in a sample held strongly on the stationary phase at the beginning of a column by reversing the flow of mobile phase after completion of an analysis.

Bonded stationary phase: Bonded stationary phase a stationary phase consisting of organic molecules bonded to a stationary phase support material (micro particulate silica gel), usually through siloxane bonds.

Buffer: A solution that maintains constant PH by resisting changes in PH from dilution o addition or small amounts of acids and bases.

Capacity Factor (k): Old term for a chromatographic parameter that measures the degree of retention. Now defined as the retention factor (k) by the international union of pure and applied chemistry(UPAC). See also retention factor for a method of calculation.

Carrier gas: Carrier gas inert mobile phase gas for gas-liquid or gas-solid chromatography, usually nitrogen is used with packed columns, helium with capillary columns.

Chromatogram: A plot of detector signal output or sample concentration versus time or elution volume during the chromatographic process.

Chromatography: Chromatography is a physical method of separation in which the components to be separated are distributed between two phases, one of which is stationary (Stationary Phase) while the other (The mobile phase) moves in a definite direction.

Column chromatography: Column chromatography mobile phase flows through a column containing the stationary phase material, either as a particulate packing or as a film on the inner wall (WCOT) of the column tubing.

Column switching: Column switching use of two or more columns to effect a separation of a sample mixture. The mobile phase can be switched between the columns to direct specific fractions from the first column onto a second column, or switched to back-flush retained components off the first column.

Counter-ion: Counter-ion in ion pair chromatography, the ion of opposite charge to the component which forms a neutral ion pair that can be separated by reverse phase HPLC, also in ion chromatography the counter ion in solution that displaces the sample ion from the stationary phase.

Dead volume: Volume of mobile phase between the injection port and detector which does not contain stationary phase causes excessive band broadening.

Degassing process: Degassing process to remove gases dissolved in an HPLC mobile phase by sparging with helium and/or using vacuum, to prevent micro bubbles forming in the detector causing signal noise.

Development separation: Development separation procedure to obtain a chromatogram, also, visualisation of separated components in TLC, Pc.

Displacement chromatography: Displacement chromatography the separation process where a sample mixture is introduced at the start of the column and is then eluted with a mobile phase that is more strongly attracted to the stationary phase than the sample components. Individual components are therefore displaced by the mobile phase and by each other, in order of retention forces.

Eluant strength: Eluant strength refers to the polarity of the eluant or mobile phase and its ability to attract and retain the solutes or components being separated. An increase in eluant polarity reduces the distribution ratio K, and the elution time for a given component. The term is most frequently used in TLC and reverse phase HPLC, remember, 'like attracts like'.

Elution: Elution descriptive term for separated components passing through and eventually emerging from the column.

Exclusion chromatography: Exclusion chromatography EC, separation of components in a mixture on the basis of molecular size. Usually a liquid mobile phase is used with a stationary phase consisting of a cross-linked polymer gel containing pores of a narrow size range. Components are retarded to varying degrees by the way the molecules partially fit into the pores, larger molecules are less retarded and elute first. EC is also known as size or steric exclusion chromatography.

Frontal chromatography: Frontal chromatography frontal analysis, form of chromatography where the sample is continuously added to the column. Only the least retained component in a sample mixture is obtained in a pure form, the other components co-elute.

Fronting asymmetric peak: Fronting asymmetric peak shape where the front part of the peak tapers forward of the main peak often caused by overloading or poor injection techniques; see asymmetry.

Gradient elution: Gradient elution refers to the progressive change of mobile phase composition (usually by increasing the proportion of organic solvent) in HPLC, that is, increasing the solvent strength and therefore progressively decreasing retention on the stationary phase and retention times.

Guard column: Guard column short column containing the same stationary phase as the main column. It is located just before the main column to protect it by collecting strongly retained components in a sample mixture that would otherwise remain on the top of the main column.

High performance liquid chromatography: High performance liquid chromatography HPLC, employs a liquid mobile phase and solid or bonded liquid stationary phase. High resolution and column efficiency is achieved by using 3-10 J-lm microporous or micropellicular column packing materials. In reverse phase chromatography an aqueous buffer containing varying amounts of a miscible organic solvent is used as the polar mobile phase, e.g. I % acetic acid/methanol, I % ammonium acetate/methanol and ODS as a non-polar stationary phase.

Hydrophilic: Hydrophilic describes the 'water loving' or polar characteristics of molecules, used to describe the 'like attracts like' features of a mobile phase or stationary phase.

Hydrophobic: Hydrophobic describes the 'water hating' or non-polar characteristics of molecules, used to describe the 'like attracts like' properties of a mobile or stationary phase.

Immobilized Phase: A stationary phase which is immobilized on the support particles, or on the inner wall of the column tubing, e.g., by in situ polymerization (cross-linking) after coating.

Ion pair chromatography: Ion pair chromatography ions in a sample are paired to form a neutral 'ion pair' by addition of an ion pair reagent to the mobile phase, e.g. heptane sulphonic acid. The alkane chain gives non polar retention characteristics to the ion pair and therefore separation can be carried out using reverse phase HPLC.

Isothermal chromatography: Isothermal chromatography chromatographic separations using the same column temperature throughout the analysis, particularly in Gc.

Isotherms plot of distribution ratios: Isotherms plot of distribution ratios K, that is, a plot of concentration of the component in the stationary phase versus concentration in the mobile phase. Isotherms represent the relative attraction of a solute for the stationary and mobile phases, the plot is linear for a symmetrical Gaussian peak, non linear isotherms lead to unsymmetrical peaks, e.g. peak fronting and tailing. Langmuir isotherms describe the equilibrium process in adsorption chromatography, Nernst isotherms relate to partition chromatography.

Linear elution chromatography: Linear elution chromatography used to describe the separation process in partition chromatography where a constant partition or distribution ratio is maintained for a given component throughout the separation. Thus the equilibrium that determines the partition of the solute between the mobile and stationary phases remains constant and the rate at which the solute moves through the column will be proportional to the time it spends in the mobile phase relative to the time spent retained on the stationary phase.

Liquid-liquid chromatography: Liquid-liquid chromatography LLC, chromatography using liquid mobile and liquid stationary phases, e.g. reverse phase HPLC and TLC; see partition chromatography.

Liquid-solid chromatography: Liquid-solid chromatography LSC, chromatography using a liquid mobile phase and solid stationary phase, e.g. HPLC, TLC; see adsorption chromatography.

Mobile phase: Mobile phase transports the solutes or components to be separated through a column or plate of stationary phase material. The solution properties of a liquid mobile phase compete with the retention forces of the stationary phase to determine the distribution ratio and hence elution time. In GC the gaseous mobile phase transports components in the vapour phase.

Non-polar dispersion: Non-polar dispersion forces used to describe the retention of non-polar solutes on a non-polar liquid stationary phase. London's dispersion forces postulate an intermolecular induced dipole mechanism to account for attraction of a component onto a non-polar stationary phase.

Normal phase chromatography: Normal phase chromatography uses a polar stationary phase and a less polar or non-polar mobile phase, e.g. in TLC or HPLC with a silica gel stationary phase and hexane or methanol/chloroform/diethyl ether mobile phase.

Packed column chromatography: Packed column chromatography where the stationary phase is in the form of uniform particles which are packed into a column, the particles may be a solid material, e.g. silica gel or alumina or particles coated with a liquid stationary phase.

Partition chromatography: Chromatographic separations which involve a liquid stationary phase immobilised on an inert support material. The solutes are distributed between the mobile phase and stationary phases according to their respective partition coefficients.

Partition retention: Partition retention forces intermolecular forces that result in attraction of solute molecules to the stationary phase. Polar retention forces include dipole-dipole attractions, van der Waal's forces and hydrogen bonding. Non-polar retention forces consist of London's dispersion forces arising from induced polarity in non-polar molecules, remember that 'like attracts like'.

Peak area: Peak area area under a peak obtained from integration of the detector signal for a given component. Peak area is proportional to the number of molecules contained in an eluted band and hence may be used to calculate the proportion of each component in a mixture; see integration, detector response factors and internal standard.

Peaks: Peaks representation of the detector signal resulting from elution of component molecules obtained by continuously plotting the signal as the analysis proceeds. The distribution of molecules within an eluted band results in a Gaussian peak; see chromatogram.

Planar chromatography: Planar chromatography type of chromatography where the stationary phase is a flat film in the form of a thin layer of material coated onto a rectangular glass or inert plastic support plate (TLC) or as a paper sheet (PC). The mobile phase moves through the stationary phase by capillary action.

Polar retention forces: Polar retention forces forces of attraction between a solute and the stationary phase due to dipole~dipole interactions, van der Waal's forces and hydrogen bonding.

Pre-columns: Pre-columns short guard columns packed with the same stationary phase material as the analytical column but with a larger particle size to avoid reducing the overall efficiency of the system. Pre-columns are used to condition the mobile phase for the stationary phase and also to clean up the sample by removing any components that are strongly retained on the stationary phase.

Sample Components: The chemically pure constituents of the sample. They may be unretained (i.e., not delayed) by the stationary phase, partially retained (i.e., eluted at different times) or retained permanently. The terms Eluite or Analyte are also acceptable for a sample component.

Solid support for stationary phases: Solid support for stationary phases robust inert material used as support for a liquid mobile phase, has a large surface area, common materials are silica gel for HPLC, processed 'brick' and celite particles for Gc.

Solute: Solute an alternative term for a sample component that is present in the mobile phase or attracted on to the stationary phase as it moves through the chromatographic system.

Stationary phase: Stationary phase solid material or liquid immobilised on an inert support, which attracts components in the mobile phase according to characteristic retention forces and thus retards their progress through the column or plate; see mobile phase.

Thin layer chromatography: Thin layer chromatography, TLC a form of planar chromatography having a coating of a solid stationary phase, e.g. silica gel or alumina, on a plate of glass or inert plastic. The liquid mobile phase moves up the plate by capillary action, reverse phase TLC plates are available where a liquid stationary phase is immobilised on the solid support. The polarity of the mobile phase is carefully chosen to obtain the best separation; see elutropic series.

Zone A: region in the chromatographic bed where one or more components of the sample are located. The term Band may also be used for it.

REFERENCES

A. Little, and J. A. Schleimer, "Application Of Paper Chromatography To Oil Field Problems."

Allen-Meares, P. and Lane, B. (1987). Grounding social work practice in theory: Ecosystems. Social Casework, 68, 515-521.

Alpert, A.J., 1983. Cation-exchange high-performance liquid chromatography of proteins on poly(aspartic acid)- silica. J. Chromatogr. A 266, 23e37.

Alpert, A.J., 1990. Hydrophilic-interaction chromatography for the separation of peptides, nucleic acids and other polar compounds. J. Chromatogr. 449, 177e196.

Alpert, A.J., 2008. Electrostatic repulsion hydrophilic interaction chromatography for isocratic separation of charged solutes and selective isolation of phosphopeptides. Anal. Chem. 80, 62e76.

Alpert, A.J., Andrews, P.C., 1988. Cation-exchange chromatography of peptides on poly(2-sulfoethyl aspartamide)- silica. J. Chromatogr. 443, 85e96.

Alpert, A.J., Hudecz, O., Mechtler, K., 2015. Anion-exchange chromatography of phosphopeptides: weak anion exchange versus strong anion exchange and anion-exchange chromatography versus electrostatic repulsionhydrophilic interaction chromatography. Anal. Chem. 87, 4704e4711.

American Petroleum Institute (2006). Manual of Petroleum Measurement Standards. Chapter 14 - Natural Gas Fluids. Measurement. Section 1 - Collecting and Handling of Natural Gas Samples for Custody Transfer.

Anderson, J. (1981). Social work methods and processes. Belmont, California: Wadsworth Publishing Company.

Appelblad, P., Abrahamsson, P., September, 2005. MS detection of homocysteine, methyl-malonic acid, and succinic acid using HILIC separation on a covalently bonded zwitterionic stationary phase. LCGC N. Am. (Suppl.), 24e25.

Aral, T., Aral, H., Zizadanogullari, B., Zizadanogullari, R., 2015. Synthesis of a mixed-model stationary phase derived from glutamine for HPLC separation of structurally different biologically active compounds: HILIC and reversed-phase applications. Talanta 131, 64e73.

Armstrong, D.W., Jin, H.L., 1989. Evaluation of the liquid chromatographic separation of monosaccharides, disaccharides, trisaccharides, tetrasaccharides, deoxysaccharides and sugar alcohols with stable cyclodextrin bonded phase columns. J. Chromatogr. 462, 219e232.

Awapara, "Application of paper chromatography to the estimation of some free amino acids in tissues of the rat," Journal of Biological Chemistry, vol. 178, pp. 113-116. 1949.

Bäckström M, Thomsson KA, Karlsson H, Hansson GC. Sensitive liquid chromatography-electrospray mass spectrometry allows for the analysis of the O-Glycosylation of immunoprecipitated proteins from cells or tissues: application to MUC1 glycosylation in cancer. J Proteome Res 2009; 8: 538-45.

Badoud, F., Grata, E., Perrenoud, L., Avois, L., Saugy, M., Rudaz, S., Veuthey, J.L., 2009. Fast analysis of doping agents in urine by ultra-high-pressure liquid chromatographyequadrupole time-of-flight mass spectrometry: I. Screening analysis. J. Chromatogr. A 1216, 4423e4433.

Beelders, T., Kalili, K.M., Joubert, E., de Beer, E., de Villiers, A., 2012. Comprehensive two-dimensional liquid chromatographic analysis of rooibos (Aspalathus linearis) phenolics. J. Sep. Sci. 35, 1808e1829.

Berg, T., Lundanes, E., Christophersen, A.S., Strand, D.H., 2009. Determination of opiates and cocaine in urine by high pH mobile phase reversed phase UPLCeMS/MS. J. Chromatogr. B 877, 421e432.

Berthod, A., Chang, S.S.C., Kullman, J.P.S., Armstrong, D.W., 1998. Practice and mechanism of HPLC oligosaccharide separation with a cyclodextrin bonded phase. Talanta 47, 1001e1012.

Bicker, W., Wu, J.Y., Lammerhofer, M., Lindner, W., 2008. Hydrophilic interaction chromatography in nonaqueous elution mode for separation of hydrophilic analytes on silica-based packings with noncharged polar bondings. J. Sep. Sci. 31, 2971e2987.

Boersema, P.J., Divecha, N., Heck, A.J.R., Mohammed, S., 2007. Evaluation and optimization of ZIC-HILIC-RP as an alternative MudPIT strategy. J. Proteome Res. 6, 937e946.

Boggs, "Two-Dimensional Paper Chromatography Using One-Dimensional Apparatus," Analytical Chemistry, vol. 24, no. 10, pp. 1673-1675. 1952.

Bones, J., Duffy, C., Macka, M., Paull, B., 2008. Using coupled monolithic rods for ultra-high peak capacity LC and LC-MS under normal LC operating pressures. Analyst 133, 180e183.

Booth, T.J., Blake, P., Nair, R.R., Jiang, D., Hill, E.W., Bangert, U., Bleloch, A., Gass, M., Novoselov, K.S., Katsnelson, M.I., 2008. Macroscopic graphene membranes and their extraordinary stiffness. Nano Lett. 8, 2442e2446.

Borch-Jensen C, Mollerup J (1996) Applications of supercritical fluid chromatography to food and natural products. Semin Food Anal 1:101–116

Borders Jr, "Descending paper chromatography of oligosaccharides. A biochemistry laboratory experiment," Journal of Chemical Education, vol. 49, no. 6, pp. 437. 1972.

Borgono CA, Grass L, Soosaipillai A, Yousef GM, Petraki CD, Howarth DHC, et al. Human kallikrein 14: A new potential biomarker for ovarian and breast cancer. Cancer Res 2003; 63:

Bouchal P, Roumeliotis T, Hrstka R, Nenutil R, Vojtesek B, Garbis SD. Biomarker discovery in low-grade breast cancer using isobaric stable isotope tags and two-dimensional liquid chromatographytandem mass spectrometry (iTRAQ-2DLC-MS/MS) based quantitative proteomic analysis. J Proteome Res 2009; 8: 362-73.

Boysen, R.A., Yang, Y., Chowdhury, J., Matyska, M.T., Pesek, J.J., Hearn, M.T.W., 2011. Simultaneous separation of hydrophobic and hydrophilic peptides with a silica hydride stationary phase using aqueous normal phase conditions. J. Chromatogr. A 1218, 8021e8026.

BRIAN, P. W., AND HEMMING, H. G. 1945 Gliotoxin, a fungistatic metabolic product of Trichoderma viride. Ann. Appl. Biol., 32, 214-220.

Brice, R.W., Zhang, X., Colon, L.A., 2009. Fused-core, sub-2 mm packings, and monolithic HPLC columns: a comparative evaluation. J. Sep. Sci. 32, 2723e2731.

Brizel DM, Schroeder T, Scher RL, Walenta S, Clough RW, Dewhirst MW, Mueller-Klieser W. Elevated tumor lactate concentrations predict for an increased risk of metastases in head-and-neck cancer. Int J Radiat Oncol Biol Phys. 2001;51:349-353. [Cited in This Article: 1]

Broeckhoven, K., Billen, J., Verstraeten, M., Choikhet, K., Dittman, M., Rozing, G., Desmet, G., 2010. Towards a solution for viscous heating in ultra-high pressure liquid chromatography using intermediate cooling. J. Chromatogr. A 1217, 2022e2031.

Buszewski, B., Noga, S., 2012. Hydrophilic interaction liquid chromatography (HILIC)da powerful separation technique. Anal. Bioanal. Chem. 402, 231e247.

C. Giddings, Dynamics of Chromatography, Part I, Principles and Theory, Marcel Dekker, New York, 1965, pp. 13–26.

Cabooter, D., Choikhet, K., Lestreau, F., Dittman, M., Desmet, G., 2014. Towards a generic variable column length method development strategy for samples with a large variety in polarity. J. Chromatogr. A 1372, 174e186.

Cabooter, D., Lestremau, F., Lynen, F., Sandra, P., Desmet, G., 2008. Kinetic plot method as a tool to design coupled column systems producing 100,000 theoretical plates in the shortest possible time. J. Chromatogr. A 1212, 23e34.

Caude M, Thiébaut D (1999) Practical supercritical fluid chromatography and extraction. Harwood Academic, Amsterdam

Caulum MM, Murphy BM, Ramsayand LM, Henry CS. Detection of cardiac biomarkers using micellar electrokinetic chromatography and a cleavable tag immunoassay. Anal Chem 2007; 79: 5249-56.

Chen J, Wang W, Lv S, Yin P, Zhao X, Lu X, Zhang F, Xu G. Metabonomics study of liver cancer based on ultra performance liquid chromatography coupled to mass spectrometry with HILIC and RPLC separations. Anal Chim Acta. 2009;650:3-9.

Chen JL, Hong J, Lu JL, Chen MX, Chen WX, Zhu JS, Chen NW, Chen GQ, Geng JG. Effect of non-anticoagulant N-desulfated heparin on expression of vascular endothelial growth factor, angiogenesis and metastasis of orthotopic implantation of human gastric carcinoma. World J Gastroenterol. 2007;13:457-461

Cheng, X.D., Peng, X.T., Yu, Q.W., Yuan, B.F., Feng, Y.Q., 2013. Preparation of a novel amino-phosphate zwitterionic stationary phase for hydrophilic interaction chromatography. Chromatographia 76, 1569e1576.

Chester TL, Pinkston JD, Raynie DE (1996) Supercritical fluid chromatography and extraction (fundamental review). Anal Chem 68:487R–514R

Cheung KL, Graves CR, Robertson JF. Tumour markermeasurements in the diagnosis and monitoring of breast cancer. Cancer Treat Rev 2000; 26: 91-102.

Chirita, R.-I., West, C., Zubrzycki, S., Finaru, A.L., Elfakir, C., 2011. Investigations on the chromatographic behaviour of zwitterionic stationary phases used in hydrophilic interaction chromatography. J. Chromatogr. A 1218, 5939e5963.

Chromatography for Breast Cancer Biomarkers Recent Patents on Biomarkers 2011, Vol. 1, No. 1 97

chromatography-tandem mass spectrometry. Anal. Bioanal. Chem. 390, 947e959.

Churms, S.C., 1996. Recent progress in carbohydrate separation by high-performance liquid chromatography based on hydrophilic interaction. J. Chromatogr. A 720, 75e91.

Comprehensive screening and quantification of veterinary drugs in milk using UPLC-ToF-MS. Anal. Bioanal. Chem. 391, 2309e2322.

Consden, A. H. Gordon, and A. J. P. Martin, Biochem. J. 38, 224–232 (1944).

Coskun, "Separation techniques: chromatography," Northern clinics of Istanbul, vol. 3, no. 2, pp. 156. 2016. © 2018 JETIR October 2018, Volume 5, Issue 10 www.jetir.org (ISSN-2349-5162)

Craig LC (1943) Identification of small amounts of organic compounds by distribution studies. Application to Atabrine. J Biol Chem 150:33–45

Cremer and F. Prior, Z. Elektrochem. 55, 66 (1951); E. Cremer and R. Muller, Z. Elektrochem. 55, 217 (1951).

Crockford, D.J., Maher, A.D., Ahmadi, K.R., Barrett, A., Plumb, R.S., Wilson, I.D., Nicholson, J.K., 2008. 1H NMR and UPLC-MSE statistical heterospectroscopy: characterization of drug metabolites (xenometabolome) in epidemiological studies. Anal. Chem. 80, 6835e6844.

Cunliffe, J.M., Adams-Hall, S.B., Maloney, T.D., 2007. Evaluation and comparison of very high pressure liquid chromatography systems for the separation and validation of pharmaceutical compounds. J. Sep. Sci. 30, 1214e1223.

Currivan, S., Jandera, P., 2014. Modification reactions applicable to polymeric monolithic columns. A review. Chromatography 1, 24e53.

Currivan, S., Maca´k, J.M., Jandera, P., 2015. Polymethacrylate monolithic columns for hydrophilic interaction liquid chromatography prepared using a secondary surface polymerization. J. Chromatogr. A 1402, 82e93.

Desmet, G., Gzil, P., Clicq, D., 2005b. Kinetic plots to directly compare the performance of LC supports. LC-GC Eur. 18, 403e409.

Destefano, J.J., Langlois, T.J., Kirkland, J.J., 2008. Characteristics of superficially-porous silica particles for fast HPLC: some performance comparisons with sub-2-mm particles. J. Chromatogr. Sci. 46, 254e260.

Dinh, N.P., Jonsson, T., Irgum, K., 2011. Probing the interaction mode in hydrophilic interaction chromatography. J. Chromatogr. A 1218, 5880e5891.

Dinh, N.P., Jonsson, T., Irgum, K., 2013.Water uptake on polar stationary phases under conditions for hydrophilic interaction chromatography and its relation to solute retention. J. Chromatogr. A 1320, 33e47.

Dmitrovsky, E., Liu, X., Freemantle, S., Sempere, L., Cole, C., Kauppinen, S., Bak, M.; Christensen, M. MicroRNA biomarkers for human breast and lung cancer. US2010010072 (2010).

Dolan, J.W., 2006. Dwell volume revisited. LC-GC N. Am. 24, 458e466.

Dolan, J.W., Snyder, L.R., 1998. Maintaining fixed band spacing when changing column dimensions in gradient elution. J. Chromatogr. A 799, 21e34.

Dolzan, M.D., Spudeit, D.A., Breitbach, Z.S., Barber,W.E., Micke, G.A., Armstrong, D.W., 2014. Comparison of superficially porous and fully porous silica supports used for a cyclofructan 6 hydrophilic interaction liquid chromatographic stationary phase. J. Chromatogr. A 1365, 124e130.

Dreyer, D.R., Park, S., Bielawski, C.W., Ruoff, R.S., 2010. The chemistry of graphene oxide. Chem. Soc. Rev. 39, 228e240.

Duffy MJ, Duggan C, Keane R, Hill AD, McDermott E, Crown J, et al. High preoperative CA 15-3 concentrations predict adverse outcome in node-negative and node-positive breast cancer: Study of 600 patients with histologically confirmed breast cancer. Clin Chem 2004; 50: 559-63.

E. Morgan, and I. Wilson, "An early description of paper chromatography?," Chromatographia, vol. 60, no. 1-2, pp. 135-136. 2004.

Echan LA, Tang HY, Ali-Khan N, Lee K, Speicher DW. Depletion of multiple high-abundance proteins improves protein profiling capacities of human serum and plasma. Proteomics 2005; 5: 3292-

Feste, A.S., Khan, I., 1992. Separation of glucooligosaccharides and polysaccharide hydrolysates by gradient elution hydrophilic interaction chromatography with pulsed amperometric detection. J. Chromatogr. 630, 129e139.

Foley, J.P., Crow, J.A., Thomas, B.A., Zamora, M., 1989. Unavoidable flow-rate errors in high-performance liquid chromatography. J. Chromatogr. A 478, 287e309.

Foo, H.C., Heaton, J., Smith, N.W., Stanley, S., 2012. Monolithic poly (SPE-co-BVPE) capillary columns as a novel hydrophilic interaction liquid chromatography stationary phase for the separation of polar analytes. Talanta 100, 344e348.

Fountain, K.J., Neue, U.D., Grumbach, E.S., Diehl, D.M., 2009. Effects of extra-column band spreading, liquid chromatography system operating pressure, and column temperature on the performance of sub-2-mm porous particles. J. Chromatogr. A 1216, 5979e5988.

Fountain, K.J., Xu, J., Diehl, D.M., Morrison, D., 2010. Influence of stationary phase chemistry and mobile-phase composition on retention, selectivity, and MS response in hydrophilic interaction chromatography. J. Sep. Sci. 33, 740e751.

Frickenschmidt A, Frhlich H, Bullinger D, Zell A, Laufer S, Gleiter CH, et al. Metabonomics in cancer diagnosis: Mass spectrometrybased profiling of urinary nucleosides from breast cancer patients. Biomarkers 2008; 13: 435-49.

Fried B, Sherma J (1999) Thin-layer chromatography, 4th edn. Marcel Dekker, New York

Fritz J.S., 1991, Principles and applications of ionexclusion chromatography, J. Chromatogr. A., 546, 111-118.

Furberg A, Jasienska G, Bjurstam N, Torjesen PA, Emaus A, Lipson SF, et al. Metabolic and hormonal profiles: HDL cholesterol as a plausible biomarker of breast cancer risk. The norwegian EBBA study. Cancer Epidemiol Biomarkers Prev 2005; 14: 33-40.

G. Cassidy, "Investigation of paper chromatography," Analytical Chemistry, vol. 24, no. 9, pp. 1415-1421. 1952.

G. Rasul, "Extraction, Isolation and Characterization of Natural Products from Medicinal Plants," International Journal of Basic Sciences and Applied Computing, vol. 2, no. 6, pp. 1-6. 2011.

Gaikwad NW, Yang L, Muti P, Meza JL, Pruthi S, Ingle JN, et al. The molecular etiology of breast cancer: Evidence from biomarkers of risk. Int J Cancer 2008; 122:1949-57.

Garbis, S.D., Melse-Boonstra, A., West, C.E., van Breemen, R.B., 2001. Determination of folates in human plasma using hydrophilic interaction chromatography e tandem mass spectrometry. Anal. Chem. 73, 5358e5364.

Garrido Frenich, A., Martinez Vidal, J., Pastor-Montoro, E., Romero-Gonzalez, R., 2008. High-throughput determination of pesticide residues in food commodities by use of ultra-performance liquid

Gika, H.G., Theodoridis, G.A., Wilson, I.D., 2008c. Hydrophilic interaction and reversed-phase ultraperformance liquid chromatography TOF-MS for metabonomic analysis of Zucker rat urine. J. Sep. Sci. 31, 1598e1608.

Gilar, M., Daly, A.E., Kele, M., Neue, U.D., Gebler, J.C., 2004. Implications of column peak capacity on the separation of complex peptide mixtures in single- and two-dimensional high-performance liquid chromatography. J. Chromatogr. A 1061, 183e192.

Gilar, M., Olivova, P., Daly, A.E., Gebler, J.C., 2005. Orthogonality of separation in two-dimensional liquid chromatography. Anal. Chem. 77, 6426e6434.

Goldknopf, I. L. Isoform specificities of blood serum proteins and their use as differentially expressed protein biomarkers for diagnosis of breast cancer. US2010129846 (2010).

Grata, E., Guillarme, D., Glauser, G., Boccard, J., Carrupt, P.A., Veuthey, J.L., Rudaz, S., Wolfender, J.L., 2009.

Greco, G., Grosse, S., Letzel, T., 2013. Serial coupling of reversed-phase and zwitterionic hydrophilic interaction LC/MS for the analysis of polar and nonpolar phenols in wine. J. Sep. Sci. 36, 1379e1388.

Greiderer, A., Trojer, L., Huck, C.W., Bonn, G.K., 2009. Influence of the polymerisation time on the porous and chromatographic properties of monolithic poly(1,2-bis(p-vinylphenyl))ethane capillary columns. J. Chromatogr. A 1216, 7747e7754.

Grinnell, R. (1973). Environmental modification casework's concern or casework's neglect? Social Service Review, 47, 208-220.

Gritti, F., Cavazzini, A., Marchetti, N., Guiochon, G., 2007. Comparison between the efficiencies of columns packed with fully and partially porous C18-bonded silica materials. J. Chromatogr. A 1157, 289e303.

Guillarme, D., Nguyen, D.T.T., Rudaz, S., Veuthey, J.L., 2007a. Method transfer for fast liquid chromatography in pharmaceutical analysis: application to short columns packed with small particles e part I, isocratic separation. Eur. J. Pharm. Biopharm. 66, 475e482.

Guillarme, D., Nguyen, D.T.T., Rudaz, S., Veuthey, J.L., 2007b. Recent developments in liquid chromatography e impact on qualitative and quantitative performance. J. Chromatogr. A 1149, 20e29.

Guy, P.A., Tavazzi, I., Bruce, S.J., Ramadan, Z., Kochhar, S., 2008. Global metabolic profiling analysis on human urine by UPLCeTOFMS: issues and method validation in nutritional metabolomics. J. Chromatogr. B 871, 253e260.

H. Peterson, and L. M. Reineke, "A paper chromatographic technique and its application to the study of new antibiotics," Journal of the American Chemical Society, vol. 72, no. 8, pp. 3598-3603. 1950.

1961.

H. Schriftman, and R. C. Shultz, "Application of paper chromatography and electrophoresis to the assay of pharmaceutical products," Journal of pharmaceutical sciences, vol. 50, no. 4, pp. 332-336. 1961.

Ha´jek, T., Jandera, P., Stankova´, M., Cesla, P., 2016. Automated dual two-dimensional liquid chromatography approach for fast acquisition of three-dimensional data using combinations of zwitterionic polymethacrylate and silica-based monolithic columns. J. Chromatogr. A 1446, 91e102.

Haddad P.R., N. Nesterenko, W. Buchberger, 2008, Recent developments and emerging directions in ion chromatography. J. Chromatogr. A, 1184, 456-473.

Hahn-Deinstrop E (2007) Applied thin-layer chromatography: best practice and avoidance of mistakes, 2nd edn. Wiley-VCH, Weinheim, Germany

Halasz, I., Endele, R., Asshauer, J., 1975. Ultimate limits in high-pressure liquid chromatography. J. Chromatogr. A 112, 37e60.

Hamann U. Hereditary breast cancer: high risk genes, genetic testing and clinical implications. Clin Lab 2000; 46: 447-61.

Hao, Z., Xiao, B., Weng, N., 2008. Impact of column temperature and mobile phase composition on selectivity of hydrophilic interaction chromatography (HILIC). J. Sep. Sci. 31, 1449e1464.

Harris DC (1999) Quantitative chemical analysis, 5th edn.W.H. Freeman, New York

Hartman, A. (1976). Finding families: An ecological approach to family assessment in adoption. Beverly Hills, California: Sage Publishing Company.

Hartmann, E., Chen, Y., Mant, C.T., Jungbauer, A., Hodges, R.S., 2003. Comparison of reversed-phase liquid chromatography and hydrophilic interaction/cation-exchange chromatography for the separation of amphipathic a-helical peptides with L- and D-amino acid substitutions in the hydrophilic face. J. Chromatogr. A 1009, 61e71.

Hartonen, K., Riekkola, M.L., 2008. Liquid chromatography at elevated temperatures with pure water as the mobile phase. Trends Anal. Chem. 27, 1e14.

Heaton, J.C., McCalley, D.V., 2014. Comparison of the kinetic performance and retentivity of sub-2 mm core-shell, hybrid and conventional bare silica phases in hydrophilic interaction chromatography. J. Chromatogr. A 1371, 106e116.

Heaton, J.C., Russell, J.J., Underwood, T., Boughtflower, R., McCalley, D.V., 2014. Comparison of peak shape in hydrophilic interaction chromatography using acidic salt buffers and simple acid solutions. J. Chromatogr. A 1347, 39e48.

Heftmann E (ed) (2004) Chromatography, 6th edn. Fundamentals and applications of chromatography and related differential migration methods. Part A: fundamentals and techniques. Part B: applications. J Chromatogr Library Ser vols 69A and 69B. Elsevier, Amsterdam

Hirayama A, Kami K, Sugimoto M, Sugawara M, Toki N, Onozuka H, Kinoshita T, Saito N, Ochiai A, Tomita M. Quantitative metabolome profiling of colon and stomach cancer microenvironment by capillary electrophoresis time-of-flight mass spectrometry. Cancer Res. 2009;69:4918-4925. [Cited in This Article: 3]

Hjerten, S., Liao, J.L., Zhang, R., 1989. High-performance liquid chromatography on continuous polymer beds. J. Chromatogr. A 473, 273e275.

Hobbs, N. (1980). An Ecologically Oriented Service-based System for the Classification of Handicapped Children. In S. Salzinger, J. Antrobus, and J. Glick (Eds.), The ecosystem of the "sick" child implications for classification and intervention for disturbed and mentally retarded children (pp. 212- 290). New York: Academic Press.

Hoque MO, Feng Q, Toure P, Dem A, Critchlow CW, Hawes SE, et al. Detection of aberrant methylation of four genes in plasma DNA for the detection of breast cancer. J Clin Oncol 2006; 24:

Horie, K., Ikegami, T., Hosoya, K., Saad, N., Fiehn, O., Tanaka, N., 2007. Highly efficient monolithic silica capillary columns modified with poly(acrylic acid) for hydrophilic interaction chromatography. J. Chromatogr. A 1164, 198e205.

Hormann, K., Tallarek, U., 2014. Mass transport properties of second-generation silica monoliths with mean mesopore size from 5 to 25 nm. J. Chromatogr. A 1365, 94e105.

Huang HL, Stasyk T, Morandell S, Dieplinger H, Falkensammer G, Griesmacher A, et al. Biomarker discovery in breast cancer serum using 2-D differential gel electrophoresis /MALDI-TOF/TOF and data validation by routine clinical assays. Electrophoresis 2006; 27: 1641-50.

Huber, J.F.K., Pawlowska, M., Markl, P., 1984. Solvent-generated liquid-liquid chromatography with aqueous ternary systems. Chromatographia 19, 19e28.

Hulka BS. In: Hulka BS, Griffith JD, Wilcosky TC, Eds. Overview of Biological Markers. Biological Markers in Epidemiology. New York: Oxford University Press 1990; 3-15.

Hydrophilic interaction chromatography on amino-silica phases complements reversed-phase highperformance liquid chromatography and capillary electrophoresis for peptide analysis. J. Chromatogr. A 724, 378e383.

Ibrahim, M.E.A., Lucy, C.A., 2012. Mixed mode HILIC/anion exchange separations on latex coated silica monolith. Talanta 100, 313e339.

Ibrahim, M.E.A., Zhou, T., Lucy, C.A., 2010. Agglomerated silica monolithic column for hydrophilic interaction LC. J. Sep. Sci. 33, 773e778.

Idborg, H., Zamani, L., Edlund, P.-O., Schuppe-Koistinen, I., Jacobsson, S.P., 2005a. Metabolic fingerprinting of rat urine by LC/MS: part 1. Analysis by hydrophilic interaction liquid chromatography-electrospray ionization mass spectrometry. J. Chromatogr. B 828, 9e13.

Idborg, H., Zamani, L., Edlund, P.-O., Schuppe-Koistinen, I., Jacobsson, S.P., 2005a. Metabolic fingerprinting of rat urine by LC/MS: part 1. Analysis by hydrophilic interaction liquid chromatography-electrospray ionization mass spectrometry. J. Chromatogr. B 828, 9e13.

Ikegami, T., Tomomatsu, K., Takubo, H., Horie, K., Tanaka, N., 2008. Separation efficiencies in hydrophilic interaction chromatography. J. Chromatogr. A 1184, 474e503.

Imai K, Ichibangase T, Saitoh R, Hoshikawa Y. A proteomics study on human breast cancer cell lines by fluorogenic derivatization-liquid chromatography/tandem mass spectrometry. Biomed Chromatogr 2008; 22: 1304-14.

Iwasaki, Y., Ishii, Y., Ito, R., Saito, K., Nakazawa, H., 2007. New approaches for analysis of metabolism compounds in hydrophilic interaction chromatography. J. Liq. Chromatogr. Rel. Technol. 30, 2117e2126.

J. E. Golay, in Gas Chromatography, V. J. Coates, H. J. Noebels, and I. S. Fagerson, eds., Academic Press, New York, 1958, pp. 1–13.

J. N. Wilson, J. Am. Chem. Soc. 62, 1583–1591 (1940).

J. P. Martin and R. L. M. Synge, Biochem. J. (Lond.) 35, 1358 (1941).

J. P. Martin, Biochem. Soc. Symp. 3, 4–15 (1949).

J. van Deemter, F. J. Zuiderweg, and A. Klinkenberg, Chem. Eng. Sci. 5, 271–289 (1956).

J. Zijp, "Application of paper chromatography to the identification of accelerators and antioxidants: IV. Detection of antioxidants," Recueil des Travaux Chimiques des Pays-Bas, vol. 75, no. 10, pp. 1129-1136. 1956.

Jandera, P., Ha´jek, T., Stankova´, M., Vynuchalova´, K., Cesla, P., 2012. Optimization of comprehensive twodimensional gradient chromatography coupling in-line hydrophilic interaction and reversed phase liquid chromatography. J. Chromatogr. A 1268, 91e101.

Jandera, P., Novotna´, K., Beldean-Galea, M.S., Jı´sa, K., 2006. Retention and selectivity tests of silica-based and metal-oxide bonded stationary phases for RP-HPLC. J. Sep. Sci. 29, 856e871.

Jandera, P., Stankova´, M., 2015. The effects of the column length on the efficiency of capillary zwitterionic organic polymer monolithic columns in HILIC chromatography. Chromatographia 79, 853e859.

Jandera, P., Stankova´, M., Ha´jek, T., 2013. New zwitterionic polymethacrylate monolithic columns for unidimensional and two-dimensional micro liquid chromatography. J. Sep. Sci. 36, 2430e2440.

Jandera, P., Urban, J., Skerı´kova´, V., Langmaier, P., Kubı´ckova´, R., Planeta, J., 2010b. Polymethacrylate monolithic and hybrid particle-monolithic columns for reversed-phase and hydrophilic interaction capillary liquid chromatography. J. Chromatogr. A 1217, 22e33.

Jandera, P., Vynuchalova´, K., Ha´jek, T., Cesla, P., Vohralı´k, G., 2008. Characterization of HPLC columns for twodimensional LC separations of phenolic acids and flavonoids. J. Chemom. 22, 203e217.

Janeckova´, L., Kalı´kova´, K., Bosa´kova´, Z., Tesarova´, E., 2010. Study of interaction mechanisms on zirconia-based polystyrene HPLC column. J. Sep. Sci. 33, 3043e3051.

Jeanes, C. Wise, and R.-J. Dimler, "Improved techniques in paper chromatography of carbohydrates," Analytical Chemistry, vol. 23, no. 3, pp. 415-420. 1951.

Jemal A, Siegel R, Ward E, Hao Y, Xu J, Thun MJ. Cancer statistics, 2009. CA Cancer J Clin. 2009;59:225-249. [Cited in This Article: 1]

Jerkovich, A.D., Mellors, J.S., Jorgenson, J.W., 2003. The use of micron-sized particles in ultrahigh-pressure liquid chromatography. LC-GC Eur. 16, 20e23.

Jerkovich, A.D., Mellors, J.S., Thompson, J.W., Jorgenson, J.W., 2005. Linear velocity surge caused by mobilephase compression as a source of band broadening in isocratic ultrahigh-pressure liquid chromatography. Anal. Chem. 77, 6292e6299.

Jesneck JL, Mukherjee S, Yurkovetsky Z, Clyde M, Marks JR, Lokshin AE, et al. Do serum biomarkers really measure breast cancer? BMC Cancer 2009; 28(9): 164.

Johann DJ Jr, Blonder J. Biomarker discovery: Tissues versus fluids versus both. Expert Rev Mol Diagn 2007; 7: 473-5.

Johnson EL, Stevenson R (1978) Basic liquid chromatography. Varian Associates, Palo Alto, CA

Kuhn, A. Wunterstein, and E. Lederer, Hoppe-Seyler's Z. Physiol. Chem. 197, 141–160 (1931).

Kuhnle GG, Joosen AM, Wood TR, Runswick SA, Griffin JL, Bingham SA. Detection and quantification of sucrose as dietary biomarker using gas chromatography and liquid chromatography

Kulsing, C., Nolvachai, Y., Marriott, P.J., Boysen, R.L., Matyska, M.T., Pesek, J.J., Hearn, M.T.W., 2015. Insights into the origin of the separation selectivity with silica hydride adsorbents. J. Phys. Chem. B 119, 3063e3069.

Kulsing, C., Yang, Y., Munerab, C., Tseb, C., Matyska, M.T., Pesek, J.J., Boysen, R.L., Hearn, M.T.W., 2014. Correlations between the zeta potentials of silica hydride-based stationary phases, analyte retention behaviour and their ionic interaction descriptors. Anal. Chim. Acta 817, 48e60.

Kumar, A., Heaton, J.C., McCalley, D.V., 2013. Practical investigation of the factors that affect the selectivity in hydrophilic interaction chromatography. J. Chromatogr. A 1276, 33e46.

L. Clegg, "Paper chromatography," Analytical Chemistry, vol. 22, no. 1, pp. 48-59. 1950.

Lammerhofer, M., Richter, M., Wu, J., Nogueira, R., Bicker, W., Lindner, W., 2008. Mixed-mode ion-exchangers and their comparative chromatographic characterization in reversed-phase and hydrophilic interaction chromatography elution modes. J. Sep. Sci. 31, 2572e2588.

Lestremau, F., Cooper, A., Szucs, R., David, F., Sandra, P., 2006. High-efficiency liquid chromatography on conventional columns and instrumentation by using temperature as a variable: I. Experiments with 25 cm 4.6 mm I.D., 5 mm ODS columns. J. Chromatogr. A 1109, 191e196.

Lestremau, F., De Villiers, A., Lynen, F., Cooper, A., Szucs, R., Sandra, P., 2007. High efficiency liquid chromatography on conventional columns and instrumentation by using temperature as a variable: kinetic plots and experimental verification. J. Chromatogr. A 1138, 120e131.

Li, F., Maguigad, J., Peizer, M., Jiang, X., Ji, Q.C., 2008. A novel "peak parking" strategy for ultra-performance liquid chromatography/tandem mass spectrometric detection for enhanced performance of bioanalytical assays. Rapid Commn. Mass Spectrom. 22, 486e494.

Li, H., Liu, C., Wang, Q., Zhou, H., Jiang, Z., 2016. The effect of charged groups on hydrophilic monolithic stationary phases on their chromatographic properties. J. Chromatogr. A 1469, 77e87.

Li, J., Hu, Y., Carr, P.W., 1997. Fast separations at elevated temperatures on polybutadiene-coated zirconia reversed-phase material. Anal. Chem. 69, 3884e3888.

Liang, T., Fu, Q., Shen, A.J., Wang, H., Jin, Y., Xin, H.X., Ke, Y.X., Guo, Z.M., Liang, X.M., 2015. Preparation and chromatographic evaluation of a newly designed steviol glycoside modified-silica stationary phase in hydrophilic interaction liquid chromatography and reversed phase liquid chromatography. J. Chromatogr. A 1388, 110e118.

Liu, X., Pohl, C.A., 2010. HILIC behavior of a reversed-phase/cationexchange/anion-exchange trimode column. J. Sep. Sci. 33, 779e786.

MacNair, J.E., Patel, K.D., Jorgenson, J.W., 1999. Ultrahigh-pressure reversed-phase capillary liquid chromatography e isocratic and gradient elution using columns packed with 1.0-mm particles. Anal. Chem. 71, 700e708.

Michalski R., 2014, Recent Development and applications of ion chromatography, Current Chromatography, 1(2), 90-99.

Michalski R., 2016, Principles and Applications of Ion Chromatography [in] Application of IC-MS and ICICP- MS in Environmental Research, Eds. Michalski R., John Wiley & Sons, Inc. Hoboken, New Jersey, 1-46.

Miyamoto, K., Hara, T., Kobayashi, H., Morisaka, H., Tokuda, D., Horie, K., Koduki, K., Makino, S., Nunez, O., Yang, C., 2008. High-efficiency liquid chromatographic separation utilizing long monolithic silica capillary columns. Anal. Chem. 80, 8741e8750.

Molina R, Barak V, van Dalen A, Duffy MJ, Einarsson R, Gion M, et al. Tumor markers in breast cancer- European Group on Tumor Markers recommendations. Tumour Biol 2005; 26: 281-93.

Moni, L., Ciogli, A., Acquarica, I.D., Dondoni, A., Gasparrini, F., Marra, A., 2010. Synthesis of sugar-based silica gels by copper-catalysed azide-alkyne cycloaddition via a single-step azido-activated silica intermediate and the use of the gels in hydrophilic interaction chromatography. Chem. Eur. J. 16, 5712e5722.

Moravcova´, D., Haapala, M., Planeta, J., Hyotylainen, T., Kostiainen, R., Wiedmer, S.K., 2014. Separation of nucleobases, nucleosides, and nucleotides using two zwitterionic silica-based monolithic capillary columns coupled with tandem mass spectrometry. J. Chromatogr. A 1373, 90e96.

Muhr D, Wagner T, Ofner PJ. Polymerase chain reaction fidelity and denaturing high performance liquid chromatography. J Chromatogr B Analyt Technol Biomed Life Sci 2002; 782: 105-10.

Nakamura T, Fidler IJ, Coombes KR. Gene expression profile of metastatic human pancreatic cancer cells depends on the organ microenvironment. Cancer Res. 2007;67:139-148. [Cited in This Article: 1]

Nordstrom, A., O'Maille, G., Qin, C., Siuzdak, G., 2006. Nonlinear data alignment for UPLCMS and HPLCMS based metabolomics: quantitative analysis of endogenous and exogenous metabolites in human serum. Anal. Chem. 78, 3289e3295.

Nova´kova´, L., Havlı´kova´, L., Vlckova´, H., 2014. Hydrophilic interaction chromatography of polar and ionizable compounds by UHPLC. Trends Anal. Chem. 63, 55e64.

Nova´kova´, L., Kaufmannova´, I., Ja´nska´, R., 2010. Evaluation of hybrid hydrophilic interaction chromatography stationary phases for ultra-HPLC in analysis of polar pteridines. J. Sep. Sci. 33, 765e772.

Peng, X., Liu, T., Ji, S., Feng, Y., 2013b. Preparation of a novel carboxyl stationary phase by "thiol-ene" click chemistry for hydrophilic interaction chromatography. J. Sep. Sci. 36, 2571e2577.

Pereira, A.D., David, F., Vanhoenacker, G., Sandra, P., 2009. The acetonitrile shortage: is reversed HILIC with water an alternative for the analysis of highly polar ionizable solutes? J. Sep. Sci. 32, 2001-2007.

Pereira, A.D., Jime´nez Giro´n, A., Admasu, E., Sandra, P., 2010. Green hydrophilic interaction chromatography using ethanol-water-carbon dioxide mixtures. J. Sep. Sci. 33, 834e837.

Persson, J., Hemstro¨m, P., Irgum, K., 2008. Preparation of a sorbitol methacrylate grafted silica as stationary phase for hydrophilic interaction chromatography. J. Sep. Sci. 31, 1504e1510.

Pesek, J.J., Matyska, M.T., 2009. Our favorite materials: silica hydride stationary phases. J. Sep. Sci. 32, 3999e4011.

Pesek, J.J., Matyska, M.T., Boysen, R.L., Yang, Y., Hearn, M.T.W., 2013. Aqueous normal-phase chromatography using silica-hydride-based stationary phases. Trends Anal. Chem. 42, 64e73.

PETERSON, D. H., AND REINEKE, L. M. 1950 A paper chromatographic technique and its application to the study of new antibiotics. J. Am. Chem. Soc., 72, 3598-3603.

Plumb, R.S., Johnson, K.A., Rainville, P., Smith, B.W., Wilson, I.D., Castro-Perez, J.M., Nicholson, J.K., 2006.

Pollari S, Käkönen SM, Edgren H, Wolf M, Kohonen P, Sara H, Guise T, Nees M, Kallioniemi O. Enhanced serine production by bone metastatic breast cancer cells stimulates osteoclastogenesis. Breast Cancer Res Treat. 2010;Epub ahead of print. [Cited in This Article: 1]

Poppe, H., 1997. Some reflections on speed and efficiency of modern chromatographic methods. J. Chromatogr. A 778, 3e21.

Qiao, L.Y., Shi, X., 2015. Preparation and evaluation of surface-bonded tricationic ionic liquid silica as stationary phases for high-performance liquid chromatography. J. Chromatogr. A 1396, 62e71.

Qin S, Ferdinand AS, Richie JP, O'Leary MP, Mok SC, Liu BC. Chromatofocusing fractionation and two-dimensional difference gel electrophoresis for low abundance serum proteins. Proteomics

Qiu, H.X., Loukotkova′, L., Sun, P., Tesarova′, E., Bosa′kova′, Z., Armstrong, D.W., 2011. Cyclofructan 6 based stationary phases for hydrophilic interaction liquid chromatography. J. Chromatogr. A 1218, 270e279.

Richardson AD, Yang C, Osterman A, Smith JW. Central carbon metabolism in the progression of mammary carcinoma. Breast Cancer Res Treat. 2008;110:297-307.

Risley, D.S., Strege, M.A., 2000. Chiral separations of polar compounds by hydrophilic interaction chromatography with evaporative light scattering detection. Anal. Chem. 72, 1736e1739.

Rodriguez-Aller, M., Gurny, R., Veuthey, J.L., Guillarme, D., 2013. Coupling UHPLC with MS(/MS): constraints and possible applications. J. Chromatogr. A 1292, 2e18.

S. El Khadem, Z. M. El-Shafei, and M. M. A. A. Rahman, "Solvents for the Paper Chromatography of Amino Acids and Sugars without Spraying," Analytical Chemistry, vol. 35, no. 11, pp. 1766-1766, 1963/10/01. 1963.

Schappler, J., Nicoli, R., Nguyen, D.T.T., Rudaz, S., Veuthey, J.L., Guillarme, D., 2009. Coupling ultra-high pressure liquid chromatography with single quadrupole mass spectrometry for the analysis of a complex drug mixture. Talanta 78, 377e387.

Schembecker, G., Winkelnkemper, T. Process for separating and/or purifying taxane, preferably paclitaxel which is useful for treating breast and lung cancer, colon cancer and melanoma, comprises carrying out at least one liquid-liquid partition chromatography step. DE102008031208 (2010).

Schiesel, S., Laemmerhofer, M., Lindner, W., 2012. Comprehensive impurity profiling of nutritional infusion solutions by multidimensional off-line reversed-phase liquid chromatography hydrophilic interaction chromatography-ion trap mass-spectrometry and charged aerosol detection with universal calibration. J. Chromatogr. A 1259, 100e110.

Schuster, G., Lindner, W., 2013. Comparative characterization of hydrophilic interaction liquid chromatography columns by linear solvation energy relationships. J. Chromatogr. A 1273, 73e94.

Scopes RK (1994) Protein purification: principles and practice, 3rd edn. Springer-Verlag, New York

SENECA, H., KANE, J. H., AND ROCKENBACH, J. 1952 Bactericidal, protozoidal and fungicidal properties of thiolutin. Antibiotics & Chemotherapy, 2, 357-360.

Shen, A.J., Guo, Z.M., Cai, X.M., Xue, X.Y., Liang, X.M., 2012. Preparation and chromatographic evaluation of a cysteine-bonded zwitterionic hydrophilic interaction liquid chromatography stationary phase. J. Chromatogr. A 1228, 175e182.

Shen, A.J., Li, X.L., Dong, X.F.,Wei, J., Guo, Z.M., Liang, X.M., 2013. Glutathione-based zwitterionic stationary phase for hydrophilic interaction/cation-exchange mixed-mode chromatography. J. Chromatogr. A 1314, 63e69.

Sheng, Q.Y., Su, X.D., Li, X.L., Ke, Y.K., Liang, X.M., 2014. A dextran-bonded stationary phase for saccharide separation. J. Chromatogr. A 1345, 57e67.

Shundo, A., Mallik, A.K., Sakurai, T., Takafuji, M., Nagaoka, S., Ihara, H., 2009. Controllable shape selectivity based on highly ordered carbonyl and methyl groups in simple beta-structural polypeptide on silica. J. Chromatogr. A 1216, 6170e6176.

Siporin, M. (1980). Ecological systems theory in social work. Journal of Sociology and Social Welfare, 7, 507-532.

Skates SJ. Pooling of case specimens to create standard serum sets for screening cancer biomarkers. Cancer Epidemiol Biomarkers Prev 2007; 16: 334-341.

Snyder, L.R., Kirkland, J.R., Dolan, J.W., 2009. Introduction to Modern Liquid Chromatography, third ed. Wiley, Hoboken, NJ. Soukup, J., Jandera, P., 2012a. Hydrosilated silica-based columns: the effects of mobile phase and temperature on dual HILIC-RP separation mechanism of phenolic acids. J. Chromatogr. A 1228, 125e134.

Snyder, L.R., Poppe, H., 1980. Mechanism of solute retention in liquid-solid chromatography and the role of the mobile phase in affecting separation: competition versus "sorption". J. Chromatogr. 184, 363e413.

Soukup, J., Jana´s, P., Jandera, P., 2013. Gradient elution in aqueous normal-phase liquid chromatography on hydrosilated silica-based stationary phases. J. Chromatogr. A 1286, 111e118.

Soukup, J., Jandera, P., 2012b. The effect of temperature and mobile phase composition on separation mechanism of flavonoid compounds on hydrosilated silica-based columns. J. Chromatogr. A 1245, 98e108.

Spagou, K., Tsoukali, H., Raikos, N., Gika, H., Wilson, I.D., Theodoridis, G., 2010. Hydrophilic interaction chromatography coupled to MS for metabonomic/metabolomic studies. J. Sep. Sci. 33, 716e727.

Spratlin JL, Serkova NJ, Eckhardt SG. Clinical applications of metabolomics in oncology: a review. Clin Cancer Res. 2009;15:431-440. [Cited in This Article: 3]

Sreekumar A, Poisson LM, Rajendiran TM, Khan AP, Cao Q, Yu J, Laxman B, Mehra R, Lonigro RJ, Li Y. Metabolomic profiles delineate potential role for sarcosine in prostate cancer progression. Nature. 2009;457:910-914.

Stankova′, M., Jandera, P., 2016. Dual retention mechanism in two-dimensional LC separations of barbiturates, sulfonamides, nucleic bases and nucleosides on polymethacrylate zwitterionic monolithic micro-columns. Chromatographia 79, 657e666.

Stankova′, M., Jandera, P., Urban, J., Skerı′kova′, V., 2013. Cross-linker effects on the separation efficiency on (poly)methacrylate capillary monolithic columns. Part II. Aqueous normal-phase liquid chromatography. J. Chromatogr. A 1289, 47e57.

Stolker, A.A.M., Rutgers, P., Oosterink, E., Lasaroms, J.J.P., Peters, R.J.B., Van Rhijn, J.A., Nielen, N.W.F., 2008.

Stoll, D.R., Li, X., Wang, X., Carr, P.W., Porter, S.E.G., Rutan, G.C., 2007. Fast, comprehensive two-dimensional liquid chromatography. J. Chromatogr. A 1168, 3e43.

Streckfus CF, Dubensky WP. Proteomic analysis of saliva for cancer diagnosis. Expert Rev Proteomics 2007; 4: 329-32.

Strege, M.A., 1998. Hydrophilic interaction chromatography-electrospray mass spectrometry analysis of polar compounds for natural product drug discovery. Anal. Chem. 70, 2439e2445.

Takegawa, Y., Deguchi, K., Keira, T., Ito, H., Nakagawa, H., Nishimura, S., 2006. Separation of isomeric 2- aminopyridine derivatized N-glycans and N-glycopeptides of human serum immunoglobulin G by using a zwitterionic type of hydrophilic-interaction chromatography. J. Chromatogr. A 1113, 177e181.

Takegawa, Y., Ito, H., Keira, T., Deguchi, K., Nakagawa, H., Nishimura, S.-I., 2008. Profiling of N- and O-glycopeptides of erythropoietin by capillary zwitterionic type of hydrophilic interaction chromatography/electrosprayionization mass spectrometry. J. Sep. Sci. 31, 1585e1593.

Takikawa M, Akiyama Y, Maruyama K, Suzuki A, Liu F, Tai S, Ohshita C, Kawaguchi Y, Bandou E, Yonemura Y. Proteomic analysis of a highly metastatic gastric cancer cell line using two-dimensional differential gel electrophoresis. Oncol Rep. 2006;16:705-711.

Unified Nomenclature for Chromatography, IUPAC, J. Pure Appl. Chem., 65(4), 819–872 (1993) (c 1993 IUPAC).

Urayama S, Zou W, Brooks K, Tolstikov V. Comprehensive mass spectrometry based metabolic profiling of blood plasma reveals potent discriminatory classifiers of pancreatic cancer. Rapid Commun Mass Spectrom. 2010;24:613-620.

Urban, J., Jandera, P., 2013. Recent advances in the design of organic polymer monoliths for reversed-phase and hydrophilic interaction chromatography separations of small molecules. Anal. Bioanal. Chem. 405, 2123e2131.

Urban, J., Skerı´kova´, V., Jandera, P., Kubı´ckova´, R., Pospı´silova´, M., 2009. Preparation and characterization of polymethacrylate monolithic capillary columns with dual hydrophilic interaction reversed-phase retention mechanism for polar compounds. J. Sep. Sci. 32, 2530e2543.

Urban, J., Svec, F., Fre´chet, J.M.J., 2010. Hypercrosslinking: new approach to porous polymer monolithic capillary columns with large surface area for the highly efficient separation of small molecules. J. Chromatogr. A 1217, 8212e8221.

Vajda, P., Felinger, A., Guiochon, G., 2013. Evaluation of surface excess isotherms in liquid chromatography. J. Chromatogr. A 1291, 41e47.

Vivilecchia, R.V., Cotter, R.L., Limpert, R.J., Thimot, N.Z., Little, J.N., 1974. Considerations of small particles in different modes of liquid chromatography. J. Chromatogr. A 99, 407e424.

W. Eigsti, "Paper Chromatography for Student Research," The American Biology Teacher, pp. 123-134. 1967.

W. Smit, I. Elliott, R. W. Peeling, D. Mabey, and P. N. Newton, "An overview of the clinical use of filter paper in the diagnosis of tropical diseases," The American journal of tropical medicine and hygiene, vol. 90, no. 2, pp. 195-210. 2014.

Wade, K.L., Garard, I.J., Fahey, J.W., 2007. Improved hydrophilic interaction chromatography method for the identification and quantification of glucosinolates. J. Chromatogr. A 1154, 469e472.

Walenta S, Schroeder T, Mueller-Klieser W. Lactate in solid malignant tumors: potential basis of a metabolic classification in clinical oncology. Curr Med Chem. 2004;11:2195-2204.

Wall PE (2005) Thin-layer chromatography: a modern practical approach. The Royal Society of Chemistry, Cambridge, UK

Walles, M., Gauvin, C., Morin, P.E., Panetta, R., Ducharme, J., 2007. Comparison of sub-2-mm particle columns for fast metabolite ID. J. Sep. Sci. 30, 1191e1199.

Walters RR (1985) Report on affinity chromatography. Anal Chem 57:1099A–1113A

Wang Q.Y., A. Mkishima, E. Nakamura, 2010, Determination of Fluorine and Chlorine by Pyrohydrolysis and Ion Chromatography: Comparison with Alkaline Fusion Digestion and Ion Chromatography, Gepst. Geoanal. Res., 34, 175.

Wang, Q., Luo, Z., Ye, M., Wang, Y., Xu, L., Shi, Z., Xu, L., 2015. Preparation, chromatographic evaluation and application of adenosine 50-monophosphate modified ZrO2/SiO2 stationary phase in hydrophilic interaction chromatography. J. Chromatogr. A 1383, 58e69.

Wang, X., Li, W., Rasmussen, H.T., 2005. Orthogonal method development using hydrophilic interaction chromatography and reversed-phase high-performance liquid chromatography for the determination of pharmaceuticals and impurities. J. Chromatogr. A 1083, 58e62.

Xue, M.Z., Huang, H.X., Ke, Y.X., Chu, C.H., 2009. "Click dipeptide": a novel stationary phase applied in twodimensional liquid chromatography. J. Chromatogr. A 1216, 8623e8629.

Yadav, M., Contractor, P., Upadhyay, V., Gupta, A., Guttikar, S., Singhal, P., Goswami, S., Shrivastav, P.S., 2008. Automated liquideliquid extraction based on 96-well plate format in conjunction with ultra-performance liquid chromatography tandem mass spectrometry (UPLCeMS/MS) for the quantitation of methoxsalen in human plasma. J. Chromatogr. B 872, 167e171.

Yang Z, Hancock WS. Approach to the comprehensive analysis of glycoproteins isolated from human serum using a multi-lectin affinity column. J Chromatogr A 2004; 1053: 79-88.

Yuan, G.X., Peng, Y.B., Liu, Z.H., Hong, J.Y., Xiao, Y., Guo, J.L., Smith, N.W., Crommen, J., Jiang, Z.J., 2013. A facile and efficient strategy to enhance hydrophilicity of zwitterionic sulfoalkylbetaine type monoliths. J. Chromatogr. A 1301, 88e97.

Z. Fu, L. Sun, J. Cai, X. Wang, X. Wang, S. Zheng, and X. Shen, "Paper Chromatographic Separation of Racemic Diphenylmethyl Alcohols Using Only Pure Water Surfactant Micellar Mobile Phase and Host—Guest Chromatography," Journal of chromatographic science, vol. 35, no. 7, pp. 309-314. 1997.

Zachariou M (2008) Affinity chromatography: methods and protocols. Humana, Totowa, NJ

Zachariou M (2008) Affinity chromatography: methods and protocols. Humana, Totowa, NJ

Zhang Z, Smith DL, Smith JB. Multiple separations facilitate identifi cation of protein variants by mass spectrometry. Proteomics 2001; 1: 1001-9.

Zhang, T., Creek, D.J., Barrett, M.P., Blackburn, B., Watson, D.G., 2012. Evaluation of coupling reversed phase, aqueous normal phase, and hydrophilic interaction liquid chromatography with orbitrap mass spectrometry for metabolomic studies of human urine. Anal. Chem. 84, 1994e2001.

Zhang, Y., Wang, X., Mukherjee, P., Petersson, P., 2009. Critical comparison of performances of superficially porous particles and sub-2 mm particles under optimized ultra-high pressure conditions. J. Chromatogr. A 1216, 4597e4605.

Zhong L, Ge K, Zu J, Zhao L, Shen W, Wang J, et al. Autoantibodies as potential biomarkers for breast cancer. Breast Cancer Res 2008; 10: R40.

Zhu K, Miller FR, Barder TJ, Lubman DM. Identification of low molecular weight proteins isolated by 2-D liquid separations. J Mass Spectrom 2004; 39: 770-80.

Zhu, B.Y., Mant, C.T., Hodges, R.S., 1991. Hydrophilic-interaction chromatography of peptides on hydrophilic and strong cation-exchange columns. J. Chromatogr. 548, 13e24.

Ziegler RG, Rossi SC, Fears TR, Bradlow HL, Adlercreutz H, Sepkovic D, *et al.* Quantifying estrogen metabolism: An evaluation of the reproducibility and validity of enzyme immunoassays for 2-hydroxyestrone and 16alpha-hydroxyestrone in urine. Environ Health Perspect 1997; 105: 607-14.

Zweig and J. Sherma, J. Chromatogr. Sci. 11, 279–283 (1973).

Zweig, and J. Sherma, "Paper Chromatography—Past, Present and Future,"*Journal of Chromatographic Science*, vol. 11, no. 6, pp. 279-283. 1973.

INDEX

www.ingramcontent.com/pod-product-compliance
Lightning Source LLC
LaVergne TN
LVHW062217200926
848484LV00006B/1420